Science and Culture

HERMANN VON HELMHOLTZ
in 1894, shortly before his death, in a sketch by Franz von Lenbach.
Courtesy of the Siemens-Forum, Munich.

Hermann von Helmholtz

Science and Culture

Popular and Philosophical Essays

Edited and with an Introduction by
DAVID CAHAN

THE UNIVERSITY OF CHICAGO PRESS

Chicago and London

DAVID CAHAN is professor of history at the University of Nebraska.

The University of Chicago Press, Chicago 60637
The University of Chicago Press, Ltd., London
© 1995 by The University of Chicago
All rights reserved. Published 1995
Printed in the United States of America
04 03 02 01 00 99 98 97 96 95   5 4 3 2 1

ISBN (cloth): 0-226-32658-6
ISBN (paper): 0-226-32659-4

Library of Congress Cataloging-in-Publication Data

Helmholtz, Hermann von, 1821–1894.
        Science and culture : popular and philosophical essays /
    Hermann von Helmholtz ; edited and with an introduction by David
    Cahan.
            p.   cm.
        Includes bibliographical references.
        ISBN 0-226-32658-6. — ISBN 0-226-32659-4 (pbk.)
        1. Science—Philosophy—History—19th century.   2. Science—
    Social aspects—History—19th century.   3. Helmholtz,
    Hermann von, 1821–1894—Knowledge—Philosophy.   I. Cahan,
    David.   II. Title.
    Q174.8.H45   1995
    500—dc20   95-12217
                                                                 CIP

⊗ The paper used in this publication meets the minimum requirements of the
American National Standard for Information Sciences—Permanence of Paper for
Printed Library Materials, ANSI Z39.48-1984.

Contents

Introduction

THE ISSUE OF the public understanding of science is virtually as old as the formation of the modern scientific enterprise itself. Galileo had not only to explain his findings concerning the moon, Jupiter, Saturn, and other heavenly objects, and about the legitimate use of his newly invented instrument, the telescope, to laypersons. He had also to convince other men of knowledge—all of whom were unfamiliar with his new instrument and some of whom (especially Aristotelian philosophers) had a vested interest in refuting his claims—of his epochmaking findings and their implications. Galileo proved to be superb at solving his problem: ever since their first publication, his *Sidereal Messenger* (1610) and *Dialogue Concerning the Two Chief World Systems* (1632) have proved to be excellent vehicles for bringing his results to scientific and lay readers alike.

Despite Galileo's success, the epistemological, psychological, and social problems of popularization have broadened and deepened over time. As the modern scientific enterprise continued to take shape during the course of the seventeenth century, Galileo's successors continued to be confronted with the problem of informing others of ever-increasing empirical findings and explaining the nature and implications of increasingly mathematical and technical developments in the exact sciences (in particular, astronomy, mechanics, optics, and statics) and natural philosophy in general. How could natural philosophers demonstrate the credibility of their knowledge to others? How could they translate it into an idiom that their colleagues and the public in general might understand? And how could they gain the public's support for their efforts and so further legitimize their activities?[1]

1. The related issues of the contributions of nonscientists to science and of the use of general culture as a resource for scientific ideas cannot be treated here, but see Richard Whitley, "Knowledge Producers and Knowledge Acquirers: Popularisation as a Relation between Scientific Fields and their Publics," in Terry Shinn and Richard Whitley, eds., *Expository Science: Forms and Functions of Popularisation* (Dordrecht: D. Reidel, 1985), pp. 3–28; and Steven Shapin, "Science and the Public," in R. C. Olby, G. N. Cantor, J. R. R. Christie, and M. J. S. Hodge, eds., *Companion to the History of Modern Science* (London and New York: Routledge, 1990), pp. 990–1007. My discussion here draws in part on Shapin's essay.

During the Enlightenment, natural philosophers, along with men and women of letters, developed a number of vehicles for addressing these epistemological, social, and political issues of science. For example, the Boyle Lectures (from 1692) in Britain, and popular science books like Voltaire's *Elements of the Philosophy of Newton* (1737), and Francesco Algarotti's *Newtonianism for the Ladies* (1735) sought to explain Newton's views to laypersons, while Leonhard Euler's enormously popular *Letters to a German Princess on Diverse Points of Physics and Philosophy* (1768) presented a non-Newtonian version of contemporary physical thought. Similarly, innumerable learned and popular books on natural theology claimed to demonstrate how science was a means to understanding God's physical universe (if not God himself); that science was, in effect, a religious activity, and hence worthy of pursuit and support. Itinerant lecturers and spectacles ranging from the preacher's pulpit to the coffeehouse also helped promulgate and legitimize scientific ideas, as did the proliferation of magazines and societies devoted to popularizing natural knowledge.

As large as these accomplishments of the early modern natural philosophers were in the popularization of science, they were greatly outdistanced by those of their nineteenth-century successors. The reasons for this are not far to seek, and we may here briefly touch on several of them.

First, the nineteenth-century witnessed an enormous and unprecedented expansion in formal schooling, from the primary level through higher and technical education. It also saw the widespread establishment of public libraries. As a result, literacy increased and access to books and journals became relatively easy. The market of potential readers expanded greatly. Second, science became widely perceived as essential to, if not the foundation of, industrialization, and hence of wealth and progress. Indeed, science and progress became virtual synonyms for almost everyone—workers and industrialists, minor civil servants and senior statesmen—interested in economic and political advance. (In fact, science was often confused with technology, and mistakenly identified as the primary cause of industrial advance.) Third, from the late Enlightenment onward the social structure of science itself changed greatly. Before then, support for science had largely been a matter of individual patrons supporting individual scientists. In the wake of the political revolution in France and the industrial revolution in England, however, science gradually became institutionalized, professionalized, and specialized. More scientific research was conducted in universities by increasing numbers of scientists working in and for universities that were largely supported as state institutions and controlled by state bureaucracies. Especially after the 1820s, specialization and professionalization increased; academic disciplines separated themselves out and formed distinct disciplinary societies. As a result, specialized scientists—it was not by accident that in 1834 the British polymath William Whewell coined the term "scientist," and that new and more specialized titles such as "biologist" and "physicist" arose during the first third of the century—became more and more insulated and isolated from one another and the lay public; some of them thus felt the need to make additional efforts to explain the state of their individual specialties to their colleagues and the public. Fourth and finally, the publication of Darwin's *Origin of Species* (1859) and his *Descent of Man* (1871) shook whatever consensus and understanding may have existed on the relations of science and religion, and undermined natural theology. From the 1860s onward, a naturalistic view of human life emerged: scientists and nonscientists began struggling toward some new vision of humankind's origins and nature. Accordingly, an ever-larger number of educated men and women turned to popularizations of Darwinism and its alleged social consequences (e.g., Social Darwinism) as a means for understanding humankind and its changed place in nature.

By the mid–nineteenth century, the German states contained the most widely schooled population; they had begun their industrial revolution; they stood at the forefront of scientific research, and hence of specialization and professionalization; and they were fully aware of the possible social, cultural, and theological implications of the Darwinian revolution. As a result, there was a large audience for popular science in Germany.

The most influential German, perhaps international, figure in the popularization of science during the first half of the nineteenth century was Alexander von Humboldt.[2] Humboldt's first major popularization was his *Aspects of Nature, in Different Lands and Different Climates; with Scientific Elucidations* (1808). Two decades later, in the winter of 1827–28, he held a series of public lectures at the University of Berlin that subsequently, though in quite different form, appeared as his five-volume *Cosmos: A Sketch of a Physical Description of the Universe* (1845–62). It became an international bestseller, being translated into ten foreign languages and read by, among many thousands of others, the Bavarian schoolboy Albert Einstein. Humboldt sought to synthesize all knowledge and to stress what he saw as the philosophical ideas behind science and the spiritual (as opposed to the socioeconomic) effect of science on people.

Humboldt's *Cosmos* represented only one type of popularization in Germany. Though its success lasted through the century, other types also began appearing around mid-century. In 1841–44, for example, Germany's

---

2. The following general discussion of the popularization of science in nineteenth-century Germany draws in part on Kurt Bayertz, "Spreading the Spirit of Science: Social Determinants of the Popularization of Science in Nineteenth-Century Germany," in Shinn and Whitley, eds., *Expository Science*, pp. 209–27.

leading chemist, Justus von Liebig, published his *Chemical Letters*, which appeared originally in installment form in a newspaper. Liebig's extremely popular *Letters* addressed the educated public and, unlike Humboldt's *Cosmos*, concentrated on one subject, explaining chemistry's empirical results, methods, and theories, and, again unlike Humboldt, arguing for his science's economic usefulness. Liebig's aim was thus not only one of education but also one of garnering greater public support for his discipline.

After 1850, most popularizers of science in Germany (and elsewhere) did not fail to argue for the potential economic or, as in the writings of the medical scientist and politician, Rudolf Virchow, medical benefit to be derived from the pursuit of science. And most believed that by helping the German states economically, science would eliminate the last remnants of the feudal order, thus leading (so the argument went) to increased industrialization, the beginnings of representative government, and to a more secular, anti-authoritarian way of life. Yet most German popularizers also tended to emphasize the cultural or philosophical contribution that they thought their science offered German and world culture—assuming, of course, that Germany's educational ministries continued, or would increase, their financial support of science and hence the social standing of German scientists.

During the later 1840s, a triumvirate of radical German scientists—Karl Vogt, Jacob Moleschott, and Ludwig Büchner—appeared on the scene of science popularization. They argued for scientific materialism. Their general message was pro-science, anti-religion, and for republican government. They considered science to be the basis of a progressive, anti-authoritarian political order. Through books such as Büchner's immensely popular *Force and Matter, or Principles of the Natural Order of the Universe* (1855), they helped shape popular understanding of science in Germany during the 1850s and after.[3] Then, with the rise of Darwinism after 1859, came the rise of Ernst Haeckel, the premier German spokesman for the Darwinian viewpoint, though one who modified Darwin's evolutionary theory and sought to extend it beyond biology. From the late 1860s onward, the prolific Haeckel published a series of popular books, essays, and lectures on science. Above all, his *The Riddle of the Universe* (1899) had an enormous impact on the public's understanding of science. Like the scientific materialists before him, the neo-romantic Haeckel sought to make science philosophical, in his case in the form of a philosophical doctrine called monism, or the unification of mind and matter.

---

3. Frederick Gregory, *Scientific Materialism in Nineteenth-Century Germany* (Dordrecht: D. Reidel, 1977).

While the popular scientific writings of all these German figures were widely read, and substantially shaped the public's understanding of science from the 1840s onward, they did not quite have the field entirely to themselves. After mid-century, Hermann von Helmholtz also entered into the culturally important, intellectually vibrant, and expanding field of the popularization of science. Helmholtz's reputation as a popularizer grew more slowly than did those of his counterparts. His popular writings did not have the sudden, dramatic impact on his contemporaries that the better-known works of other German popularizers had, in part because his efforts took the form of individual lectures, beginning in 1853, in response to specific requests from various cultural or scientific groups. Moreover, his thought and his sentiments stood somewhere in the great intellectual, cultural, and political middle of the spectrum of German scientific and philosophical life. Neither materialist nor idealist; neither altogether democratic, nor aristocratic; and certainly neither socialist nor monarchist, Helmholtz's voice was the voice of reason and moderation. After the appearance of the first collection of his popular scientific lectures in 1865, however, it became a voice increasingly eminent throughout Germany. With the translation of this first collection into English in 1873, and with the subsequent appearance in the 1870s, 1880s, and early 1890s of new essays, a substantial two-volume set of essays (appearing in English as well as in German) eventually took shape. During these politically unstable decades in Bismarckian Germany, Helmholtz's name became increasingly known to many in the educated world within and beyond Germany's borders, and his popular scientific writings further increased his standing as one of Germany's and the world's spokesmen of science.

Helmholtz's life (1821–94) spanned much of the nineteenth century. He thus participated in and observed firsthand much of science's emergence from natural philosophy and medicine into its specialized disciplinary and professional formations, and much of the concomitant rise of German science from its secondary position behind France and Britain early in the century to one of leadership by the last third of the century.

Helmholtz received an excellent primary and secondary education in the schools of his native Potsdam, as well as from his father, who taught in the local gymnasium. As a youth, Helmholtz wanted to become a physicist, but his father's insufficient financial resources led Hermann to study medicine at a Prussian army medical institute associated with the University of Berlin. In return for a four-year medical education, Helmholtz agreed to serve eight years in the army as a medical officer. He spent most of the years from 1843 to 1848 serving as a staff surgeon in the army in Potsdam while simultaneously managing to produce a series of important publications on physiological topics and a bold, pathbreaking essay on the conservation of force (i.e., energy). The young scientist's work was greatly encouraged by his much-admired mentor, the anatomist and physiologist Johannes Müller, who helped him get an early discharge from the military, obtain a temporary academic position as an instructor at the Art Academy and assistant at the Anatomy Museum in Berlin (1848–49), and enter permanently into an academic career in 1849 with a position as associate professor of physiology at the University of Königsberg.[4]

In late 1850, as part of his preparation for his lectures on general pathology and physiology, Helmholtz invented the ophthalmoscope. As he says in his "Autobiographical Sketch," the ophthalmoscope was probably "the most popular of my scientific performances." Indeed it was. For despite the decisive scientific contributions that he had already made and the credit that he had already received for his work in heat and muscle physiology and for measuring the propagation rate of the nerve impulse, and despite the later recognition he would receive for his co-discovery of the conservation of force, it was the invention of the device for observing the living retina that brought him immediate and widespread recognition among his professional colleagues, the Prussian ministry of education (which promoted him to full professor in 1851), and the public at large. More important for present purposes, it was chiefly the recognition resulting from the invention of the ophthalmoscope that paved the way to his becoming a popularizer of science. Less than two years later, in the spring of 1853, Helmholtz gave his first popular scientific lecture, "On Goethe's Scientific Research," before the German Society in Königsberg.

Helmholtz remained in Königsberg until 1855, when he accepted a professorship of anatomy and physiology at the University of Bonn. But Bonn proved not to his liking: he remained there for only three years. In 1858, he accepted a new position as professor of physiology at the University of Heidelberg. Helmholtz thus spent much of the 1850s and 1860s at these three universities, conducting research in various aspects of physiology—nerve physiology, electrophysiology, color theory, and physiological acoustics—and, intimately related to this research, conducting experiments in and reflecting on the physiology, philosophy, and psychology of perception. During these decades, moreover, he was not only a prolific author of numerous physiological papers; he also researched and wrote two books: his gigantic and fundamental reference work, the three-part *Handbook of Physiological Optics* (1856–67); and a long scientific and scholarly study on the physics, physiology, and aesthetics of music, *On the Sensations of Tone as a Physiological Basis for the Theory of Music* (1863), an epochmaking work which influenced musicologists into the twentieth century.

---

4. For a detailed discussion of Helmholtz's early life and career see the introduction to David Cahan, ed., *Letters of Hermann von Helmholtz to His Parents, 1837–1846* (Stuttgart: Franz Steiner Verlag, 1993), pp. 1–30.

In 1871, Helmholtz moved to the University of Berlin, as professor of physics, and remained there in a full-time capacity until 1888. Although his pre-1871 work in physics—on energy conservation, electrical-current distribution, hydrodynamics, acoustics, and electrodynamics—had been important enough to gain him a reputation as one of Europe's leading physicists, it had nonetheless taken second place to physiology in his formal academic career. During the last twenty-three years of his career, however, Helmholtz devoted his academic research almost entirely to physics, above all to investigating problems of electrodynamics, thermodynamics, and mechanics, while also making a series of important contributions to microscopy, physical chemistry, and meteorology.[5]

In addition to research in physiology and physics, Helmholtz was predominant in German and world science through another aspect of professional activity. He was the director of no fewer than four university scientific institutes (Königsberg, Bonn, Heidelberg, and Berlin); the last two of these he had built to his specifications. He also helped plan and served as the founding president (1888–94) of the first national institute devoted to physics and technology, the Physikalisch-Technische Reichsanstalt in Berlin.[6] He held academic office as the prorector of the University of Heidelberg (1862–63) and as the rector of the University of Berlin (1877–78). During the final dozen years of his life, he represented German interests by leading several German scientific and technical delegations at international meetings.

The popular scientific lectures represented another dimension of Helmholtz's scientific leadership. Because the lectures originated in response to requests from various groups, and because his scientific interests were so extensive, the subject matter and general intellectual approach of the lectures varied widely. They range from discussions of fundamental concepts, laws, and results in physics and physiology to issues of epistemology, aesthetics, and the philosophy of science as well as the relations of science, society, and culture broadly construed. In contrast to the other German popularizers we have mentioned, his lectures are thus not dominated by any one recurring issue, or concept, or point of view. On the other hand, despite their diversity, the totality of his popular scientific lectures published over a thirty-year period suggests a theme that I have elsewhere described as the civilizing power of science.[7] Broadly speaking,

---

5. For analysis of Helmholtz's scientific work, see the pertinent essays in David Cahan, ed., *Hermann von Helmholtz and the Foundations of Nineteenth-Century Science* (Berkeley and Los Angeles: University of California Press, 1994).

6. David Cahan, *An Institute for an Empire: The Physikalisch-Technische Reichsanstalt 1871–1918* (Cambridge: Cambridge University Press, 1989).

7. For analysis of this theme see my "Helmholtz and the Civilizing Power of Science," *Helmholtz and the Foundations of Nineteenth-Century Science*, pp. 559–601.

Helmholtz's lectures illumine his conviction, at points explicit and at points implicit, that science enables us to understand the natural world and humanity's place in it; that it empowers us to control that world; that it provides the foundations for aesthetic life; and that it helps unite individuals into social and political communities, national and international. In propounding these views in lectures, and subsequently in published form, Helmholtz addressed over the course of his professional lifetime some of the deepest questions of material, social, and spiritual well-being. And for many in the educated world during the second half of the nineteenth century, science and its relations with philosophy, culture, and society was what Helmholtz claimed it to be.

For some years now, Helmholtz's essays have been out of print or available only in costly editions devoted to a single aspect of his multifaceted work and writings (e.g., to his epistemology or physiology). In selecting and assembling this collection of Helmholtz's essays, I have therefore sought to offer in an inexpensive paperback edition a broad selection of his most important essays on popular science, on culture, and on philosophy. In so doing, I have had three particular, partially overlapping categories of potential readers in mind who, I believe, may welcome the appearance of such an edition.

First, I have sought to provide the increasing number of scholars specializing in the study of Helmholtz's work and its place in nineteenth-century thought and culture with a convenient edition of his essays that might prove useful for their own work. I have, secondly, sought to supply historians of science, philosophers (especially of science), scientists with an interest in the history and philosophy of science, scholars of modern European intellectual and cultural history, scholars of German literature, and humanists, as well as the educated public at large, with a volume of Helmholtz's essays which they may want for their bookshelves. Third and above all, I have sought to provide college and university teachers with a collection of Helmholtz's essays for classroom use. In my view, Helmholtz's essays merit being readily available to students, both undergraduates and graduates. There is a growing movement in the United States and elsewhere today to inculcate "scientific literacy"—above all among students who do not intend to become scientists or who seek to avoid the study of as much science as possible. For such students, Helmholtz's essays offer relatively easy access to one of the most important visions of nineteenth-century science and to some of its results, while at the same time presenting his conception of science's relationships to philosophy and to society. As the titles of the essays in this volume suggest, they raise important issues about the nature of science, its purposes, methods, and its intellectual and social contexts; and they concern the scientific influences on art and on the nature of academic life. For more critical students, the essays should also raise questions about Helmholtz's ideological attitudes toward science and toward its relationships with philosophy and society; as such, the essays offer excellent discussion material for the undergraduate classroom and the graduate seminar.

I have chosen to present fifteen of Helmholtz's most important essays. Thirteen of those are reproduced essentially unchanged from the two volumes of his *Popular Lectures on Scientific Subjects* (New York: D. Appleton and Company, 1873 and 1881). Selections from the first volume are reproduced from the 1897 printing; for the second series, I relied on an edition published in London in 1898 by Longmans, Green, and Company, that is a reissue of Appleton's 1881 volume, with the addition of Helmholtz's autobiographical sketch of 1891. Both English-language volumes were based on Helmholtz's *Populäre wissenschaftliche Vorträge* (2 vols., Braunschweig: Friedrich Vieweg and Sohn, 1865 and 1871; 2d ed. in 3 vols., 1876).

Because the large type and small page format of the English-language originals make facsimile reproduction impractical, I decided to maximize the number of essays I might offer readers while minimizing their cost by scanning and resetting them electronically. The scanned essays had no fewer than seven different translators, working over a period of many years, and contain minor stylistic inconsistencies which I have not attempted to regularize. The nineteenth-century flavor of the translations remains, along with some British spellings or turns of phrase. The reader may care to know that the abbreviation "Tr." used in several footnotes stands for "Translator," and the initials "J.T." in notes to the essay "On the Interaction of the Natural Forces," refer to John Tyndall, who introduced the 1897 Appleton edition.

The two remaining essays in this collection, "The Facts in Perception" and "Goethe's Presentiments of Coming Scientific Ideas," I have translated anew from the German-language originals appearing in the 1903 fifth edition of Helmholtz's *Vorträge und Reden* (an expanded edition of the *Populäre wissenschaftliche Vorträge*), first published in 1884.

In translating these two essays, I have naturally sought to remain as true as possible to Helmholtz's presumed meaning and style. For translations of Goethe's and Schiller's verse, I turned to nineteenth-century translators of their poetry. However, in one instance I altered Bayard Taylor's translation of *Faust* (see pp. 362 and 409), where, in my view, Taylor's translation was seriously misleading with respect to the meaning of a key term ("*Gleichniss*") that is crucial to Helmholtz's epistemology. Moreover, because Helmholtz (significantly) misquoted Schiller's verse from "Der Spaziergang," I have provided my own, unpoetical translation. Finally, I changed Helmholtz's internal cross-references to these two essays as they appeared in his *Vorträge und Reden*.

I could not have achieved the goal of scanning in and resetting thirteen of Helmholtz's essays without the help of several assistants. I thank Jamal Mashlab, who did the actual scanning of the essays; Gena Bomberger and Peter Dahm, who assisted in the first round of correcting the scanned essays; Kent Krause, who assisted me in reading and correcting the penultimate draft of the corrected essays against the original English-language essays; and Kenneth Jensen, who is responsible for electronically reproducing the graphic illustrations in Helmholtz's essay "On the Physiological Causes of Harmony in Music." I thank, in addition, Joe Alderfer and Claudia Rex of the University of Chicago Press for their advice in technical matters, and my editor, Susan Abrams, for her interest in and encouragement of this project. Above all, I am deeply grateful to my wife, Jean Axelrad Cahan, for her careful review of my translations against Helmholtz's German-language texts.

David Cahan

Selected Further Readings

The following readings are intended only as an introduction to the literature on Helmholtz. The emphasis here is, for the most part, on the more recent, English-language literature that is most pertinent to the essays in the present edition. For an extensive bibliography of writings on Helmholtz, see the edited volume (1994) by David Cahan cited below.

Barnouw, Jeffrey. "Goethe and Helmholtz: Science and Sensation." Pp. 45–82 in Frederick Amrine, Francis J. Zucker, and Harvey Wheeler, eds. *Goethe and the Sciences: A Reappraisal.* Dordrecht: D. Reidel, 1987.

Cahan, David. *An Institute for an Empire: The Physikalisch-Technische Reichsanstalt 1871–1918.* Cambridge: Cambridge University Press, 1989.

_____, ed. *Letters of Hermann von Helmholtz to His Parents: The Medical Education of a German Scientist, 1837–1846.* Stuttgart: Franz Steiner Verlag, 1993.

_____, ed. *Hermann von Helmholtz and the Foundations of Nineteenth-Century Science.* Berkeley and Los Angeles: University of California Press, 1994.

Cohen, Robert S., and Yehuda Elkana, eds. *Hermann von Helmholtz. Epistemological Writings.* Trans. Malcom F. Lowe. Dordrecht: D. Reidel, 1977.

Elkana, Yehuda. "Helmholtz' 'Kraft': An Illustration of Concepts in Flux." *Historical Studies in the Physical Sciences* 2 (1970): 263–98.

Hatfield, Gary. *The Natural and the Normative: Theories of Spatial Perception from Kant to Helmholtz.* Cambridge, Mass.: MIT Press, 1990.

Heimann, Peter. "Helmholtz and Kant: The Metaphysical Foundations of *Über die Erhaltung der Kraft.*" *Studies in History and Philosophy of Science* 5 (1974): 205–38.

Holmes, Frederic L., and Kathryn M. Olesko. "The Images of Precision: Helmholtz and the Graphical Method in Physiology." Pp. 198–221 in M. Norton Wise, ed. *The Values of Precision.* Princeton, N.J.: Princeton University Press, 1995.

Kahl, Russell, ed. *Selected Writings of Hermann von Helmholtz.* Middletown, Conn.: Wesleyan University Press, 1971.

Kirsten, Christa, et al., eds. *Dokumente einer Freundschaft. Briefwechsel zwischen Hermann von Helmholtz und Emil du Bois-Reymond 1846–1894.* Berlin: Akademie Verlag, 1986.

Koenigsberger, Leo. *Hermann von Helmholtz*. 3 vols. Braunschweig: Friedrich Vieweg und Sohn, 1902–3.

_____. *Hermann von Helmholtz*. Trans. Frances A. Welby. Oxford: Clarendon Press, 1906; reprinted New York: Dover, 1965.

Kremer, Richard, ed. *Letters of Hermann von Helmholtz to His Wife, 1847–1859*. Stuttgart: Franz Steiner Verlag, 1990.

Krüger, Lorenz, ed. *Universalgenie Helmholtz: Rückblick nach 100 Jahren*. Berlin: Akademie Verlag, 1994.

Kuhn, Thomas. "Energy Conservation as an Example of Simultaneous Discovery." Reprinted in Kuhn's *The Essential Tension: Selected Studies in Scientific Tradition and Change*, pp. 66–104. Chicago: University of Chicago Press, 1977.

Lenoir, Timothy. *The Strategy of Life: Teleology and Mechanics in Nineteenth-Century German Biology*. Reprint. Chicago: University of Chicago Press, 1989.

_____. "Helmholtz and the Materialities of Communication." Pp. 184–207 in Albert Van Helden and Thomas L. Hankins, eds. *Instruments*. (= *Osiris* 9 [1994]). Chicago: University of Chicago Press, 1994.

Pastore, Nicholas. "Helmholtz's 'Popular Lectures on Vision.'" *Journal of the History of the Behavioral Sciences* 9 (1973): 190–202.

Turner, R. Steven. "Helmholtz, Hermann von." *Dictionary of Scientific Biography*, 6:241–53. Ed. Charles Coulston Gillispie. 16 vols. New York: Charles Scribner's Sons, 1970–80.

_____. "Hermann von Helmholtz and the Empiricist Vision." *Journal of the History of the Behavioral Sciences* 13 (1977): 48–58.

_____. *In the Eye's Mind: Vision and the Helmholtz-Hering Controversy*. Princeton, N.J.: Princeton University Press, 1994.

1

On Goethe's Scientific Researches

IT could not but be that Goethe, whose comprehensive genius was most strikingly apparent in that sober clearness with which he grasped and reproduced with lifelike freshness the realities of nature and human life in their minutest details, should, by those very qualities of his mind, be drawn towards the study of physical science. And in that department, he was not content with acquiring what others could teach him, but he soon attempted, as so original a mind was sure to do, to strike out an independent and a very characteristic line of thought. He directed his energies, not only to the descriptive, but also to the experimental sciences; the chief results being his botanical and osteological treatises on the one hand, and his theory of colour on the other. The first germs of these researches belong for the most part to the last decade of the eighteenth century, though some of them were not completed nor published till later. Since that time science has not only made great progress, but has widely extended its range. It has assumed in some respects an entirely new aspect, it has opened out new fields of research and undergone many changes in its theoretical views. I shall attempt in the following Lecture to sketch the relation of Goethe's researches to the present stand-point of science, and to bring out the guiding idea that is common to them all.

The peculiar character of the descriptive sciences—botany, zoology, anatomy, and the like—is a necessary result of the work imposed upon them. They undertake to collect and sift an enormous mass of facts, and, above all, to bring them into a logical order or system. Up to this point their work is only the dry task of a lexicographer; their system is nothing more than a muniment-room in which the accumulation of papers is so arranged that any one can find what he wants at any moment. The more intellectual part of their work and their real interest only begins when they attempt to feel after the scattered traces of law and order in the disjointed, heterogeneous mass, and out of it to construct for themselves an orderly system, accessible at a glance, in which every detail has its due place, and gains additional interest from its connection with the whole.

In such studies, both the organising capacity and the insight of our poet found a congenial sphere—the epoch was moreover propitious to him. He found ready to his hand a sufficient store of logically arranged materials in botany and comparative anatomy, copious and systematic enough to admit of a comprehensive view, and to indicate the way to some happy glimpse of an all-pervading law; while his contemporaries, if they made any efforts in this direction, wandered without a compass, or else they were so absorbed in the dry registration of facts, that they scarcely ventured to think of anything beyond. It was reserved for Goethe to introduce two ideas of infinite fruitfulness.

The first was the conception that the differences in the anatomy of different animals are to be looked upon as variations from a common phase or type, induced by differences of habit, locality, or food. The observation which led him to this fertile conception was by no means a striking one; it is to be found in a monograph on the intermaxillary bone, written as early as 1786. It was known that in most vertebrate animals (that is, mammalia, birds, amphibia, and fishes) the upper jaw consists of two bones, the upper jaw-bone and the intermaxillary bone. The former always contains in the mammalia the molar and the canine teeth, the latter the incisors. Man, who is distinguished from all other animals by the absence of the projecting snout, has, on the contrary, on each side only one bone, the upper jaw-bone, containing all the teeth. This being so, Goethe discovered in the *human* skull faint traces of the sutures, which in animals unite the upper and middle jaw-bones, and concluded from it that man had originally possessed an intermaxillary bone, which had subsequently coalesced with the upper jaw-bone. This obscure fact opened up to him a source of the most intense interest in the field of osteology, generally so much decried as the driest of studies. That details of structures should be the same in man and in animals when the parts continue to perform similar functions had involved nothing extraordinary. In fact, Camper had already attempted, on this principle, to trace similarities of structure even between man and fishes. But the persistence of this similarity, at least in a rudimentary form, even in a case when it evidently does not correspond to any of the requirements of the complete human structure, and consequently needs to be adapted to them by the coalescence of two parts originally separate, was what struck Goethe's far-seeing eye, and suggested to him a far more comprehensive view than had hitherto been taken. Further studies soon convinced him of the universality of his newly-discovered principle, so that in 1795 and 1796 he was able to define more clearly the idea that had struck him in 1786, and to commit it to writing in his 'Sketch of a General Introduction to Comparative Anatomy.' He there lays down with the utmost confidence and precision, that all differences in the structure of animals must be looked upon as variations of a singly primitive type, induced by the coalescence, the alteration, the increase, the diminution, or even the complete removal of single parts of the structure; the very principle, in fact, which has become the leading idea of comparative anatomy in its present stage. Nowhere has it been better or more clearly expressed than in Goethe's writings. Subsequent authorities have made but few essential alterations in his theory. The most important of these is, that we no longer undertake to construct a common type for the whole animal kingdom, but are content with one for each of Cuvier's great divisions. The industry of Goethe's successors has accumulated a well-sifted stock of facts, infinitely more copious than what he could command, and has followed up successfully into the minutest details what he could only indicate in a general way.

The second leading conception which science owes to Goethe enunciated the existence of an analogy between the different parts of one and the same organic being, similar to that which we have just pointed out as subsisting between corresponding parts of different species. In most organisms we see a great repetition of single parts. This is most striking in the vegetable kingdom; each plant has a great number of similar stem leaves, similar petals, similar stamens, and so on. According to Goethe's own account, the idea first occurred to him while looking at a fan-palm at Padua. He was struck by the immense variety of changes of form which the successively-developed stem-leaves exhibit, by the way in which the first simple root leaflets are replaced by a series of more and more divided leaves, till we come to the most complicated.

He afterwards succeeded in discovering the transformation of stem-leaves into sepals and petals, and of sepals and petals into stamens, nectaries, and ovaries, and thus he was led to the doctrine of the metamorphosis of plants, which he published in 1790. Just as the anterior extremity of vertebrate animals takes different forms, becoming in man and in apes an arm, in other animals a paw with claws, or a forefoot with a hoof, or a fin, or a wing, but always retains the same divisions, the same position, and the same connection with the trunk, so the leaf appears as a cotyledon, stem-leaf, sepal, petal, stamen, nectary, ovary, &c., all resembling each other to a certain extent in origin and composition, and even capable, under certain unusual conditions, of passing from one form into the other, as, for example, may be seen by any one who looks carefully at a full-blown rose, where some of the stamens are completely, some of them partially, changed into petals. This view of Goethe's, like the other, is now completely adopted into science, and enjoys the universal assent of botanists, though of course some details are still matters of controversy, as, for instance, whether the bud is a single leaf or a branch.

In the animal kingdom, the composition of an individual out of several similar parts is very striking in the great sub-kingdom of the articulata—for example, in insects and worms. The larva of an insect, or the caterpillar of a butterfly, consists of a number of perfectly similar segments; only the first and last of them differ, and that but slightly, from the others. After their transformation into perfect insects, they furnish clear and simple exemplifications of the view which Goethe had grasped in his doctrine of the metamorphosis of plants, the development, namely, of apparently very dissimilar forms from parts originally alike. The posterior segments retain their original simple form; those of the breast-plate are drawn closely together, and develop feet and wings; while those of the head develop jaws and feelers; so that in the perfect insect, the original segments are recognised only in the posterior part of the body. In the vertebrata, again, a repetition of similar parts is suggested by the vertebral column, but has ceased to be observable in the external form. A fortunate glance at a broken sheep's skull, which Goethe found by accident on the sand of the Lido at Venice, suggested to him that the skull itself consisted of a series of very much altered vertebrae. At first sight, no two things can be more unlike than the broad uniform cranial cavity of the mammalia, enclosed by smooth plates, and the narrow cylindrical tube of the spinal marrow, composed of short, massy, jagged bones. It was a bright idea to detect the transformation in the skull of a mammal; the similarity is more striking in the amphibia and fishes. It should be added that Goethe left this idea unpublished for a long time, apparently because he was not quite sure how it would be received. Meantime, in 1806, the same idea occurred to Oken, who introduced it to the scientific world, and afterwards disputed with Goethe the priority of discovery. In fact, Goethe had waited till 1817, when the opinion had begun to find adherents, and then declared that he had had it in his mind for thirty years. Up to the present day, the number and composition of the vertebrae of the skull are a subject of controversy, but the principle has maintained its ground.

Goethe's views, however, on the existence of a common type in the animal kingdom do not seem to have exercised any direct influence on the progress of science. The doctrine of the metamorphosis of plants was introduced into botany as his distinct and recognised property; but his views on osteology were at first disputed by anatomists, and only subsequently attracted attention when the science had, apparently on independent grounds, found its way to the same discovery. He himself complains that his first ideas of a common type had encountered nothing but contradiction and skepticism at the time when he was working them out in his own mind, and that even men of the freshest and most original intellect, like the two Von Humboldts, had listened to them with something like impatience. But it is almost a matter of course that in any natural or physical science, theoretical ideas attract the attention of its cultivators only when they are advanced in connection with the whole of the evidence on which they rest, and thus justify their title to recognition. Be that as it may, Goethe is entitled to the credit of having caught the first glimpse of the guiding ideas to which the sciences of botany and anatomy were tending, and by which their present form is determined.

But great as is the respect which Goethe has secured by his achievements in the descriptive natural sciences, the denunciation heaped by all physicists on his researches in their department, and especially on his 'theory of colour,' is at least as uncompromising. This is not the place to plunge into the controversy that raged on the subject, and so I shall only attempt to state clearly the points at issue, and to explain what principle was involved, and what is the latent significance of the dispute.

To this end it is of some importance to go back to the history of the origin of the theory, and to its simplest form, because at that stage of the controversy the points at issue are obvious, and admit of easy and distinct statement, unencumbered by disputes about the correctness of detached facts and complicated theories.

Goethe himself describes very gracefully, in the confession at the end of his 'Theory of Colour,' how he came to take up the subject. Finding himself unable to grasp the aesthetic principles involved in effects of colour, he resolved to resume the study of the physical theory, which he had been taught at the university, and to repeat for himself the experiments connected with it. With that view he borrowed a prism of Hofrath Büttner, of Jena, but was prevented by other occupations from carrying out his plan, and kept it by him for a long time unused. The owner of the prism, a very orderly man, after several times asking in vain, sent a messenger with instructions to bring it back directly. Goethe took it out of the case, and thought he would take one more peep through it. To make certain of seeing something, he turned it towards a long white wall, under the impression that as there was plenty of light there he could not fail to see a brilliant example of the resolution of light into different colours; a supposition, by the way, which shows how little Newton's theory of the phenomena was then present to his mind. Of course he was disappointed. On the white wall he saw no colours; they only appeared where it was bounded by darker objects. Accordingly he made the observation—which, it should be added, is fully accounted for by Newton's theory—that colour can only be seen through a prism where a dark object and a bright one have the same boundary. Struck by this observation, which was quite new to him, and convinced that it was irreconcilable with Newton's theory, he induced the owner of the prism to relent, and devoted himself to the question with the utmost zeal and interest. He prepared sheets of paper with black and white spaces, and studied the phenomena under every variety of condition, until he thought he had sufficiently proved his rules. He next attempted to explain his supposed discovery to a neighbour, who was a physicist, and was disagreeably surprised to be assured by him that the experiments were well known, and fully accounted for in Newton's theory. Every other natural philosopher whom he consulted told him exactly the same, including even the brilliant Lichtenberg, who he tried for a long time to convert, but in vain. He studied Newton's writings, and fancied he had found some fallacies in them which accounted for the error. Unable to convince any of his acquaintances, he at last resolved to appear before the bar of public opinion, and in 1791 and 1792 published the first and second parts of his 'Contributions to Physical Optics.'

In that work he describes the appearances presented by white discs on a black ground, black discs on a white ground, and coloured discs on a black or white ground, when examined through a prism. As to the results of the experiments there is no dispute whatever between him and the physicists. He describes the phenomena he saw with great truth to nature; the style is lively, and the arrangement such as to make a conspectus of them easy and inviting; in short, in this as in all other cases where facts are to be described, he proves himself a master. At the same time he expresses his conviction that the facts he has adduced are calculated to refute Newton's theory. There are two points especially which he considers fatal to it: first, that the centre of a broad white surface remains white when seen through a prism; and secondly, that even a black streak on a white ground can be entirely decomposed into colours.

Newton's theory is based on the hypothesis that there exists light of different kinds, distinguished from one another by the sensation of colour which they produce in the eye. Thus there is red, orange, yellow, green, blue, and violet light, and light of all intermediate colours. Different kinds of light, or differently coloured lights, produce, when mixed, derived colours, which to a certain extent resemble the original colours from which they are derived; to a certain extent form new tints. White is a mixture of all the before-named colours in certain definite proportions. But the primitive colours can always be reproduced by analysis from derived colours, or from white, while themselves incapable of analysis or change. The cause of the colours of transparent and opaque bodies is, that when white light falls upon them they destroy some of its constituents and send to the eye other constituents, but no longer mixed in the right proportions to produce white light. Thus a piece of red glass looks red, because it transmits only red rays. Consequently all colour is derived solely from a change in the proportions in which light is mixed, and is, therefore, a property of light, not of the coloured bodies, which only furnish an occasion for its manifestation.

A prism refracts transmitted light; that is to say, deflects it so that it makes a certain angle with its original direction; the rays of simple light of different colours have, according to Newton, different refrangibilities, and therefore, after refraction in the prism, pursue different courses and separate from each other. Accordingly a luminous point of infinitely small dimensions appears, when seen through the prism, to be first displaced, and secondly, extended into a coloured line, the so-called prismatic spectrum, which shows what are called the primary colours in the order above-named. If, however, you look at a broader luminous surface, the spectra of the points near the middle are superposed, as may be seen from a simple geometrical investigation, in such proportions as to give white light, except at the edges, where certain of the colours are free. This white surface appears displaced, as the luminous point did; but instead of being coloured throughout, it has on one side a margin of blue and violet, on the other a margin of red and yellow. A black patch between two bright surfaces may be entirely covered by their coloured edges; and when these spectra meet in the middle, the red of the one and the violet of the other combine to form purple. Thus the colours into which, at first sight, it seems as if the black were analysed are in reality due, not to the black strip, but to the white on each side of it.

It is evident that at the first moment Goethe did not recollect Newton's theory well enough to be able to find out the physical explanation of the facts I have just glanced at. It was afterwards laid before him again and again, and that in a thoroughly intelligible form, for he speaks about it several times in terms that show he understood it quite correctly. But he is still so dissatisfied with it, that he persists in his assertion that the facts just cited are of a nature to convince any one who observes them of the absolute incorrectness of Newton's theory. Neither here nor in his later controversial writings does he ever clearly state in what he conceives the insufficiency of the explanation to consist. He merely repeats again and again that it is quite absurd. And yet I cannot see how any one, whatever his views about colour, can deny that the theory is perfectly consistent with itself; and that if the hypothesis from which it starts be granted, it explains the observed facts completely and even simply. Newton himself mentions these spurious spectra in several passages of his optical works, without going into any special elucidation of the point, considering, of course, that the explanation follows at once from his hypothesis. And he seems to have had good reason to think so; for Goethe no sooner began to call attention of his scientific friends to the phenomena, than all with one accord, as he himself tells us, met his difficulties with this explanation from Newton's principles, which, though not actually in his writings, instantly suggested itself to every one who knew them.

A reader who tries to realise attentively and thoroughly every step in this part of the controversy is apt to experience at this point an uncomfortable, almost a painful feeling to see a man of extraordinary abilities persistently declaring that there is an obvious absurdity lurking in a few inferences apparently quite clear and simple. He searches and searches, and at last unable, with all his efforts, to find any such absurdity, or even the appearance of it, he gets into a state of mind in which his own ideas are, so to speak, crystallised. But it is just this obvious, flat contradiction that makes Goethe's point of view in 1792 so interesting and so important. At this point he has not as yet developed any theory of his own; there is nothing under discussion but a few easily-grasped facts, as to the correctness of which both parties are agreed, and yet both hold distinctly opposite views; neither of them even understands what his opponent is driving at. On the one side are a number of physicists, who, by a long series of the ablest investigations, the most elaborate calculations, and the most ingenious inventions, have brought optics to such perfection, that it, and it alone, among the physical sciences, was beginning almost to rival astronomy in accuracy. Some of them have made the phenomena the subject of direct investigation; all of them, thanks to the accuracy with which it is possible to calculate beforehand the result of every variety in the construction and combination of instruments, have had the opportunity of putting the inferences deduced from Newton's views to the test of experiment, and all, without exception, agree in accepting them. On the other side is a man whose remarkable mental endowments, and whose singular talent for seeing through whatever obscures reality, we have had occasion to recognise, not only in poetry, but also in the descriptive parts of the natural sciences; and this man assures us with the utmost zeal that the physicists are wrong: he is so convinced of the correctness of his own view, that he cannot explain the contradiction except by assuming narrowness or malice on their part, and finally declares that he cannot help looking upon his own achievement in the theory of colour as far more valuable than anything he has accomplished in poetry.[1]

So flat a contradiction leads us to suspect that there must be behind some deeper antagonism of principle, some difference of organisation between his mind and theirs, to prevent them from understanding each other. I will try to indicate in the following pages what I conceive to be the grounds of this antagonism.

Goethe, though he exercised his powers in many spheres of intellectual activity, is nevertheless, *par excellence*, a poet. Now in poetry, as in every other art, the essential thing is to make the material of the art, be it words, or music, or colour, the direct vehicle of an idea. In a perfect work of art, the idea must be present and dominate the whole, almost unknown to the poet himself, not as the result of a long intellectual process, but as inspired by a direct intuition of the inner eye, or by an outburst of excited feeling.

An idea thus embodied in a work of art, and dressed in the garb of reality, does indeed make a vivid impression by appealing directly to the senses, but loses, of course, that universality and that intelligibility which

---

1. See Eckermann's *Conversations*.

it would have had if presented in the form of an abstract notion. The poet, feeling how the charm of his works is involved in an intellectual process of this type, seeks to apply it to other materials. Instead of trying to arrange the phenomena of nature under definite conceptions, independent of intuition, he sits down to contemplate them as he would a work of art, complete in itself, and certain to yield up its central idea, sooner or later, to a sufficiently susceptible student. Accordingly, when he sees the skull on the Lido, which suggests to him the vertebral theory of the cranium, he remarks that it serves to revive his old belief, already confirmed by experience, that Nature has no secrets from the attentive observer. So again in his first conversation with Schiller on the 'Metamorphosis of Plants.' To Schiller, as a follower to Kant, the idea is the goal, ever to be sought, but ever unattainable, and therefore never to be exhibited as realised in a phenomenon. Goethe, on the other hand, as a genuine poet, conceives that he finds in the phenomenon the direct expression of the idea. He himself tells us that nothing brought out more sharply the separation between himself and Schiller. This, too, is the secret of his affinity with the natural philosophy of Schelling and Hegel, which likewise proceeds from the assumption that Nature shows us by direct intuition the several steps by which a conception is developed. Hence, too, the ardour with which Hegel and his school defended Goethe's scientific views. Moreover this view of Nature accounts for the war which Goethe continued to wage against complicated experimental researches. Just as a genuine work of art cannot bear retouching by a strange hand, so he would have us believe Nature resists the interference of the experimenter who tortures her and disturbs her; and in revenge, misleads the impertinent kill-joy by a distorted image of herself.

Accordingly, in his attack upon Newton he often sneers at spectra, tortured through a number of narrow slits and glasses, and commends the experiments that can be made in the open air under a bright sun, not merely as particularly easy and particularly enchanting, but also as particularly convincing! The poetic turn of mind is very marked even in his morphological researches. If we only examine what has really been accomplished by the help of the ideas which he contributed to science, we shall be struck by the very singular relation which they bear to it. No one will refuse to be convinced if you lay before him the series of transformations by which a leaf passes into a stamen, an arm into a fin or a wing, a vertebra into the occipital bone. The idea that all the parts of a flower are modified leaves, reveals a connecting law, which surprises us into acquiescence. But now try and define the leaf-like organ, determine its essential characteristics, so as to include all the forms that we have named. You will find yourself in a difficulty, for all distinctive marks vanish, and you have nothing left, except that a leaf in the wider sense of the term is a lateral appendage of the axis of a plant. Try then to express the proposition 'the parts of the flower are modified leaves' in the language of scientific definition, and it reads, 'the parts of the flower are lateral appendages of the axis.' To see this does not require a Goethe. So again it has been objected, and not unjustly, to the vertebral theory, that it must extend the notion of a vertebra so much that nothing is left but the bare fact—a vertebra is a bone. We are equally perplexed if we try to express in clear scientific language what we mean by saying that such and such part of one animal corresponds to such and such a part of another. We do not mean that their physiological use is the same, for the same piece which in a bird serves as the lower jaw, becomes in mammals a tiny tympanal bone. Nor would the shape, the position, or the connection of the part in question with other parts, serve to identify it in all cases. But yet it has been found possible in most cases, by following the intermediate steps, to determine with tolerable certainty which parts correspond to each other. Goethe himself said this very clearly: he says, in speaking of the vertebral theory of the skull, 'Such an *aperçu*, such an intuition, conception, representation, notion, idea, or whatever you choose to call it, always retains something esoteric and indefinable, struggle as you will against it; as a general principle, it may be enunciated, but cannot be proved; in detail it may be exhibited, but can never be put in a cut and dry form.' And so, or nearly so, the problem stands to this day. The difference may be brought out still more clearly if we consider how physiology, which investigates the relations of vital processes as cause and effect, would have to treat this idea of a common type of animal structure. The science might ask, Is it, on the one hand, a correct view, that during the geological periods that have passed over the earth, one species has been developed from another, so that, for example, the breast-fin of the fish has gradually changed into an arm or a wing? Or again, shall we say that the different species of animals were created equally perfect—that the points of resemblance between them are to be ascribed to the fact, that in all vertebrate animals the first steps in development from the egg can only be effected by Nature in one way, almost identical in all cases, and that the later analogies of structure are determined by these features, common to all embryos? Probably the majority of observers incline to the latter view,[2] for the agreement between the embryos of different vertebrate animals, in the earlier stages, is very striking. Thus even young mammals have occasionally rudimentary gills on the side of the neck, like fishes. It seems, in fact, that what are in the mature animals corresponding parts, originate in the same way during the process of development, so the scientific men have lately begun to make use of embryology as a sort of check on the theoretical views of comparative anatomy. It is evident that by the application of the physiological views just suggested, the idea of a common

2. This was written before the appearance of Darwin's *Origin of Species*.

type would acquire definiteness and meaning as a distinct scientific conception. Goethe did much: he saw by a happy intuition that there was a law, and he followed up the indications of it with great shrewdness. But what law it was, he did not see; nor did he even try to find it out. That was not in his line. Moreover, even in the present condition of science, a definite view on the question is impossible; the very form in which it should be proposed is scarcely yet settled. And therefore we readily admit that in this department Goethe did all that was possible at the time when he lived. I said just now that he treated nature like a work of art. In his studies on morphology, he reminds one of a spectator at a play, with strong artistic sympathies. His delicate instinct makes him feel how all the details fall into their places, and work harmoniously together, and how some common purpose governs the whole; and yet, while this exquisite order and symmetry give him intense pleasure, he cannot formulate the dominant idea. That is reserved for the scientific critic of the drama, while the artistic spectator feels perhaps, as Goethe did in the presence of natural phenomena, an antipathy to such dissection, fearing, though without reason, that his pleasure may be spoilt by it.

Goethe's point of view in the Theory of Colour is much the same. We have seen that he rebels against the physical theory just at the point where it gives complete and consistent explanations from principles once accepted. Evidently it is not the insufficiency of the theory to explain individual cases that is a stumbling-block to him. He takes offence at the assumption made for the sake of explaining the phenomena, which seem to him so absurd, that he looks upon the interpretation as no interpretation at all. Above all, the idea that white light could be composed of coloured light seems to have been quite inconceivable to him; at the very beginning of the controversy, he rails at the disgusting Newtonian white of the natural philosophers, an expression which seems to show that this was the assumption that most annoyed him.

Again, in his later attacks on Newton, which were not published till after his Theory of Colour was completed, he rather strives to show that Newton's facts might be explained on his own hypothesis, and that therefore Newton's hypothesis was not fully proved, than attempts to prove that hypothesis inconsistent with itself or with the facts. Nay, he seems to consider the obviousness of his own hypothesis so overwhelming, that it need only be brought forward to upset Newton's entirely. There are only a few passages where he disputes the experiments described by Newton. Some of them, apparently, he could not succeed in refuting, because the result is not equally easy to observe in all positions of the lenses used, and because he was unacquainted with the geometrical relations by which the most favourable positions of them are determined. In other experiments on the separation of simple coloured lights by means of prisms alone, Goethe's objections are not quite groundless, inasmuch as the isolation of single colours cannot by this means be so effectually carried out, that after refraction through another prism there are no traces of other tints at the edges. A complete isolation of light of one colour can only be effected by very carefully arranged apparatus, consisting of combined prisms and lenses, a set of experiments which Goethe postponed to a supplement, and finally left unnoticed. When he complains of the complication of these contrivances, we need only think of the laborious and roundabout methods which chemists must often adopt to obtain certain elementary bodies in a pure form; and we need not be surprised to find that it is impossible to solve a similar problem in the case of light in the open air in a garden, and with a single prism in one's hand.[3] Goethe must, consistently with his theory, deny *in toto* the possibility of isolating pure light of one colour. Whether he ever experimented with the proper apparatus to solve the problem remains doubtful, as the supplement in which he promised to detail these experiments was never published.

To give some idea of the passionate way in which Goethe, usually so temperate and even courtier-like, attacks Newton, I quote from a few pages of the controversial part of his work the following expressions, which he applies to the propositions of this consummate thinker in physical and astronomical science—'incredibly impudent;' 'mere twaddle;' 'ludicrous explanation;' 'admirable for school-children in a go-cart;' 'but I see nothing will do but lying, and plenty of it.'[4]

Thus, in the theory of colour, Goethe remains faithful to his principle, that Nature must reveal her secrets of her own free will; that she is but the transparent representation of the ideal world. Accordingly, he demands as a preliminary to the investigation of physical phenomenon that the observed facts shall be so arranged that one explains to the other, and that thus we may attain an insight into their connection without ever having to trust to any thing but our senses. This demand of his looks most attractive, but is essentially wrong in principle. For a natural phenomenon is not considered in physical science to be fully explained until you have traced it back to the ultimate forces which are concerned in its production and its maintenance. Now, as we can never become cognizant of forces *qua* forces, but only of their effects, we are compelled in every explanation of natural phenomena to leave the sphere of sense, and to pass to things which are not objects of

---

3. I venture to add that I am acquainted with the impossibility of decomposing or changing simple coloured light, the two principles which form the basis of Newton's theory, not merely by hearsay, but from actual observation, having been under the necessity in one of my own researches of obtaining light of one colour in a state of the greatest possible purity. (See Poggendorff's *Annalen*, vol. lxxxvi. p. 501, on Sir D. Brewster's *New Analysis of Sunlight*.)

4. Something parallel to this extraordinary proceeding of Goethe's may be found in Hobbes's attack on Wallis.—TR.

sense, and are defined only by abstract conceptions. When we find a stove warm, and then observe that a fire is burning in it, we say, though somewhat inaccurately, that the former sensation is explained by the latter. But in reality this is equivalent to saying, we are always accustomed to find heat where fire is burning; now, a fire is burning in the stove, therefore we shall find heat there. Accordingly, we bring our single fact under a more general, better known fact, rest satisfied with it, and call it falsely an explanation. Evidently, however, the generality of the observation does not necessarily imply an insight into causes; such an insight is only obtained when we can make out what forces are at work in the fire, and how the effects depend upon them.

But this step into the region of abstract conceptions, which must necessarily be taken, if we wish to penetrate to the causes of phenomena, scares the poet away. In writing a poem he has been accustomed to look, as it were, right into the subject, and to reproduce his intuition without formulating any of the steps that led him to it. And his success is proportionate to the vividness of the intuition. Such is the fashion in which he would have Nature attacked. But the natural philosopher insists on transporting him into a world of invisible atoms and movements, of attractive and repulsive forces, whose intricate actions and reactions, though governed by strict laws, can scarcely be taken in at a glance. To him the impressions of sense are not an irrefragable authority; he examines what claim they have to be trusted; he asks whether things which they pronounce alike are really alike, and whether things which they pronounce different are really different; and often finds that he must answer, no! The result of such examination, as at present understood, is that the organs of sense do indeed give us information about external effects produced on them, but convey those effects to our consciousness in a totally different form, so that the character of a sensuous perception depends not so much on the properties of the object perceived as on those of the organ by which we receive the information. All that the optic nerve conveys to us, it conveys under the form of a sensation of light, whether it be the rays of the sun, or a blow in the eye, or an electric current passing through it. Again, the auditory nerve translates everything into phenomena of sound, the nerves of the skin into sensations of temperature or touch. The same electric current whose existence is indicated by the optic nerve as a flash of light, or by the organ of taste as an acid flavour, excites in the nerves of the skin the sensation of burning. The same ray of sunshine, which is called light when it falls on the eye, we call heat when it falls on the skin. But on the other hand, in spite of their different effects upon our organisation, the daylight which enters through our windows, and the heat radiated by an iron stove, do not in reality differ more or less from each other than the red and blue constituents of light. In fact, just as in the Undulatory Theory, the red rays are distinguished from the blue rays only by their longer period of vibration, and their smaller refrangibility, so the dark heat rays of the stove have a still longer period and still smaller refrangibility than the red rays of light, but are in every other respect exactly similar to them. All these rays, whether luminous or non-luminous, have heating properties, but only a certain number of them, to which for that reason we give the name of light, can penetrate through the transparent part of the eye to the optic nerve, and excite a sensation of light. Perhaps the relation between our senses and the external world may be best enunciated as follows: our sensations are for us only *symbols* of the objects of the external world, and correspond to them only in some such way as written characters or articulate words to the things they denote. They give us, it is true, information respecting the properties of things without us, but no better information than we give a blind man about colour by verbal descriptions.

We see that science has arrived at an estimate of the senses very different from that which was present to the poet's mind. And Newton's assertion that white was composed of all the colours of the spectrum was the first germ of the scientific view which has subsequently been developed. For at that time there were none of those galvanic observations which paved the way to a knowledge of the functions of the nerves in the production of sensations. Natural philosophers asserted that white, to the eye the simplest and purest of all our sensations of colour, was compounded of less pure and complex materials. It seems to have flashed upon the poet's mind that all his principles were unsettled by the results of this assertion, and that is why the hypothesis seems to him so unthinkable, so ineffably absurd. We must look upon his theory of colour as a forlorn hope, as a desperate attempt to rescue from the attacks of science the belief in the direct truth of our sensations. And this will account for the enthusiasm with which he strives to elaborate and to defend his theory, for the passionate irritability with which he attacks his opponent, for the overweening importance which he attaches to these researches in comparison with his other achievements, and for his inaccessibility to conviction or compromise.

If we now turn to Goethe's own theories on the subject, we must, on the grounds above stated, expect to find that he cannot, without being untrue to his own principle, give us anything deserving to be called a scientific explanation of the phenomena, and that is exactly what happens. He starts with the proposition that all colours are darker than white, that they have something of shade in them (on the physical theory, white compounded of all colours must necessarily be brighter than any of its constituents). The direct mixture of dark and light, of black and white, gives grey; the colours must therefore owe their existence to some form of the co-operation of light and shade. Goethe imagines he has discovered it in the phenomena presented by slightly opaque or hazy media. Such media usually look blue when the light falls on them, and they are seen in front of a dark object, but yellow when a bright object is looked at through them. Thus in the day time the air looks blue against the dark background of the sky, and the sun, when viewed, as is the case at sunset, through a thick and hazy stratum of air, appears yellow. The physical explanation of this phenomenon, which, however, is not exhibited by all such media, as, for instance, by plates of unpolished glass, would lead us too far from the subject. According to Goethe, the semi-opaque medium imparts to the light something corporeal, something of the nature of shade, such as is requisite, he would say, for the formation of colour. This conception alone is enough to perplex anyone who looks upon it as a physical explanation. Does he mean to say that material particles mingle with the light and fly away with it? But this is Goethe's fundamental experiment, this is the typical phenomenon under which he tries to reduce all the phenomena of colour, especially those connected with the prismatic spectrum. He looks upon all transparent bodies as slightly hazy, and assumes that the prism imparts to the image which it shows to an observer something of its own opacity. Here, again, it is hard to get a definite conception of what is meant. Goethe seems to have thought that a prism never gives perfectly defined images, but only indistinct, half-obliterated ones, for he puts them all in the same class with the double images which are exhibited by parallel plates of glass and by Iceland spar. The images formed by a prism are, it is true, indistinct in compound light, but they are perfectly defined when simple light is used. If you examine, he says, a bright surface on a dark ground through a prism, the image is displaced and blurred by the prism. The anterior edge is pushed forward over the dark background, and consequently a hazy light on a dark ground appears blue, while the other edge is covered by the image of the black surface which comes after it, and, consequently, being a light image behind a hazy dark colour, appears yellowish-red. But why the anterior edge appears in front of the ground, the posterior edge behind it, and not *vice versa*, he does not explain. Let us analyse this explanation, and try to grasp clearly the conception of an optical image. When I see a bright object reflected in a mirror, the reason is that the light which proceeds from it is thrown back exactly as if it came from an object of the same kind behind the mirror. The eye of the observer receives the impression accordingly, and therefore he imagines he really sees the object. Everyone knows there is nothing real behind the mirror to correspond to the image—that no light can penetrate thither, but that what is called the image is simple a geometrical point, in which the reflected rays, if produced backwards, would intersect. And, accordingly, no one expects the image to produce any real effect behind the mirror. In the same way the prism shows us images of objects which occupy a different position from the objects themselves; that is to say, the light which an object sends to the prism is refracted by it, so that it appears to come from an object lying to one side, called the image. This image, again, is not real; it is, as in the case of reflection, the geometrical point in which the refracted rays intersect when produced backwards. And yet, according to Goethe, this image is to produce real effects by its displacement; the displaced patch of light makes, he says, the dark space behind it appear blue, just as an imperfectly transparent body would, and so again the displaced dark patch makes the bright space behind appear reddish-yellow. That Goethe really treats the image as an actual object in the place it appears to occupy is obvious enough, especially as he is compelled to assume, in the course of his explanation, that the blue and red edges of the bright space are respectively before and behind the dark image which, like it, is displaced by the prism. He does, in fact, remain loyal to the appearance presented to the senses, and treats a geometrical locus as if it were a material object. Again, he does not scruple at one time to make red and blue destroy each other, as, for example, in the blue edge of a red surface seen through the prism, and at another to construct out of them a beautiful purple, as when the blue and red edges of two neighbouring white surfaces meet in a black ground. And when he comes to Newton's more complicated experiments, he is driven to still more marvelous expedients. As long as you treat his explanations as a pictorial way of representing the physical processes, you may acquiesce in them, and even frequently find them vivid and characteristic, but as physical elucidations of the phenomena they are absolutely irrational.

In conclusion, it must be obvious to everyone that the theoretical part of the Theory of Colour is not natural philosophy at all; at the same time we can, to a certain extent, see that the poet wanted to introduce a totally different method into the study of Nature, and more or less understand how he came to do so. Poetry is concerned solely with the 'beautiful show' which makes it possible to contemplate the ideal; how that show is produced is a matter of indifference. Even Nature is, in the poet's eyes, but the sensible expression of the spiritual. The natural philosopher, on the other hand, tries to discover the levers, the cords, and the pulleys, which work behind the scenes, and shift them. Of course the sight of the machinery spoils the beautiful show, and therefore the poet would gladly talk it out of existence, and ignoring cords and pulleys as the chimeras of a pedant's brain, he would have us believe that the scenes shift themselves, or are governed by the idea of the drama. And it is just characteristic of Goethe, that he, and he alone among poets, must needs break a lance with natural philosophers. Other poets are either so entirely carried away by the fire of their enthusiasm that they do not trouble themselves about the disturbing influences of the outer world, or else they rejoice in the triumphs of mind over matter, even on that unpropitious battlefield. But Goethe, whom no intensity of subjective feeling could blind to the realities around him, cannot rest satisfied until he has stamped reality itself with the image and superscription of poetry. This constitutes the peculiar beauty of his poetry, and at the same time fully accounts for his resolute hostility to the machinery that every moment threatens to disturb his poetic repose, and for his determination to attack the enemy in his own camp.

But we cannot triumph over the machinery of matter by ignoring it; we can triumph over it only by subordinating it to the aims of our moral intelligence. We must familiarise ourselves with its levers and pulleys, fatal though it be to poetic contemplation, in order to be able to govern them after our own will, and therein lies the complete justification of physical investigation, and its vast importance for the advance of human civilisation.

From what I have said it will be apparent that Goethe did follow the same line of thought in all his contributions to science, but that the problems he encountered were of diametrically opposite characters. And, perhaps, when it is understood how the self-same characteristic of his intellect, which in one branch of science won for him immortal renown, entailed upon him egregious failure in the other, it will tend to dissipate, in the minds of many worshippers of the great poet, a lingering prejudice against natural philosophers, whom they suspect of being blinded by narrow professional pride to the loftiest inspirations of genius.

A Lecture Delivered before the German Society of Königsberg, in the Spring of 1853.

2

On the Interaction of the Natural Forces

A NEW conquest of very general interest has been recently made by natural philosophy. In the following pages I will endeavour to give an idea of the nature of this conquest. It has reference to a new and universal natural law, which rules the action of natural forces in their mutual relations towards each other, and is as influential on our theoretic views of natural processes as it is important in their technical applications.

Among the practical arts which owe their progress to the development of the natural sciences, from the conclusion of the middle ages downwards, practical mechanics, aided by the mathematical science which bears the same name, was one of the most prominent. The character of the art was, at the time referred to, naturally very different from its present one. Surprised and stimulated by its own success, it thought no problem beyond its power, and immediately attacked some of the most difficult and complicated. Thus it was attempted to build automaton figures which should perform the functions of men and animals. The marvel of the last century was Vaucanson's duck, which fed and digested its food; the flute-player of the same artist, which moved all its fingers correctly; the writing-boy of the elder, and the pianoforte-player of the younger Droz; which latter, when performing, followed its hands with its eyes, and at the conclusion of the piece bowed courteously to the audience. That men like those mentioned, whose talent might bear comparison with the most inventive heads of the present age, should spend so much time in the construction of these figures which we at present regard as the merest trifles, would be incomprehensible, if they had not hoped in solemn earnest to solve a great problem. The writing-boy of the elder Droz was publicly exhibited in Germany some years ago. Its wheelwork is so complicated, that no ordinary head would be sufficient to decipher its manner of action. When, however, we are informed that this boy and its constructor, being suspected of the black art, lay for a time in the Spanish Inquisition, and with difficulty obtained their freedom, we may infer that in those days even such a toy appeared great enough to excite doubts as to its natural origin. And though these artists may not have hoped to breathe into the creature of their ingenuity a soul gifted with moral completeness, still there were many who would be willing to dispense with the moral qualities of their servants, if at the same time their immoral qualities could also be got rid of; and to accept, instead of the mutability of flesh and bones, services which should combine the regularity of a machine with the durability of brass and steel.

The object, therefore, which the inventive genius of the past century placed before it with the fullest earnestness, and not as a piece of amusement merely, was boldly chosen, and was followed up with an expenditure of sagacity which has contributed not a little to enrich the mechanical experience which a later time knew how to take advantage of. We no longer seek to build machines which shall fulfill the thousand services required of *one* man, but desire, on the contrary, that a machine shall perform *one* service, and shall occupy in doing it the place of a thousand men.

From these efforts to imitate living creatures, another idea, also by a misunderstanding, seems to have developed itself, and which, as it were, formed the new philosopher's stone of the seventeenth and eighteenth centuries. It was now the endeavour to construct a perpetual motion. Under this term was understood a machine, which, without being wound up, without consuming in the working of it falling water, wind, or any other natural force, should still continue in motion, the motive power being perpetually supplied by the machine itself. Beasts and human beings seemed to correspond to the idea of such an apparatus, for they moved themselves energetically and incessantly as long as they lived, and were never wound up; nobody set them in motion. A connexion between the supply of nourishment and the development of force did not make itself apparent. The nourishment seemed only necessary to grease, as it were, the wheelwork of the animal machine, to replace what was used up, and to renew the old. The development of force out of itself seemed to be the essential peculiarity, the real quintessence of organic life. If, therefore, men were to be constructed, a perpetual motion must first be found.

Another hope also seemed to take up incidentally the second place, which in our wiser age would certainly have claimed the first rank in the thoughts of men. The perpetual motion was to produce work inexhaustibly without corresponding consumption, that is to say, out of nothing. Work, however, is money. Here, therefore, the great practical problem which the cunning heads of all centuries have followed in the most diverse ways, namely, to fabricate money out of nothing, invited solution. The similarity with the philosopher's stone sought by the ancient chemists was complete. That also was thought to contain the quintessence of organic life, and to be capable of producing gold.

The spur which drove men to inquiry was sharp, and the talent of some of the seekers must not be estimated as small. The nature of the problem was quite calculated to entice poring brains, to lead them round a circle for years, deceiving ever with new expectations which vanished upon nearer approach, and finally reducing these dupes of hope to open insanity. The phantom could not be grasped. It would be impossible to give a history of these efforts, as the clearer heads, among whom the elder Droz must be ranked, convinced themselves of the futility of their experiments, and were naturally not inclined to speak much about them. Bewildered intellects, however, proclaimed often enough that they had discovered the grand secret; and as the incorrectness of their proceedings was always speedily manifest, the matter fell into bad repute, and the opinion strengthened itself more and more that the problem was not capable of solution; one difficulty after another was brought under the dominion of mathematical mechanics, and finally a point was reached where it could be proved, that at least by the use of pure mechanical forces no perpetual motion could be generated.

We have here arrived at the idea of the driving force or power of a machine, and shall have much to do with it in future. I must therefore give an explanation of it. The idea of work is evidently transferred to machines by comparing their performances with those of men and animals, to replace which they were applied. We still reckon the work of steam-engines according to horse-power. The value of manual labour is determined partly by the force which is expended in it (a strong labourer is valued more highly than a weak one), partly, however, by the skill which is brought into action. Skilled workmen are not to be had in any quantity at a moment's notice; they must have both talent and instruction, their education requires both time and trouble. A machine, on the contrary, which executes work skilfully, can always be multiplied to any extent; hence its skill has not the high value of human skill in domains where the latter cannot be supplied by machines. Thus the idea of the quantity of work in the case of machines has been limited to the consideration of the expenditure of force; this was the more important, as indeed most machines are constructed for the express purpose of exceeding, by the magnitude of their effects, the powers of men and animals. Hence, in a mechanical sense, the idea of work has become identical with that of the expenditure of force, and in this way I will apply it in the following pages.

How, then, can we measure this expenditure, and compare it in the case of different machines?

I must here conduct you a portion of the way—as short a portion as possible—over the uninviting field of mathematico-mechanical ideas, in order to bring you to a point of view from which a more rewarding prospect will open. And though the example which I will here choose, namely, that of a water-mill with iron hammer, appears to be tolerably romantic, still, alas! I must leave the dark forest valley, the foaming brook, the spark-emitting anvil, and the black Cyclops wholly out of sight, and beg a moment's attention for the less poetic side of the question, namely, the machinery. This is driven by a water-wheel, which in its turn is set in motion by the falling water. The axle of the water-wheel has at certain places small projections, thumbs, which, during the rotation, lift the heavy hammer and permit it to fall again. The falling hammer belabours the mass of metal, which is introduced beneath it. The work therefore done by the machine consists, in this case, in the lifting of the hammer, to do which the gravity of the latter must be overcome. The expenditure of force will in the first place, other circumstances being equal, be proportional to the weight of the hammer; it will, for example, be double when the weight of the hammer is doubled. But the action of the hammer depends not upon its weight alone, but also upon the height from which it falls. If it falls through two feet, it will produce a greater effect than if it falls through only one foot. It is, however, clear that if the machine, with a certain expenditure of force, lifts the hammer a foot in height, the same amount of force must be expended to raise it a second foot in height. The work is therefore not only doubled when the weight of the hammer is increased twofold, but also when the space through which it falls is doubled. From this it is easy to see that the work must be measured by the product of the weight into the space through which it ascends. And in this way, indeed, we measure in mechanics. The unit of work is a foot-pound, that is, a pound weight raised to the height of one foot.

While the work in this case consists in the raising of the heavy hammer-head, the driving force which sets the latter in motion is generated by falling water. It is not necessary that the water should fall vertically, it can also flow in a moderately inclined bed; but it must always, where it has water-mills to set in motion, move from a higher to a lower position. Experiment and theory concur in teaching, that when a hammer of a hundredweight is to be raised one foot, to accomplish this at least a hundredweight of water must fall through the space of one foot; or what is equivalent to this, two hundredweight must fall half a foot, or four hundredweight a quarter of a foot, &c. In short, if we multiply the weight of the falling water by the height through which it falls, and regard, as before, the product as the measure of the work, then the work performed by the machine in raising the hammer can, in the most favourable case, be only equal to the number of foot-pounds of water which have fallen in the same time. In practice, indeed, this ratio is by no means attained: a great portion of the work of the falling water escapes unused, inasmuch as part of the force is willingly sacrificed for the sake of obtaining greater speed.

I will further remark, that this relation remains unchanged whether the hammer is driven immediately by the axle of the wheel, or whether—by the intervention of wheelwork, endless screws, pulleys, ropes—the motion is transferred to the hammer. We may, indeed, by such arrangements succeed in raising a hammer of ten hundredweight, when by the first simple arrangement the elevation of a hammer of one hundredweight might alone be possible; but either this heavier hammer is raised to only one-tenth of the height, or tenfold the time is required to raise it to the same height; so that, however we may alter, by the interposition of machinery, the intensity of the acting force, still in a certain time, during which the mill-stream furnishes us with a definite quantity of water, a certain definite quantity of work, and no more, can be performed.

Our machinery, therefore, has in the first place done nothing more than make use of the gravity of the falling water in order to overpower the gravity of the hammer, and to raise the latter. When it has lifted the hammer to the necessary height, it again liberates it, and the hammer falls upon the metal mass which is pushed beneath it. But why does the falling hammer here exercise a greater force than when it is permitted simply to press with its own weight on the mass of metal? Why is its power greater as the height from which it falls is increased, and the greater therefore the velocity of its fall? We find, in fact, that the work performed by the hammer is determined by its velocity. In other cases, also, the velocity of moving masses is a means of producing great effects. I only remind you of the destructive effects of musket-bullets, which in a state of rest are the most harmless things in the world. I remind you of the windmill, which derives its force from the moving air. It may appear surprising that motion, which we are accustomed to regard as a non-essential and transitory endowment of bodies, can produce such great effects. But the fact is, that motion appears to us, under ordinary circumstances, transitory, because the movement of all terrestrial bodies is resisted perpetually by other forces, friction, resistance of the air, &c., so that the motion is incessantly weakened and finally arrested. A body, however, which is opposed by no resisting force, when once set in motion moves onward eternally with undiminished velocity. Thus we know that the planetary bodies have moved without change through space for thousands of years. Only by resisting forces can motion be diminished or destroyed. A moving body, such as the hammer or the musket-ball, when it strikes against another, presses the latter together, or penetrates it, until the sum of the resisting forces presented by the body struck to pressure, or to the separation of its particles, is sufficiently great to destroy the motion of the hammer or of the bullet. The motion of a mass regarded as taking the place of working force is called the living force (*vis viva*) of the mass. The word 'living' has of course here no reference whatever to living beings, but is intended to represent solely the force of the motion as distinguished from the state of unchanged rest—from the gravity of a motionless body, for example, which produces an incessant pressure against the surface which supports it, but does not produce any motion.

In the case before us, therefore, we had first power in the form of a falling mass of water, then in the form of a lifted hammer, and thirdly in the form of the living force of the falling hammer. We should transform the third form into the second, if we, for example, permitted the hammer to fall upon a highly elastic steel beam strong enough to resist the shock. The hammer would rebound, and in the most favourable case would reach a height equal to that from which it fell, but would never rise higher. In this way its mass would ascend; and at the moment when its highest point has been attained it would represent the same number of raised foot-pounds as before it fell, never a greater number; that is to say, living force can generate the same amount of work as that expended in its production. It is therefore equivalent to this quantity of work.

Our clocks are driven by means of sinking weights, and our watches by means of the tension springs. A weight which lies on the ground, an elastic spring which is without tension, can produce no effects: to obtain such we must first raise the weight or impart tension to the spring, which is accomplished when we wind up our clocks and watches. The man who winds the clock or watch communicates to the weight or to the spring a certain amount of power, and exactly so much as is thus communicated is gradually given out again during the following twenty-four hours, the original force being thus slowly consumed to overcome the friction of the wheels and the resistance which the pendulum encounters from the air. The wheelwork of the clock therefore develops no working force, which was not previously communicated to it, but simply distributes the force given to it uniformly over a longer time.

Into the chamber of an air-gun we squeeze, by means of a condensing air-pump, a great quantity of air. When we afterwards open the cock of the gun and admit the compressed air into the barrel, the ball is driven out of the latter with a force similar to that exerted by ignited powder. Now we may determine the work consumed in the pumping-in of the air, and the living force which, upon firing, is communicated to the ball, but we shall never find the latter greater than the former. The compressed air has generated no working force, but simply gives to the bullet that which has been previously communicated to it. And while we have pumped for perhaps a quarter of an hour to charge the gun, the force is expended in a few seconds when the bullet is discharged; but because the action is compressed into so short a time, a much greater velocity is imparted to the ball than would be possible to communicate to it by the unaided effort of the arm in throwing it.

From these examples you observe, and the mathematical theory has corroborated this for all purely mechanical, that is to say, for moving forces, that all our machinery and apparatus generate no force, but simply yield up the power communicated to them by natural forces,—falling water, moving wind, or by the muscles of men and animals. After this law had been established by the great mathematicians of the last century, a perpetual motion, which should make use solely of pure mechanical forces, such as gravity, elasticity, pressure of liquids and gases, could only be sought after by bewildered and ill-instructed people. But there are still other natural forces which are not reckoned among the purely moving forces,—heat, electricity, magnetism, light, chemical forces, all of which nevertheless stand in manifold relation to mechanical processes. There is hardly a natural process to be found which is not accompanied by mechanical actions, or from which mechanical work may not be derived. Here the question of a perpetual motion remained open; the decision of this question marks the progress of modern physics, regarding which I promised to address you.

In the case of the air-gun, the work to be accomplished in the propulsion of the ball was given by the arm of the man who pumped in the air. In ordinary firearms, the condensed mass of air which propels the bullet is obtained in a totally different manner, namely, by the combustion of the powder. Gunpowder is transformed by combustion for the most part into gaseous products, which endeavour to occupy a much greater space than that previously taken up by the volume of the powder. Thus you see that, by the use of gunpowder, the work which the human arm must accomplish in the case of the air-gun is spared.

In the mightiest of our machines, the steam-engine, it is a strongly compressed aëriform body, water vapour, which, by its effort to expand, sets the machine in motion. Here also we do not condense the steam by means of an external mechanical force, but by communicating heat to a mass of water in a closed boiler, we change this water into steam, which, in consequence of the limits of the space, is developed under strong pressure. In this case, therefore, it is the heat communicated which generates the mechanical force. The heat thus necessary for the machine we might obtain in many ways: the ordinary method is to procure it from the combustion of coal.

Combustion is a chemical process. A particular constituent of our atmosphere, oxygen, possesses a strong force of attraction, or, as is said in chemistry, a strong affinity for the constituents of the combustible body, which affinity, however, in most cases can only exert itself at high temperatures. As soon as a portion of the combustible body, for example the coal, is sufficiently heated, the carbon unites itself with great violence to the oxygen of the atmosphere and forms a peculiar gas, carbonic acid, the same that we see foaming from beer and champagne. By this combination light and heat are generated; heat is generally developed by any combination of two bodies of strong affinity for each other; and when the heat is intense enough, light appears. Hence in the steam-engine it is chemical processes and chemical forces which produce the astonishing work of these machines.

In like manner the combustion of gunpowder is a chemical process, which in the barrel of the gun communicates living force to the bullet. While now the steam-engine develops for us mechanical work out of heat, we can conversely generate heat by mechanical forces. Each impact, each act of friction does it. A skilful blacksmith can render an iron wedge red-hot by hammering. The axles of our carriages must be protected by careful greasing from ignition through friction. Even lately this property has been applied on a large scale. In some factories, where a surplus of water power is at hand, this surplus is applied to cause a strong iron plate to rotate rapidly upon another, so that they become strongly heated by the friction. The heat so obtained warms the room, and thus a stove without fuel is provided. Now could not the heat generated by the plates be applied to a small steam-engine, which in its turn should be able to keep the rubbing plates in motion? The perpetual motion would thus be at length found. This question might be asked, and could not be decided by the older mathematico-mechanical investigations. I will remark beforehand, that the general law which I will lay before you answers the question in the negative.

By a similar plan, however, a speculative American set some time ago the industrial world of Europe in excitement. The magneto-electric machines often made use of in the case of rheumatic disorders are well known to the public. By imparting a swift rotation to the magnet of such a machine we obtain powerful currents of electricity. If those be conducted through water, the latter will be resolved into its two components, oxygen and hydrogen. By the combustion of hydrogen, water is again generated. If this combustion takes place, not in atmospheric air, of which oxygen only constitutes a fifth part, but in pure oxygen, and if a bit of chalk be placed in the flame, the chalk will be raised to its white heat, and give us the sun-like Drummond's light. At the same time the flame develops a considerable quantity of heat. Our American proposed to utilise in this way the gases obtained from electrolytic decomposition, and asserted, that by the combustion a sufficient amount of heat was generated to keep a small steam-engine in action, which again drove his magneto-electric machine, decomposed the water, and thus continually prepared its own fuel. This would certainly have been the most splendid of all discoveries; a perpetual motion which, besides the force that kept it going, generated light like the sun, and warmed all around it. The matter was by no means badly thought out. Each practical step in the affair was known to be possible; but those who at that time were acquainted with the physical investigations which bear upon this subject, could have affirmed, on first hearing the report, that the matter was to be numbered among the numerous stories of the fable-rich America; and indeed a fable it remained.

It is not necessary to multiply examples further. You will infer from those given in what immediate connection heat, electricity, magnetism, light, and chemical affinity, stand with mechanical forces.

Starting from each of these different manifestations of natural forces, we can set every other in motion, for the most part not in one way merely, but in many ways. It is here as with the weaver's web,—

> Where a step stirs a thousand threads,
> The shuttles shoot from side to side,
> The fibres flow unseen,
> And one shock strikes a thousand combinations.

Now it is clear that if by any means we could succeed, as the above American professed to have done, by mechanical forces, in exciting chemical, electrical, or other natural processes, which, by any circuit whatever, and without altering permanently the active masses in the machine, could produce mechanical force in greater quantity than that at first applied, a portion of the work thus gained might be made use of to keep the machine in motion, while the rest of the work might be applied to any other purpose whatever. The problem was to find, in the complicated net of reciprocal actions, a track through chemical, electrical, magnetical, and thermic processes, back to mechanical actions, which might be followed with a final gain of mechanical work: thus would the perpetual motion be found.

But, warned by the futility of former experiments, the public had become wiser. On the whole, people did not seek much after combinations which promised to furnish a perpetual motion, but the question was inverted. It was no more asked, How can I make use of the known and unknown relations of natural forces so as to construct a perpetual motion? but it was asked, If a perpetual motion be impossible, what are the relations which must subsist between natural forces? Everything was gained by this inversion of the question. The relations of natural forces rendered necessary by the above assumption, might be easily and completely stated. It was found that all known relations of forces harmonise with the consequences of that assumption, and a series of unknown relations were discovered at the same time, the correctness of which remained to be proved. If a single one of them could be proved false, then a perpetual motion would be possible.

The first who endeavoured to travel this way was a Frenchman named Carnot, in the year 1824. In spite of a too limited conception of his subject, and an incorrect view as to the nature of heat, which led him to some erroneous conclusions, his experiment was not quite unsuccessful. He discovered a law which now bears his name, and to which I will return further on.

His labours remained for a long time without notice, and it was not till eighteen years afterwards, that is in 1842, that different investigators in different countries, and independent of Carnot, laid hold of the same thought. The first who saw truly the general law here referred to, and expressed it correctly, was a German physician, J. R. Mayer of Heilbronn, in the year 1842. A little later, in 1843, a Dane named Colding presented a memoir to the Academy of Copenhagen, in which the same law found utterance, and some experiments were described for its further corroboration. In England, Joule began about the same time to make experiments having reference to the same subject. We often find, in the case of questions to the solution of which the development of science points, that several heads, quite independent of each other, generate exactly the same series of reflections.

I myself, without being acquainted with either Mayer or Colding, and having first made the acquaintance of Joule's experiments at the end of my investigation, followed the same path. I endeavoured to ascertain all the relations between the different natural processes, which followed from our regarding them from the above point of view. My inquiry was made public in 1847, in a small pamphlet bearing the title, 'On the Conservation of Force.'[1]

Since that time the interest of the scientific public for this subject has gradually augmented, particularly in England, of which I had an opportunity of convincing myself during a visit last summer. A great number of the essential consequences of the above manner of viewing the subject, the proof of which was wanting when the first theoretic notions were published, have since been confirmed by experiment, particularly by those of Joule; and during the last year the most eminent physicist of France, Regnault, has adopted the new mode of regarding the question, and by fresh investigations on the specific heat of gases has contributed much to its support. For some important consequences the experimental proof is still wanting, but the number of confirmations is so predominant, that I have not deemed it premature to bring the subject before even a non-scientific audience.

How the question has been decided you may already infer from what has been stated. In the series of natural processes there is no circuit to be found, by which mechanical force can be gained without a corresponding consumption. The perpetual motion remains impossible. Our reflections, however, gain thereby a higher interest.

We have thus far regarded the development of force by natural processes, only in its relation to its usefulness to man, as mechanical force. You now see that we have arrived at a general law, which holds good wholly independent of the application which man makes of natural forces; we must

---

1. There is a translation of this important Essay in the *Scientific Memoirs*, New Series, p. 114.—J.T.

therefore make the expression of our law correspond to this more general significance. It is in the first place clear, that the work which, by any natural process whatever, is performed under favourable conditions by a machine, and which may be measured in the way already indicated, may be used as a measure of force common to all. Further, the important question arises, If the quantity of force cannot be augmented except by corresponding consumption, can it be diminished or lost? For the purposes of our machines it certainly can, if we neglect the opportunity to convert natural processes to use, but as investigation has proved, not for nature as a whole.

In the collision and friction of bodies against each other, the mechanics of former years assumed simply that living force was lost. But I have already stated that each collision and each act of friction generates heat; and, moreover, Joule has established by experiment the important law, that for every foot-pound of force which is lost a definite quantity of heat is always generated, and that when work is performed by the consumption of heat, for each foot-pound thus gained a definite quantity of heat disappears. The quantity of heat necessary to raise the temperature of a pound of water a degree of the Centigrade thermometer, corresponds to a mechanical force by which a pound weight would be raised to the height of 1,350 feet: we name this quantity the mechanical equivalent of heat. I may mention here that these facts conduct of necessity to the conclusion, that heat is not, as was formerly imagined, a fine imponderable substance, but that, like light, it is a peculiar shivering motion of the ultimate particles of bodies. In collision and friction, according to this manner of viewing the subject, the motion of the mass of a body which is apparently lost is converted into a motion of the ultimate particles of the body; and conversely, when mechanical force is generated by heat, the motion of the ultimate particles is converted into a motion of the mass.

Chemical combinations generate heat, and the quantity of this heat is totally independent of the time and steps through which the combination has been effected, provided that other actions are not at the same time brought into play. If, however, mechanical work is at the same time accomplished, as in the case of the steam-engine, we obtain as much less heat as is equivalent to this work. The quantity of work produced by chemical force is in general very great. A pound of the purest coal gives, when burnt, sufficient heat to raise the temperature of 8,086 pounds of water one degree of the Centigrade thermometer; from this we can calculate that the magnitude of the chemical force of attraction between the particles of a pound of coal and the quantity of oxygen that corresponds to it, is capable of lifting a weight of 100 pounds to a height of twenty miles. Unfortunately, in our steam-engines we have hitherto been able to gain only the smallest portion of this work, the greater part is lost in the shape of heat. The best expansive engines give back as mechanical work only 18 per cent. of the heat generated by the fuel.

From a similar investigation of all the other known physical and chemical processes, we arrive at the conclusion that Nature as a whole possesses a store of force which cannot in any way be either increased or diminished, and that therefore the quantity of force in Nature is just as eternal and unalterable as the quantity of matter. Expressed in this form, I have named the general law 'The Principle of the Conservation of Force.'

We cannot create mechanical force, but we may help ourselves from the general storehouse of Nature. The brook and the wind, which drive our mills, the forest and the coal-bed, which supply our steam-engines and warm our rooms, are to us the bearers of a small portion of the great natural supply which we draw upon for our purposes, and the actions of which we can apply as we think fit. The possessor of a mill claims the gravity of the descending rivulet, or the living force of the moving wind, as his possession. These portions of the store of Nature are what give his property its chief value.

Further, from the fact that no portion of force can be absolutely lost, it does not follow that a portion may not be inapplicable to human purposes. In this respect the inferences drawn by William Thomson from the law of Carnot are of importance. This law, which was discovered by Carnot during his endeavours to ascertain the relations between heat and mechanical force, which, however, by no means belongs to the necessary consequences of the conservation of force, and which Clausius was the first to modify in such a manner that it no longer contradicted the above general law, expresses a certain relation between the compressibility, the capacity for heat, and the expansion by heat of all bodies. It is not yet completely proved in all directions, but some remarkable deductions having been drawn from it, and afterwards proved to be facts by experiment, it has attained thereby the highest degree of probability. Besides the mathematical form in which the law was first expressed by Carnot, we can give it the following more general expression:—'Only when heat passes from a warmer to a colder body, and even then only partially, can it be converted into mechanical work.'

The heat of a body which we cannot cool further, cannot be changed into another form of force—into electric or chemical force for example. Thus in our steam-engines we convert a portion of the heat of the glowing coal into work, by permitting it to pass to the less warm water of the boiler. If, however, all the bodies in Nature had the same temperature, it would be impossible to convert any portion of their heat into mechanical work. According to this we can divide the total force store of the universe into two parts, one of which is heat, and must continue to be such; the other, to which a portion of the heat of the warmer bodies, and the total supply of chemical, mechanical, electrical, and magnetical forces belong, is capable of the most varied changes of form, and constitutes the whole wealth of change which takes place in Nature.

But the heat of the warmer bodies strives perpetually to pass to bodies less warm by radiation and conduction, and thus to establish an equilibrium of temperature. At each motion of a terrestrial body a portion of mechanical force passes by friction or collision into heat, of which only a part can be converted back again into mechanical force. This is also generally the case in every electrical and chemical process. From this it follows that the first portion of the store of force, the unchangeable heat, is augmented by every natural process, while the second portion, mechanical, electrical, and chemical force, must be diminished; so that if the universe be delivered over to the undisturbed action of its physical processes, all force will finally pass into the form of heat, and all heat come into a state of equilibrium. Then all possibility of a further change would be at an end, and the complete cessation of all natural processes must set in. The life of men, animals, and plants could not of course continue if the sun had lost his high temperature, and with it his light,—if all the components of the earth's surface had closed those combinations which their affinities demand. In short, the universe from that time forward would be condemned to a state of eternal rest.

These consequences of the law of Carnot are, of course, only valid provided that the law, when sufficiently tested, proves to be universally correct. In the mean time there is little prospect of the law being proved incorrect. At all events, we must admire the sagacity of Thomson, who, in the letters of a long-known little mathematical formula, which only speaks of heat, volume, and pressure of bodies, was able to discern consequences which threatened the universe, though certainly after an infinite period of time, with eternal death.

I have already given you notice that our path lay through a thorny and unrefreshing field of mathematico-mechanical developments. We have now left this portion of our road behind us. The general principle which I have sought to lay before you has conducted us to a point from which our view is a wide one; and aided by this principle, we can now at pleasure regard this or the other side of the surrounding world according as our interest in the matter leads us. A glance into the narrow laboratory of the physicist, with its small appliances and complicated abstractions, will not be so attractive as a glance at the wide heaven above us, the clouds, the rivers, the woods, and the living beings around us. While regarding the laws which have been deduced from the physical processes of terrestrial bodies as applicable also to the heavenly bodies, let me remind you that the same force which, acting at the earth's surface, we call gravity (*Schwere*), acts as gravitation in the celestial spaces, and also manifests its power in the motion of the immeasurably distant double stars, which are governed by exactly the same laws as those subsisting between the earth and moon; that therefore the light and heat of terrestrial bodies do not in any way differ essentially from those of the sun or of the most distant fixed star; that the meteoric stones which sometimes fall from external space upon the earth are composed of exactly the same simple chemical substances as those with which we are acquainted. We need, therefore, feel no scruple in granting that general laws to which all terrestrial natural processes are subject are also valid for other bodies than the earth. We will, therefore, make use of our law to glance over the household of the universe with respect to the store of force, capable of action, which it possesses.

A number of singular peculiarities in the structure of our planetary system indicate that it was once a connected mass, with a uniform motion of rotation. Without such an assumption it is impossible to explain why all the planets move in the same direction round the sun, why they all rotate in the same direction round their axes, why the planes of their orbits and those of their satellites and rings all nearly coincide, why all their orbits differ but little from circles, and much besides. From these remaining indications of a former state astronomers have shaped an hypothesis regarding the formation of our planetary system, which, although from the nature of the case it must ever remain an hypothesis, still in its special traits is so well supported by analogy, that it certainly deserves our attention; and the more so, as this notion in our own home, and within the walls of this town,[2] first found utterance. It was Kant who, feeling great interest in the physical description of the earth and the planetary system, undertook the labour of studying the works of Newton; and, as an evidence of the depth to which he had penetrated into the fundamental ideas of Newton, seized the notion that the same attractive force of all ponderable matter which now supports the motion of the planets must also aforetime have been able to form from matter loosely scattered in space the planetary system. Afterwards, and independent of Kant, Laplace, the great author of the 'Mécanique céleste,' laid hold of the same thought, and introduced it among astronomers.

The commencement of our planetary system, including the sun, must, according to this, be regarded as an immense nebulous mass which filled the portion of space now occupied by our system far beyond the limits of Neptune, our most distant planet. Even now we discern in distant regions of the firmament nebulous patches the light of which, as spectrum analysis teaches, is the light of ignited gases; and in their spectra we see more especially those bright lines which are produced by ignited hydrogen and by ignited nitrogen. Within our system, also, comets, the crowds of shooting stars, and the zodiacal light exhibit distinct traces of matter dispersed like powder, which moves, however, according to the law of gravitation, and is,

2. Königsberg.

at all events, partially retarded by the larger bodies and incorporated in them. The latter undoubtedly happens with the shooting stars and meteoric stones which come within the range of our atmosphere.

If we calculate the density of the mass of our planetary system, according to the above assumption, for the time when it was a nebulous sphere, which reached to the path of the outermost planet, we should find that it would require several millions of cubic miles of such matter to weigh a single grain.

The general attractive force of all matter must, however, impel these masses to approach each other, and to condense, so that the nebulous sphere became incessantly smaller, by which, according to mechanical laws, a motion of rotation originally slow, and the existence of which must be assumed, would gradually become quicker and quicker. By the centrifugal force, which must act most energetically in the neighbourhood of the equator of the nebulous sphere, masses could from time to time be torn away, which afterwards would continue their courses separate from the main mass, forming themselves into single planets, or, similar to the great original sphere, into planets with satellites and rings, until finally the principle mass condensed itself into the sun. With regard to the origin of heat and light this theory originally gave no information.

When the nebulous chaos separated itself from other fixed star masses it must not only have contained all kinds of matter which was to constitute the future planetary system, but also, in accordance with our new law, the whole store of force which at a future time ought to unfold therein its wealth of actions. Indeed, in this respect an immense dower was bestowed in the shape of the general attraction of all the particles for each other. This force, which on the earth exerts itself as gravity, acts in the heavenly spaces as gravitation. As terrestrial gravity when it draws a weight downwards performs work and generates *vis viva*, so also the heavenly bodies do the same when they draw two portions of matter from distant regions of space towards each other.

The chemical forces must have been also present, ready to act; but as these forces can only come into operation by the most intimate contact of the different masses, condensation must have taken place before the play of chemical forces began.

Whether a still further supply of force in the shape of heat was present at the commencement we do not know. At all events, by aid of the law of the equivalence of heat and work, we find in the mechanical forces existing at the time to which we refer such a rich source of heat and light, that there is no necessity whatever to take refuge in the idea of a store of these forces originally existing. When, through condensation of the masses, their particles came into collision and clung to each other, the *vis viva* of their motion would be thereby annihilated, and must reappear as heat. Already in old theories it has been calculated that cosmical masses must generate heat by their collision, but it was far from anybody's thought to make even a guess at the amount of heat to be generated in this way. At present we can give definite numerical values with certainty.

Let us make this addition to our assumption—that, at the commencement, the density of the nebulous matter was a vanishing quantity as compared with the present density of the sun and planets: we can then calculate how much work has been performed by the condensation; we can further calculate how much of this work still exists in the form of mechanical force, as attraction of the planets towards the sun, and as *vis viva* of their motion, and find by this how much of the force has been converted into heat.

The result of this calculation[3] is, that only about the 454th part of the original mechanical force remains as such, and that the remainder, converted into heat, would be sufficient to raise a mass of water equal to the sun and planets taken together, not less than twenty-eight million of degrees of the Centigrade scale. For the sake of comparison, I will mention that the highest temperature which we can produce by the oxyhydrogen blowpipe, which is sufficient to fuse and vaporise even platinum, and which but few bodies can endure without melting, is estimated at about 2,000 degrees. Of the action of a temperature of twenty-eight millions of such degrees we can form no notion. If the mass of our entire system were pure coal, by the combustion of the whole of it only the 3,500th part of the above quantity would be generated. This is also clear, that such a great development of heat must have presented the greatest obstacle to the speedy union of the masses; that the greater part of the heat must have been diffused by radiation into space, before the masses could form bodies possessing the present density of the sun and planets, and that these bodies must once have been in a state of fiery fluidity. This notion is corroborated by the geological phenomena of our planet; and with regard to the other planetary bodies, the flattened form of the sphere, which is the form of equilibrium of a fluid of a mass, is indicative of a former state of fluidity. If I thus permit an immense quantity of heat to disappear without compensation from our system, the principle of the conservation of force is not thereby invaded. Certainly for our planet it is lost, but not for the universe. It has proceeded outwards, and daily proceeds outwards into infinite space; and we know not whether the medium which transmits the undulations of light and heat possesses an end where the rays must return, or whether they eternally pursue their way through infinitude.

The store of force at present possessed by our system is also equivalent to immense quantities of heat. If our earth were by a sudden shock brought

3. See note on page 43.

to rest in her orbit—which is not to be feared in the existing arrangement of our system—by such a shock a quantity of heat would be generated equal to that produced by the combustion of fourteen such earths of solid coal. Making the most unfavourable assumption as to its capacity for heat—that is, placing it equal to that of water—the mass of the earth would thereby be heated 11,200 degrees; it would, therefore, be quite fused, and for the most part converted into vapour. If, then, the earth, after having been thus brought to rest, should fall into the sun—which, of course, would be the case—the quantity of heat developed by the shock would be 400 times greater.

Even now from time to time such a process is repeated on a small scale. There can hardly be a doubt that meteors, fireballs, and meteoric stones are masses which belong to the universe, and before coming into the domain of our earth, moved like the planets round the sun. Only when they enter our atmosphere do they become visible and fall sometimes to the earth. In order to explain the emission of light by these bodies, and the fact that for some time after their descent they are very hot, the friction was long ago thought of which they experience in passing through the air. We can now calculate that a velocity of 3,000 feet a second, supposing the whole of the friction to be expended in heating the solid mass, would raise a piece of meteoric iron 1,000° C. in temperature, or, in other words, to a vivid red heat. Now the average velocity of the meteors seems to be thirty to fifty times the above amount. To compensate this, however, the greater portion of the heat is doubtless carried away by the condensed mass of air which the meteor drives before it. It is known that bright meteors generally leave a luminous trail behind them, which probably consists of severed portions of the red-hot surfaces. Meteoric masses which fall to the earth often burst with a violent explosion, which may be regarded as a result of the quick heating. The newly-fallen pieces have been for the most part found hot, but not red-hot, which is easily explainable by the circumstance, that during the short time occupied by the meteor in passing through the atmosphere, only a thin superficial layer is heated to redness, while but a small quantity of heat has been able to penetrate to the interior of the mass. For this reason the red heat can speedily disappear.

Thus has the falling of the meteoric stone, the minute remnant of processes which seem to have played an important part in the formation of the heavenly bodies, conducted us to the present time, where we pass from the darkness of hypothetical views to the brightness of knowledge. In what we have said, however, all that is hypothetical is the assumption of Kant and Laplace, that the masses of our system were once distributed as nebulae in space.

On account of the rarity of the case, we will still further remark in what close coincidence the results of science here stand with the earlier legends of the human family, and the forebodings of poetic fancy. The cosmogony of ancient nations generally commences with chaos and darkness. Thus for example Mephistopheles says:—

Part of the Part am I, once All, in primal night,
Part of the Darkness which brought forth the Light,
The haughty Light, which now disputes the space,
And claims of Mother Night her ancient place.

Neither is the Mosaic tradition very divergent, particularly when we remember that that which Moses names heaven, is different from the blue dome above us, and is synonymous with space, and that the unformed earth and the waters of the great deep, which were afterwards divided into waters above the firmament and waters below the firmament, resembled the chaotic components of the world:—

'In the beginning God created the heaven and the earth.

'And the earth was without form, and void; and darkness was upon the face of the deep. And the spirit of God moved upon the face of the waters.'

And just as in nebulous sphere, just become luminous, and in the new red-hot liquid earth of our modern cosmogony light was not yet divided into sun and stars, nor time into day and night, as it was after the earth had cooled.

'And God divided the light from the darkness.

'And God called the light day, and the darkness He called night. And the evening and the morning were the first day.'

And now, first, after the waters had been gathered together into the sea, and the earth had been laid dry, could plants and animals be formed.

Our earth bears still the unmistakable traces of its old fiery fluid condition. The granite formations of her mountains exhibit a structure, which can only be produced by the crystallisation of fused masses. Investigation still shows that the temperature in mines and borings increases as we descend; and if this increase is uniform, at the depth of fifty miles a heat exists sufficient to fuse all our minerals. Even now our volcanoes project from time to time mighty masses of fused rocks from their interior, as a testimony of the heat which exists there. But the cooled crust of the earth has already become so thick, that, as may be shown by calculations of its conductive power, the heat coming to the surface from within, in comparison with that reaching the earth from the sun, is exceedingly small, and increases the temperature of the surface only about $\frac{1}{30}th$ of a degree Centigrade; so that the remnant of the old store of force which is enclosed heat within the bowels of the earth has a sensible influence upon the processes at the earth's surface only through the instrumentality of volcanic phenomena. Those processes owe their power almost wholly to the action of other heavenly bodies, particularly to the light and heat of the sun, and partly also, in the case of the tides, to the attraction of the sun and moon.

Most varied and numerous are the changes which we owe to the light and heat of the sun. The sun heats our atmosphere irregularly, the warm rarefied air ascends, while fresh cool air flows from the sides to supply its place: in this way winds are generated. This action is most powerful at the equator, the warm air of which incessantly flows in the upper regions of the atmosphere towards the poles; while just as persistently at the earth's surface, the trade-wind carries new and cool air to the equator. Without the heat of the sun, all winds must of necessity cease. Similar currents are produced by the same cause in the waters of the sea. Their power may be inferred from the influence which in some cases they exert upon climate. By them the warm water of the Antilles is carried to the British Isles, and confers upon them a mild uniform warmth, and rich moisture; while, through similar causes, the floating ice of the North Pole is carried to the coast of Newfoundland and produces raw cold. Further, by the heat of the sun a portion of the water is converted into vapour, which rises in the atmosphere, is condensed to clouds, or falls in rain and snow upon the earth, collects in the form of springs, brooks, and rivers, and finally reaches the sea again, after having gnawed the rocks, carried away light earth, and thus performed its part in the geologic changes of the earth; perhaps besides all this it has driven our water-mill upon its way. If the heat of the sun were withdrawn, there would remain only a single motion of water, namely, the tides, which are produced by the attraction of the sun and moon.

How is it, now, with the motions and the work of organic beings? To the builders of the automata of the last century, men and animals appeared as clockwork which was never wound up, and created the force which they exerted out of nothing. They did not know how to establish a connexion between the nutriment consumed and the work generated. Since, however, we have learned to discern in the steam-engine this origin of mechanical force, we must inquire whether something similar does not hold good with regard to men. Indeed, the continuation of life is dependent on the consumption of nutritive materials: these are combustible substances, which, after digestion and being passed into the blood, actually undergo a slow combustion, and finally enter into almost the same combinations with the oxygen of the atmosphere that are produced in an open fire. As the quantity of heat generated by combustion is independent of the duration of the combustion and the steps in which it occurs, we can calculate from the mass of the consumed material how much heat, or its equivalent work, is thereby generated in an animal body. Unfortunately, the difficulty of the experiments is still very great; but within those limits of accuracy which have been as yet attainable, the experiments show that the heat generated in the animal body corresponds to the amount which would be generated by the chemical processes. The animal body therefore does not differ from the steam-engine as regards the manner in which it obtains heat and force, but does differ from it in the manner in which the force gained is to be made use of. The body is, besides, more limited than the machine in the choice of its fuel; the latter could be heated with sugar, with starch-flour, and butter, just as well as with coal or wood; the animal body must dissolve its materials artificially, and distribute them through its system; it must, further, perpetually renew the used-up materials of its organs, and as it cannot itself create the matter necessary for this, the matter must come from without. Liebig was the first to point out these various uses of the consumed nutriment. As material for the perpetual renewal of the body, it seems that certain definite albuminous substances which appear in plants, and form the chief mass of the animal body, can alone be used. They form only a portion of the mass of nutriment taken daily; the remainder, sugar, starch, fat, are really only materials for warming, and are perhaps not to be superseded by coal, simply because the latter does not permit itself to be dissolved.

If, then, the processes in the animal body are not in this respect to be distinguished from inorganic processes, the question arises, whence comes the nutriment which constitutes the source of the body's force? The answer is, from the vegetable kingdom; for only the material of plants, or the flesh of herbivorous animals, can be made use of for food. The animals which live on plants occupy a mean position between carnivorous animals, in which we reckon man, and vegetables, which the former could not make use of immediately as nutriment. In hay and grass the same nutritive substances are present as in meal and flour, but in less quantity. As, however, the digestive organs of man are not in a condition to extract the small quantity of the useful from the great excess of the insoluble, we submit, in the first place, these substances to the powerful digestion of the ox, permit the nourishment to store itself in the animal's body, in order in the end to gain it for ourselves in a more agreeable and useful form. In answer to our question, therefore, we are referred to the vegetable world. Now when what plants take in and what they give out are made the subjects of investigation, we find that the principle part of the former consists in the products of combustion which are generated by the animal. They take the consumed carbon given off in respiration, as carbonic acid, from the air, the consumed hydrogen as water, the nitrogen in its simplest and closest combination as ammonia; and from these materials, with the assistance of small ingredients which they take from the soil, they generate anew the compound combustible substances, albumen, sugar, oil, on which the animal subsists. Here, therefore, is a circuit which appears to be a perpetual store of force. Plants prepare fuel and nutriment, animals consume these, burn them slowly in their lungs, and from the products of combustion the plants again derive their nutriment. The latter is an eternal source of chemical, the former of mechanical forces. Would not the combination of both organic kingdoms produce the perpetual motion? We must not conclude hastily: further inquiry shows, that plants are capable of producing combustible substances only when they are under the influence of the sun. A portion of the sun's rays exhibits a remarkable relation to chemical forces,—it can produce and destroy chemical combinations; and these rays, which for the most part are blue or violet, are called therefore chemical rays. We make use of their action in the production of photographs. Here compounds of silver are decomposed at the place where the sun's rays strike them. The same rays overpower in the green leaves of plants the strong chemical affinity of the carbon of the carbonic acid for oxygen, give back the latter free to the atmosphere, and accumulate the other, in combination with other bodies, as woody fibre, starch, oil, or resin. These chemically active rays of the sun disappear completely as soon as they encounter the green portion of the plants, and hence it is that in Daguerreo-type images the green leaves of plants appear uniformly black. Inasmuch as the light coming from them does not contain the chemical rays, it is unable to act upon the silver compounds. But besides the blue and violet, the yellow rays play an important part in the growth of plants. They also are comparatively strongly absorbed by the leaves.

Hence a certain portion of force disappears from the sunlight, while combustible substances are generated and accumulated in plants; and we can assume it as very probable, that the former is the cause of the latter. I must indeed remark, that we are in possession of no experiments from which we might determine whether the *vis viva* of the sun's rays which have disappeared corresponds to the chemical forces accumulated during the same time; and as long as these experiments are wanting, we cannot regard the stated relation as a certainty. If this view should prove correct, we derive from it the flattering result, that all force, by means of which our bodies live and move, finds its source in the purest sunlight; and hence we are all, in point of nobility, not behind the race of the great monarch of China, who heretofore alone called himself Son of the Sun. But it must also be conceded, that our lower fellow-beings, the frog and leech, share the same aethereal origin, as also the whole vegetable world, and even the fuel which comes to us from the ages past, as well as the youngest offspring of the forest with which we heat our stoves and set our machines in motion.

You see, then, that the immense wealth of ever-changing meteorological, climatic, geological, and organic processes of our earth are almost wholly preserved in action by the light- and heat-giving rays of the sun; and you see in this a remarkable example, how Proteus-like the effects of a single cause, under altered external conditions, may exhibit itself in nature. Besides these, the earth experiences an action of another kind from its central luminary, as well as from its satellite the moon, which exhibits itself in the remarkable phenomenon of the ebb and flow of the tide.

Each of these bodies excites, by its attraction upon the waters of the sea, two gigantic waves, which flow in the same direction round the world, as the attracting bodies themselves apparently do. The two waves of the moon, on account of her greater nearness, are about 3½ times as large as those excited by the sun. One of these waves has its crest on the quarter of the earth's surface which is turned towards the moon, the other is at the opposite side. Both these quarters possess the flow of the tide, while the regions which lie between have the ebb. Although in the open sea the height of the tide amounts to only about three feet, and only in certain narrow channels, where the moving water is squeezed together, rises to thirty feet, the might of the phaenomenon is nevertheless manifest from the calculation of Bessel, according to which a quarter of the earth covered by the sea possesses, during the flow of the tide, about 22,000 cubic miles of water more than during the ebb, and that therefore such a mass of water must, in 6¼ hours, flow from one quarter of the earth to the other.

The phaenomenon of the ebb and flow, as already recognised by Mayer, combined with the law of the conservation of force, stands in remarkable connexion with the question of the stability of our planetary system. The mechanical theory of the planetary motions discovered by Newton teaches, that if a solid body in absolute *vacuo*, attracted by the sun, move around him in the same manner as the planets, this motion will endure unchanged through all eternity.

Now we have actually not only one, but several such planets, which move around the sun, and by their mutual attraction create little changes and disturbances in each other's paths. Nevertheless Laplace, in his great work, the 'Mécanique céleste,' has proved that in our planetary system all these disturbances increase and diminish periodically, and can never exceed certain limits, so that by this cause the eternal existence of the planetary system is unendangered.

But I have already named two assumptions which must be made: first, that the celestial spaces must be absolutely empty; and secondly, that the sun and planets must be solid bodies. The first is at least the case as far as astronomical observations reach, for they have never been able to detect any retardation of the planets, such as would occur if they moved in a resisting medium. But on a body of less mass, the comet of Encke, changes are observed of such a nature: this comet describes ellipses round the sun which are becoming gradually smaller. If this kind of motion, which certainly corresponds to that through a resisting medium, be actually due to the existence of such a medium, a time will come when the comet will strike the sun; and a similar end threatens all the planets, although after a time, the length of which baffles our imagination to conceive of it. But even should the existence of a resisting medium appear doubtful to us, there is no doubt that the planets are not wholly composed of solid materials which are inseparably bound together. Signs of the existence of an atmosphere are observed on the Sun, on Venus, Mars, Jupiter, and Saturn. Signs of water and ice upon Mars; and our earth has undoubtedly a fluid portion on its surface, and perhaps a still greater portion of fluid within it. The motions of the tides, however, produce friction, all friction destroys *vis viva*, and the loss in this case can only affect the *vis viva* of the planetary system. We come thereby to the unavoidable conclusion, that every tide, although with infinite slowness, still with certainty diminishes the store of mechanical force of the system; and as a consequence of this, the rotation of the planets in question round their axes must become more slow. The recent careful investigations of the moon's motion made by Hansen, Adams, and Delaunay, have proved that the earth does experience such a retardation. According to the former, the length of each sidereal day has increased since the time of Hipparchus by the $\frac{1}{81}$ part of a second, and the duration of a century by half a quarter of an hour; according to Adams and Sir W. Thomson, the increase has been almost twice as great. A clock which went right at the beginning of a century, would be twenty-two seconds in advance of the earth at the end of the century. Laplace had denied the existence of such a retardation in the case of the earth; to ascertain the amount, the theory of lunar motion required a greater development than was possible in his time. The final consequence would be, but after millions of years, if in the mean time the ocean did not become frozen, that one side of the earth would be constantly turned towards the sun, and enjoy a perpetual day, whereas the opposite side would be involved in eternal night. Such a position we observe in our moon with regard to the earth, and also in the case of the satellites as regards their planets; it is, perhaps, due to the action of the mighty ebb and flow to which these bodies, in the time of their fiery fluid condition, were subjected.

I would not have brought forward these conclusions, which again plunge us in the most distant future, if they were not unavoidable. Physico-mechanical laws are, as it were, the telescopes of our spiritual eye, which can penetrate into the deepest night of time, past and to come.

Another essential question as regards the future of our planetary system has reference to its future temperature and illumination. As the internal heat of the earth has but little influence on the temperature of the surface, the heat of the sun is the only thing which essentially affects the question. The quantity of heat falling from the sun during a given time upon a given portion of the earth's surface may be measured, and from this it can be calculated how much heat in a given time is sent out from the entire sun. Such measurements have been made by the French physicist Pouillet, and it has been found that the sun gives out a quantity of heat per hour equal to that which a layer of the densest coal 10 feet thick would give out by its combustion; and hence in a year a quantity equal to the combustion of a layer of 17 miles. If this heat were drawn uniformly from the entire mass of the sun, its temperature would only be diminished thereby 1⅓ of a degree Centigrade per year, assuming its capacity for heat to be equal to that of water. These results can give us an idea of the magnitude of the emission, in relation to the surface and mass of the sun; but they cannot inform us whether the sun radiates heat as a glowing body, which since its formation has its heat accumulated within it, or whether a new generation of heat by chemical processes is continually taking place at the sun's surface. At all events, the law of the conservation of force teaches us that no process analogous to those known at the surface of the earth can supply for eternity an inexhaustible amount of light and heat to the sun. But the same law also teaches that the store of force at present existing, as heat, or as what may become heat, is sufficient for an immeasurable time. With regard to the store of chemical force in the sun, we can form no conjecture, and the store of heat there existing can only be determined by very uncertain estimations. If, however, we adopt the very probable view, that the remarkably small density of so large a body is cause by its high temperature, and may become greater in time, it may be calculated that if the diameter of the sun were diminished only the ten-thousandth part of its present length, by this act a sufficient quantity of heat would be generated to cover the total emission for 2,100 years. So small a change it would be difficult to detect even by the finest astronomical observations.

Indeed, from the commencement of the period during which we possess historic accounts, that is, for a period of about 4,000 years, the temperature of the earth has not sensibly diminished. From these old ages we have certainly no thermometric observations, but we have information regarding the distribution of certain cultivated plants, the vine, the olive tree, which are very sensitive to changes of the mean annual temperature, and we find that these plants at the present moment have the same limits of distribution that they had in the times of Abraham and Homer; from which we may infer backwards the constancy of the climate.

In opposition to this it has been urged, that here in Prussia the German knights in former times cultivated the vine, cellared their own wine and drank it, which is no longer possible. From this the conclusion has been drawn, that the heat of our climate has diminished since the time referred to. Against this, however, Dove has cited the reports of ancient chroniclers, according to which, in some peculiarly hot years, the Prussian grape possessed somewhat less than its usual quantity of acid. The fact also speaks not so much for the climate of the country as for the throats of the German drinkers.

But even though the force store of our planetary system is so immensely great, that by the incessant emission which has occurred during the period of human history it has not been sensibly diminished, even though the length of the time which must flow by before a sensible change in the state of our planetary system occurs is totally incapable of measurement, still the inexorable laws of mechanics indicate that this store of force, which can only suffer loss and not gain, must be finally exhausted. Shall we terrify ourselves by this thought? Men are in the habit of measuring the greatness and the wisdom of the universe by the duration and the profit which it promises to their own race; but the past history of the earth already shows what an insignificant moment the duration of the existence of our race upon it constitutes. A Nineveh vessel, a Roman sword, awake in us the conception of grey antiquity. What the museums of Europe show us of the remains of Egypt and Assyria we gaze upon with silent astonishment, and despair of being able to carry our thoughts back to a period so remote. Still must the human race have existed for ages, and multiplied itself before the Pyramids or Nineveh could have been erected. We estimate the duration of human history at 6,000 years; but immeasurable as this time may appear to us, what is it in comparison with the time during which the earth carried successive series of rank plants and mighty animals, and no men; during which in our neighbourhood the amber-tree bloomed, and dropped its costly gum on the earth and in the sea; when in Siberia, Europe, and North America groves of tropical palms flourished; where gigantic lizards, and after them elephants, whose mighty remains we still find buried in the earth, found a home? Different geologists, proceeding from different premises, have sought to estimate the duration of the above-named creative period, and vary from a million to nine million years. The time during which the earth generated organic beings is again small when compared with the ages during which the world was a ball of fused rocks. For the duration of its cooling from 2,000° to 200° Centigrade the experiments of Bishop upon basalt show that about 350 millions of years would be necessary. And with regard to the time during which the first nebulous mass condensed into our planetary system, our most daring conjectures must cease. The history of man, therefore, is but a short ripple in the ocean of time. For a much longer series of years than that during which he has already occupied this world, the existence of the present state of inorganic nature favourable to the duration of man seems to be secured, so that for ourselves and for long generations after us we have nothing to fear. But the same forces of air and water, and of the volcanic interior, which produced former geological revolutions, and buried one series of living forms after another, act still upon the earth's crust. They more probably will bring about the last day of the human race than those distant cosmical alterations of which we have spoken, forcing us perhaps to make way for new and more complete living forms, as the lizards and the mammoth have given place to us and our fellow-creatures which now exist.

Thus the thread which was spun in darkness by those who sought a perpetual motion has conducted us to a universal law of nature, which radiates light into the distant nights of the beginning and of the end of the history of the universe. To our own race it permits a long but not an endless existence; it threatens it with a day of judgment, the dawn of which is still happily obscured. As each of us singly must endure the thought of his death, the race must endure the same. But above the forms of life gone by, the human race has higher moral problems before it, the bearer of which it is, and in the completion of which it fulfills its destiny.

NOTE TO PAGE 33

I must here explain the calculation of the heat which must be produced by the assumed condensation of the bodies of our system from scattered nebulous matter. The other calculations, the results of which I have mentioned, are to be found partly in J. R. Mayer's papers, partly in Joule's communications, and partly by aid of the known facts and method of science: they are easily performed.

The measure of the work performed by the condensation of the mass from a state of infinitely small density is the potential of the condensed mass upon itself. For a sphere of uniform density of the mass M, and the radius R, the potential upon itself V—if we call the mass of the earth m, its radius r, and the intensity of gravity at its surface g—has the value

$$V = \frac{3}{5} \cdot \frac{r^2 M^2}{Rm} \cdot g.$$

Let us regard the bodies of our system as such spheres, then the total work of condensation is equal to the sum of all their potentials on themselves. As, however, these potentials for different spheres are to each other as the quantity $\frac{M^2}{R}$, they all vanish in comparison with the sun; even that of the greatest planet, Jupiter, is only about the one hundred-thousandth part of that of the sun; in the calculation, therefore, it is only necessary to introduce the latter.

To elevate the temperature of a mass M of the specific heat σ, t degrees, we need a quantity of heat equal to $M\sigma t$; this corresponds, when Ag represents the mechanical equivalent of the unit of heat, to the work $AgM\sigma t$. To find the elevation of temperature produced by the condensation of the mass of the sun, let us set

$$AgM\sigma t = V;$$

we have then

$$t = \frac{3}{5} \cdot \frac{r^2 M}{A \cdot R \cdot m \cdot \sigma} .$$

For a mass of water equal to the sun we have $\sigma = 1$; then the calculation with the known values of A, M, R, m, and r, gives

$$t = 28611000° \; Cent.$$

The mass of the sun is 738 times greater than that of all the planets taken together; if, therefore, we desire to make the water mass equal to that of the entire system, we must multiply the value of t by the fraction $\frac{738}{739}$, which makes hardly a sensible alteration in the result.

When a spherical mass of the radius R condenses more and more to the radius R_1, the elevation of temperature thereby produced is

$$\theta = \frac{3}{5} \cdot \frac{r^2 M}{A \cdot m\sigma} \{ \frac{1}{R_1} - \frac{1}{R_0} \},$$

or

$$= \frac{3}{5} \cdot \frac{r^2 M}{A R_1 m\sigma} \{ 1 - \frac{R_1}{R_0} \}.$$

Supposing, then, the mass of the planetary system to be at the commencement, not a sphere of infinite radius, but limited, say of the radius of the path of Neptune, which is six thousand times greater than the radius of the sun, the magnitude $\frac{R_1}{R_0}$ will then be equal to $\frac{1}{6000}$, and the above value of t would have to be diminished by this inconsiderable amount.

From the same formula we can deduce that a diminution of $\frac{1}{10000}$ of the radius of the sun would generate work in a water mass equal to the sun, equivalent to 2,861 degrees Centigrade. And as, according to Pouillet, a quantity of heat corresponding to 1¼ degree is lost annually in such a mass, the condensation referred to would cover the loss for 2,289 years.

If the sun, as seems probable, be not everywhere of the same density, but is denser at the centre than near the surface, the potential of its mass and the corresponding quantity of heat will be still greater.

Of the now remaining mechanical forces, the *vis viva* of the rotation of the heavenly bodies round their own axes is, in comparison with the other quantities, very small, and may be neglected. The *vis viva* of the motion of revolution round the sum, if μ be the mass of a planet, and ρ its distance from the sun, is

$$L = \frac{gr^2 M\mu}{m}\left\{\frac{1}{R} - \frac{1}{2\rho}\right\}.$$

Omitting the quantity $\frac{1}{2\rho}$ as very small compared with $\frac{1}{R}$, and dividing by the above value of V, we obtain

$$\frac{L}{V} = \frac{5}{3}\frac{\mu}{M}.$$

The mass of all the planets together is $\frac{1}{738}$ of the mass of the sun; hence the value of L for the entire system is

$$L = \frac{1}{453} \cdot V.$$

A Lecture Delivered February 7, 1854, at Königsberg, in Prussia.

3

On the Physiological Causes of Harmony in Music

LADIES AND GENTLEMEN,—In the native town of Beethoven, the mightiest among the heroes of harmony, no subject seemed to me better adapted for a popular audience than music itself. Following, therefore, the direction of my researches during the last few years, I will endeavour to explain to you what physics and physiology have to say regarding the most cherished art of the Rhenish land—music and musical relations. Music has hitherto withdrawn itself from scientific treatment more than any other art. Poetry, painting, and sculpture borrow at least the material for their delineations from the world of experience. They portray nature and man. Not only can their material be critically investigated in respect to its correctness and truth to nature, but scientific art-criticism, however much enthusiasts may have disputed its right to do so, has actually succeeded in making some progress in investigating the causes of that aesthetic pleasure which it is the intention of these arts to excite. In music, on the other hand, it seems at first sight as if those were still in the right who reject all 'anatomisation of pleasurable sensations.' This art, borrowing no part of its material from the experience of our senses; not attempting to describe, and only exceptionally to imitate the outer world, necessarily withdraws from scientific consideration the chief points of attack which other arts present, and hence seems to be as incomprehensible and wonderful as it is certainly powerful in its effects. We are, therefore, obliged, and we purpose, to confine ourselves, in the first place, to a consideration of the material of the art, musical sounds or sensations. It always struck me as a wonderful and peculiarly interesting mystery, that in the theory of musical sounds, in the physical and technical foundations of music, which above all other arts seems in its action on the mind as the most immaterial, evanescent, and tender creator of incalculable and indescribable states of consciousness, that here in especial the science of purest and strictest thought—mathematics—should prove preeminently fertile. Thorough bass is a kind of applied mathematics. In considering musical intervals, divisions of time, and so forth, numerical fractions, and sometimes even logarithms, play a prominent part. Mathematics and music!

the most glaring possible opposites of human thought! and yet connected, mutually sustained! It is as if they would demonstrate the hidden consensus of all the actions of our mind, which in the revelations of genius makes us forefeel unconscious utterances of a mysteriously active intelligence.

When I considered physical acoustics from a physiological point of view, and thus more closely followed up the part which the ear plays in the perception of musical sounds, much became clear of which the connection had not been previously evident. I will attempt to inspire you with some of the interest which these questions have awakened in my own mind, by endeavouring to exhibit a few of the results of physical and physiological acoustics.

The short space of time at my disposal obliges me to confine my attention to one particular point; but I shall select the most important of all, which will best show you the significance and results of scientific investigation in this field; I mean the foundation of concord. It is an acknowledged fact that the numbers of the vibrations of concordant tones bear to each other ratios expressible by small whole numbers. But why? What have the ratios of small whole numbers to do with concord? This is an old riddle, propounded by Pythagoras, and hitherto unsolved. Let us see whether the means at the command of modern science will furnish the answer.

First of all, what is a musical tone? Common experience teaches us that all sounding bodies are in a state of vibration. This vibration can be seen and felt; and in the case of loud sounds we feel the trembling of the air even without touching the sounding bodies. Physical science has ascertained that any series of impulses which produce a vibration of the air will, if repeated with sufficient rapidity, generate sound.

This sound becomes a *musical* tone, when such rapid impulses recur with perfect regularity and in precisely equal times. Irregular agitation of the air generates only noise. The *pitch* of a musical tone depends on the number of impulses which take place in a given time; the more there are in the same time the higher or sharper is the tone. And, as before remarked, there is found to be a close relationship between the well-known harmonious musical intervals and the number of the vibrations of the air. If twice as many vibrations are performed in the same time for one tone as for another, the first is the octave above the second. If the numbers of vibrations in the same time are as 2 to 3, the two tones form a fifth; if they are as 4 to 5, the two tones form a major third.

If you observe that the numbers of the vibrations which generate the tones of the major chord C E G c are in the ratio of the numbers 4 : 5 : 6 : 8, you can deduce from these all other relations of musical tones, by imagining a new major chord, having the same relations of the numbers of vibrations, to be formed upon each of the above-named tones. The numbers of vibrations within the limits of audible tones which would be obtained by executing the calculation thus indicated, are extraordinarily different. Since the octave above any tone has twice as many vibrations as the tone itself, the second octave above will have four times, the third has eight times as many. Our modern pianofortes have seven octaves. Their highest tones, therefore, perform 128 vibrations in the time that their lowest tone makes one single vibration.

The deepest C_1 which our pianos usually possess, answers to the sixteen-foot open pipe of the organ—musicians call it the 'contra-C'—and makes thirty-three vibrations in one second of time. This is very nearly the limit of audibility. You will have observed that these tones have a dull, bad quality of sound on the piano, and that it is difficult to determine their pitch and the accuracy of their tuning. On the organ the contra-C is somewhat more powerful than on the piano, but even here some uncertainty is felt in judging of its pitch. On larger organs there is a whole octave of tones below the contra-C, reaching to the next lower C, with 16½ vibrations in a second. But the ear can scarcely separate these tones from an obscure drone; and the deeper they are the more plainly can it distinguish the separate impulses of the air to which they are due. Hence they are used solely in conjunction with the next higher octaves, to strengthen their notes, and produce an impression of greater depth.

With the exception of the organ, all musical instruments, however diverse the methods in which their sounds are produced, have their limit of depth at about the same point in the scale as the piano; not because it would be impossible to produce slower impulses of the air of sufficient power, but because the *ear* refuses its office, and hears slower impulses separately, without gathering them up into single tones.

The often repeated assertion of the French physicist Savart, that he heard tones of eight vibrations in a second, upon a peculiarly constructed instrument, seems due to an error.

Ascending the scale from the contra-C, pianofortes usually have a compass of seven octaves, up to the so-called five-accented *c*, which has 4,224 vibrations in a second. Among orchestral instruments it is only the piccolo flute which can reach as high, and this will give even one tone higher. The violin usually mounts no higher than the *e* below, which has 2,640 vibrations—of course we except the gymnastics of heaven-scaling *virtuosi*, who are ever striving to excruciate their audience by some new impossibility. Such performers may aspire to three whole octaves lying above the five-accented *c*, and very painful to the ear, for their existence has been established by Despretz, who, by exciting small tuning-forks with a violin bow, obtained and heard the eight-accented *c*, having 32,770 vibrations in a second. Here the sensation of tone seemed to have reached its upper limit, and the intervals were really undistinguishable in the later octaves.

The musical pitch of a tone depends entirely on the number of vibrations of the air in a second, and not at all upon the mode in which they are produced. It is quite indifferent whether they are generated by the vibrating strings of a piano or violin, the vocal chords of the human larynx, the metal tongues of the harmonium, the reeds of the clarionet, oboe and bassoon, the trembling lips of the trumpeter, or the air cut by a sharp edge in organ pipes and flutes.

A tone of the same number of vibrations has always the same pitch, by whichever one of these instruments it is produced. That which distinguishes the note A of a piano for example, from the equally high A of the violin, flute, clarionet, or trumpet, is called the *quality of the tone*, and to this we shall have to recur presently.

As an interesting example of these assertions, I beg to show you a peculiar physical instrument for producing musical tones, called the *siren*, Fig. 1, which is especially adapted to establish the properties resulting from the ratios of the numbers of vibrations.

In order to produce tones upon this instrument, the portvents g_0 and g_1 are connected by means of flexible tubes with a bellows. The air enters into round brass boxes, a_0 and a_1, and escapes by the perforated covers of these boxes at c_0 and c_1. But the holes for the escape of air are not perfectly free. Immediately before the covers of both boxes there are two other perforated discs, fastened to a perpendicular axis k, which turns with great readiness. In the figure, only the perforated disc can be seen at c_0, and immediately below it is the similarly perforated cover of the box. In the upper box, c_1, only the edge of the disc is visible. If then the holes of the disc are precisely opposite to those of the cover, the air can escape freely. But if the disc is made to revolve, so that some of its unperforated portions stand before the holes of the box, the air cannot escape at all. On turning the disc rapidly, the vent-holes of the box are alternately opened and closed. During the opening, air escapes; during the closure, no air can pass. Hence the continuous stream of air from the bellows is converted into a series of discontinuous puffs, which, when they follow one another with sufficient rapidity, gather themselves together into a tone.

Each of the revolving discs of this instrument (which is more complicated in its construction than any one of the kind hitherto made, and hence admits of a much greater number of combinations of tone) has four concentric circles of holes, the lower set having 8, 10, 12, 18, and the upper set 9, 12, 15, and 16, holes respectively. The series of holes in the covers of the boxes are precisely the same as those in the discs, but under each of them lies a perforated ring, which can be so arranged, by means of the stops i i i i, that the corresponding holes of the cover can either communicate freely with the inside of the box, or are entirely cut off from it. We are thus enabled to use any one of the eight series of holes singly, or combined two and two, or three and three together, in any arbitrary manner.

The round boxes, h_0 h_0 and h_1 h_1, of which halves only are drawn in the figure, serve by their resonance to soften the harshness of the tone.

Fig. 1.

The holes in the boxes and discs are cut obliquely, so that when the air enters the boxes through one or more of the series of holes, the wind itself drives the discs round with a perpetually increasing velocity.

On beginning to blow the instrument, we first hear separate impulses of the air, escaping as puffs, as often as the holes of the disc pass in front of those of the box. These puffs of air follow one another more and more quickly, as the velocity of the revolving discs increases, just like the puffs of steam of a locomotive on beginning to move with the train. They next produce a whirring and whizzing, which constantly becomes more rapid. At last we hear a dull drone, which, as the velocity further increases, gradually gains in pitch and strength.

Suppose that the discs have been brought to a velocity of 33 revolutions in a second, and that the series with 8 holes has been opened. At each revolution of the disc all these 8 holes will pass before each separate hole of the cover. Hence there will be 8 puffs for each revolution of the disc, or 8 times 33, that is, 264 puffs in a second. This gives us the once-accented *c* of our musical scale, [that is 'middle *c*,' written on the leger line between the bass and treble staves.] But on opening the series of 16 holes instead, we have twice as many, or 16 times 33, that is, 528 vibrations in a second. We hear exactly the octave above the first *c'*, that is the twice-accented *c"*, [or *c* on the third space of the treble staff.] By opening both the series of 8 and 16 holes at once, we have both *c'* and *c"* at once, and can convince ourselves that we have the absolutely pure concord of the octave. By taking 8 and 12 holes, which give numbers of vibrations in the ratio of 2 to 3, we have the concord of a perfect fifth. Similarly 12 and 16 or 9 and 12 give fourths, 12 and 15 give a major third, and so on.

The upper box is furnished with a contrivance for slightly sharpening or flattening the tones which it produces. This box is movable upon an axis, and connected with a toothed wheel, which is worked by the driver attached to the handle d. By turning the handle slowly while one of the series of holes in the upper box is in use, the tone will be sharper or flatter, according as the box moves in the opposite direction to the disc, or in the same direction as the disc. When the motion is in the opposite direction, the holes meet those of the disc a little sooner than they otherwise would, the time of vibration of the tone is shortened, and the tone becomes sharper. The contrary ensues in the other case.

Now, on blowing through 8 holes below and 16 above, we have a perfect octave, as long as the upper box is still; but when it is in motion, the pitch of the upper tone is slightly altered, and the octave becomes false.

On blowing through 12 holes above and 18 below, the result is a perfect fifth as long as the upper box is at rest, but if it moves the concord is perceptibly injured.

These experiments with the siren show us, therefore:—

1. That a series of puffs following one another with sufficient rapidity, produce a musical tone.

2. That the more rapidly they follow one another, the sharper is the tone.

3. That when the ratio of the number of vibrations is exactly 1 to 2, the result is a perfect octave; when it is 2 to 3, a perfect fifth; when it is 3 to 4, a pure fourth, and so on. The slightest alteration in these ratios destroys the purity of the concord.

You will perceive, from what has been hitherto adduced, that the human ear is affected by vibrations of the air, within certain degrees of rapidity—viz. from about 20 to about 32,000 in a second—and that the sensation of musical tone arises from this affection.

That the sensation thus excited is a sensation of musical tone, does not depend in any way upon the peculiar manner in which the air is agitated, but solely on the peculiar powers of sensation possessed by our ears and auditory nerves. I remarked, a little while ago, that when the tones are loud the agitation of the air is perceptible to the skin. In this way deaf mutes can perceive the motion of the air, which we call sound. But they do not hear, that is, they have no sensation of tone in the ear. They feel the motion by the nerves of the skin, producing that peculiar description of sensation called whirring. The limits of the rapidity of vibration within which the ear feels an agitation of the air to be sound, depend also wholly upon the peculiar constitution of the ear.

When the siren is turned slowly, and hence the puffs of air succeed each other slowly, you hear no musical sound. By the continually increasing rapidity of its revolution, no essential change is produced in the kind of vibration of the air. Nothing new happens externally to the ear. The only new result is the sensation experienced by the ear, which then for the first time begins to be affected by the agitation of the air. Hence the more rapid vibrations receive a new name, and are called Sound. If you admire paradoxes, you may say that aerial vibrations do not become sound until they fall upon a hearing ear.

I must now describe the propagation of sound through the atmosphere. The motion of a mass of air through which a tone passes, belongs to the so-called wave motions—a class of motions of great importance in physics. Light, as well as sound, is one of these motions.

The name is derived from the analogy of waves on the surface of water, and these will best illustrate the peculiarity of this description of motion.

When a point in a surface of still water is agitated—as by throwing in a stone—the motion thus caused is propagated in the form of waves, which spread in rings over the surface of the water. The circles of waves continue to increase even after rest has been restored at the point first affected. At the same time the waves become continually lower, the further they are removed from the centre of motion, and gradually disappear. On each wave-ring we distinguish ridges or crests, and hollows or troughs.

Crest and trough together form a wave, and we measure its length from one crest to the next.

While the wave passes over the surface of the fluid, the particles of the water which form it do not move on with it. This is easily seen, by floating a chip of straw on the water. When the waves reach the chip, they raise or depress it, but when they have passed over it, the position of the chip is not perceptibly changed.

Now a light floating chip has no motion different from that of the adjacent particles of water. Hence we conclude that these particles do not follow the wave, but, after some pitching up and down, remain in their original position. That which really advances as a wave is, consequently, not the particles of water themselves, but only a superficial form, which continues to be built up by fresh particles of water. The paths of the separate particles of water are more nearly vertical circles, in which they revolve with a tolerably uniform velocity, as long as the waves pass over them.

In Fig. 2 the dark wave-line, A B C, represents a section of the surface of the water, over which waves are running in the direction of the arrows above a and c. The three circles, a, b, and c, represent the paths of particular particles of water at the surface of the wave. The particle which revolves in the circle b, is supposed at the time that the surface of the water presents the form A B C, to be at its highest point B, and the particles revolving in the circles a and c to be simultaneously in their lowest positions.

The respective particles of water revolve in these circles in the direction marked by the arrows. The dotted curves represent other positions of the passing waves, at equal intervals of time, partly before the assumption of the A B C position (as for the crests between a and b), and partly after the same (for the crests between b and c). The positions of the crests are marked with figures. The same figures in the three circles, show where the respective revolving particle would be, at the moment the wave assumed the corresponding form. It will be noticed that the particles advance by equal arcs of the circles, as the crest of the wave advances by equal distances parallel to the water level.

In the circle b it will be further seen, that the particle of water in its positions 1, 2, 3, hastens to meet the approaching wave-crests, 1, 2, 3, rises on its left hand side, is then carried on by the crest from 4 to 7 in the direction of its advance, afterwards halts behind it, sinks down again on the right side, and finally reaches its original position at 13. (In the Lecture itself, Fig. 2 was replaced by a working model, in which the movable particles, connected by threads, really revolved in circles, while connecting elastic threads represented the surface of the water.)

All particles at the surface of the water, as you see by this drawing, describe equal circles. The particles of water at different depths move in the same way, but as the depths increase, the diameters of their circles of revolution rapidly diminish.

Fɪɢ. 2.

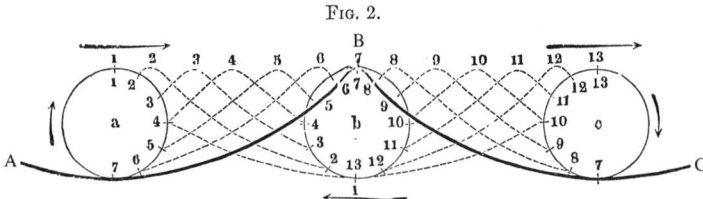

Fig. 2.

In this way, then, arises the appearance of a progressive motion along the surface of the water, while in reality the moving particles of water do not advance with the wave, but perpetually revolve in their small circular orbits.

To return from waves of water to waves of sound. Imagine an elastic fluid like air to replace the water, and the waves of this replaced water to be compressed by an inflexible plate laid on their surface, the fluid being prevented from escaping laterally from the pressure. Then on the waves being thus flattened out, the ridges where the fluid had been heaped up will produce much greater density than the hollows, from which the fluid had been removed to form the ridges. Hence the ridges are replaced by condensed strata of air, and the hollows by rarefied strata. Now further imagine that these compressed waves are propagated by the same law as before, and that also the vertical circular orbits of the several particles of water are compressed into horizontal straight lines. Then the waves of sound will retain the peculiarity of having the particles of air only oscillating backwards and forwards in a straight line, while the wave itself remains merely a progressive form of motion, continually composed of fresh particles of air. The immediate result then would be waves of sound spreading out horizontally from their origin.

But the expansion of waves of sound is not limited, like those of water, to a horizontal surface. They can spread out in any direction whatsoever. Suppose the circles generated by a stone thrown into the water to extend in all directions of space, and you will have the spherical waves of air by which sound is propagated.

Hence we can continue to illustrate the peculiarities of the motion of sound, by the well-known visible motions of waves of water.

The length of a wave of water, measured from crest to crest, is extremely different. A falling drop, or a breath of air, gently curls the surface of the water. The waves in the wake of a steamboat toss the swimmer or skiff severely. But the waves of a stormy ocean can find room in their hollows for the keel of a ship of the line, and their ridges can scarcely be overlooked from the masthead. The waves of sound present similar differences. The little curls of water with short lengths of wave correspond to high tones, the giant ocean billows to deep tones. Thus the contrabass C has a wave thirty-five feet long, its higher octave a wave of half the length, while the highest tones of a piano have waves of only three inches in length.[1]

You perceive that the pitch of the tone corresponds to the length of the

---

1. The exact lengths of waves corresponding to certain *notes*, or symbols of tone, depend upon the standard pitch assigned to one particular note, and this differs in different countries. Hence the figures of the author have been left unreduced. They are sufficiently near to those usually adopted in England, to occasion no difficulty to the reader in these general remarks.—TR.

wave. To this we should add that the height of the ridges, or, transferred to air, the degree of alternate condensation and rarefaction, corresponds to the loudness and intensity of the tone. But waves of the same height may have different forms. The crest of the ridge, for example, may be rounded off or pointed. Corresponding varieties also occur in waves of sound of the same pitch and loudness. The so-called *timbre* or quality of tone is what corresponds to the *form* of the waves of water. The conception of form is transferred from waves of water to waves of sound. Supposing waves of water of *different* forms to be pressed flat as before, the surface, having been levelled, will of course display no differences of form, but, in the interior of the mass of water, we shall have different distributions of pressure, and hence of density, which exactly correspond to the differences of form in the still uncompressed surface. In this sense then we can continue to speak of the form of waves of sound, and can represent it geometrically. We make the curve rise where the pressure, and hence density, increases, and fall where it diminishes—just as if we had a compressed fluid beneath the curve, which would expand to the height of the curve in order to regain its natural density.

Unfortunately, the form of waves of sound, on which depends the quality of the tones produced by various sounding bodies, can at present be assigned in only a very few cases.

Among the forms of waves of sound which we are able to determine with more exactness, is one of great importance, here termed the *simple* or *pure* wave-form, and represented in Fig. 3.

Fig. 3.

It can be seen in waves of water only when their height is small in comparison with their length, and they run over a smooth surface without external disturbance, or without any action of wind. Ridge and hollow are gently rounded off, equally broad and symmetrical, so that, if we inverted the curve, the ridges would exactly fit into the hollows, and conversely. This form of wave would be more precisely defined by saying that the particles of water describe exactly circular orbits of small diameters, with exactly uniform velocities. To this simple wave-form corresponds a peculiar species of tone, which, from reasons to be hereafter assigned, depending upon its relation to quality, we will term a *simple* tone. Such tones are produced by striking a tuning-fork, and holding it before the opening of a properly-tuned resonance tube. The tone of tuneful human voices, singing the vowel *oo* in *too*, in the middle positions of their register, appears not to differ materially from this form of wave.

We also know the laws of the motion of strings with sufficient accuracy to assign in some cases the form of motion which they impart to the air. Thus Fig. 4 represents the forms successively assumed by a string struck, as in the German *Zither*, by a pointed style, [the *plectrum* of the ancient

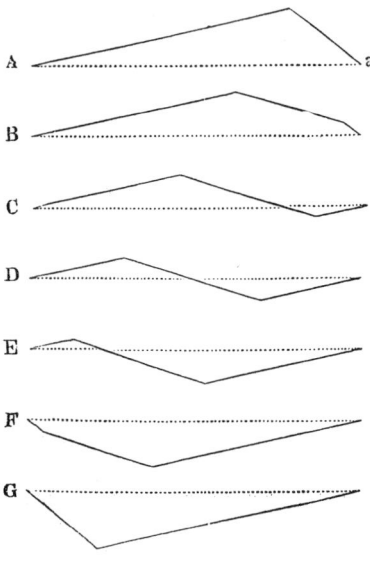

Fig. 4.

lyra, or the quill of the old harpsichord, which may be easily imitated on a guitar]. A a represents the form assumed by the string at the moment of percussion. Then, at equal intervals of time, follow the forms B, C, D, E, F, G; and then in inverse order, F, E, D, C, B, A, and so on in perpetual repetition. The form of motion which such a string, by means of an attached sounding board, imparts to the surrounding air, probably corresponds to the broken line in Fig. 5, where h h indicates the position of equilibrium, and the letters a b c d e f g show the line of the wave which is produced by the action of several forms of string marked by the corresponding capital letters in Fig. 4. It is easily seen how greatly this form of wave (which of course could not occur in water) differs from that of Fig. 3 (independently of magnitude), as the string only imparts to the air a series of short impulses, alternately directed to opposite sides.[2]

The waves of air produced by the tone of a violin would, on the same principle,

---

2. It is here assumed that the sounding-board and air in contact with it immediately obey the impulse given by the end of the string without exercising a perceptible reaction on the motion of the string.

be represented by Fig. 6. During each period of vibration the pressure increases uniformly, and at the end falls back suddenly to its minimum.

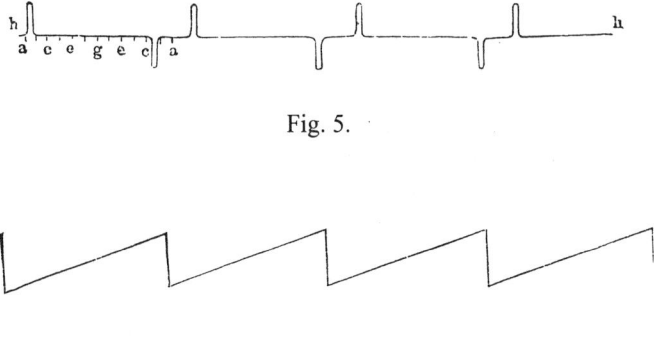

Fig. 5.

Fig. 6.

It is to such differences in the forms of the waves of sound that the variety of quality in musical tones is due. We may even carry the analogy further. The more uniformly rounded the form of wave, the softer and milder is the quality of tone. The more jerking and angular the wave-form, the more piercing the quality. Tuning-forks, with their rounded forms of wave (Fig. 3), have an extraordinarily soft quality; and the qualities of tone generated by the zither and violin resemble in harshness the angularity of their wave-forms. (Figs. 5 and 6.)

Finally, I would direct your attention to an instructive spectacle, which I have never been able to view without a certain degree of physico-scientific delight, because it displays to the bodily eye, on the surface of water, what otherwise could only be recognised by the mind's eye of the mathematical thinker in a mass of air traversed in all directions by waves of sound. I allude to the composition of many different systems of waves, as they pass over one another, each undisturbedly pursuing its own path. We can watch it from the parapet of any bridge spanning a river, but it is most complete and sublime when viewed from a cliff beside the sea. It is then rare not to see innumerable systems of waves, of various length, propagated in various directions. The longest come from the deep sea and dash against the shore. Where the boiling breakers burst shorter waves arise, and run back again towards the sea. Perhaps a bird of prey darting after a fish gives rise to a system of circular waves, which, rocking over the undulating surface, are propagated with the same regularity as on the mirror of an inland lake. And thus, from the distant horizon, where white lines of foam on the steel-blue surface betray the coming trains of wave, down to the sand beneath our feet, where the impression of their arcs remains, there is unfolded before our eyes a sublime image of immeasurable power and unceasing variety, which, as the eye at once recognises its pervading order and law, enchains and exalts without confusing the mind.

Now, just in the same way you must conceive the air of a concert-hall or ballroom traversed in every direction, and not merely on the surface, by a variegated crowd of intersecting wave-systems. From the mouths of the male singers proceed waves of six to twelve feet in length; from the lips of the songstresses dart shorter waves, from eighteen to thirty-six inches long. The rustling of silken skirts excites little curls in the air, each instrument in the orchestra emits its peculiar waves, and all these systems expand spherically from their respective centres, dart through each other, are reflected from the walls of the room, and thus rush backwards and forwards, until they succumb to the greater force of newly generated tones.

Although this spectacle is veiled from the material eye, we have another bodily organ, the ear, specially adapted to reveal it to us. This analyses the interdigitation of the waves, which in such cases would be far more confused than the intersection of the water undulations, separates the several tones which compose it, and distinguishes the voices of men and women—nay, even of individuals—the peculiar qualities of tone given out by each instrument, the rustling of the dresses, the footfalls of the walkers, and so on.

It is necessary to examine the circumstances with greater minuteness. When a bird of prey dips into the sea, rings of waves arise, which are propagated as slowly and regularly upon the moving surface as upon a surface at rest. These rings are cut into the curved surface of the waves in precisely the same way as they would have been into the still surface of a lake. The form of the external surface of the water is determined in this, as in other more complicated cases, by taking the height of each point to be the height of all the ridges of the waves which coincide at this point at one time, after deducting the sum of all similarly simultaneously coincident hollows. Such a sum of positive magnitudes (the ridges) and negative magnitudes (the hollows), where the latter have to be subtracted instead of being added, is called an *algebraical sum*. Using this term, then, we may say that *the height of every point of the surface of the water is equal to the algebraical sum of all the portions of the waves which at that moment there concur.*

It is the same with the waves of sound. They, too, are added together at every point of the mass of air, as well as in contact with the listener's ear. For them also the degree of condensation and the velocity of the particles of air in the passages of the organ of hearing are equal to the algebraical sums of the separate degrees of condensation and of the velocities of the waves of sound, considered apart. This single motion of the air produced by the simultaneous action of various sounding bodies, has now to be analyzed by the ear into the separate parts which correspond to their separate effects. For doing this the ear is much more unfavourably situated than the eye. The latter surveys the whole undulating surface at a glance. But the ear can, of course, only perceive the motion of the particles of air which impinge upon it. And yet the ear solves its problem with the greatest exactness, certainty, and determinacy. This power of the ear is of supreme importance for hearing. Were it not present it would be impossible to distinguish different tones.

Some recent anatomical discoveries appear to give a clue to the explanation of this important power of the ear.

You will all have observed the phenomena of the sympathetic production of tones in musical instruments, especially stringed instruments. The string of a pianoforte when the damper is raised begins to vibrate as soon as its proper tone is produced in its neighbourhood with sufficient force by some other means. When this foreign tone ceases the tone of the string will be heard to continue some little time longer. If we put little paper riders on the string they will be jerked off when its tone is thus produced in the neighbourhood. This sympathetic action of the string depends on the impact of the vibrating particles of air against the string and its sounding-board.

Each *separate* wave-crest (or condensation) of air which passes by the string is, of course, too weak to produce a sensible motion in it. But when a long series of wave-crests (or condensations) strike the string in such a manner that each succeeding one increases the slight tremour which resulted from the action of its predecessors, the effect finally becomes sensible. It is a process of exactly the same nature as the swinging of a heavy bell. A powerful man can scarcely move it sensibly by a single impulse. A boy, by pulling the rope at regular intervals corresponding to the time of its oscillations, can gradually bring it into violent motion.

This peculiar reinforcement of vibration depends entirely on the rhythmical application of the impulse. When the bell has been once made to vibrate as a pendulum in a very small arc, and the boy always pulls the rope as it falls, and at a time that his pull augments the existing velocity of the bell, this velocity, increasing slightly at each pull, will gradually become considerable. But if the boy apply his power at irregular intervals, sometimes increasing and sometimes diminishing the motion of the bell, he will produce no sensible effect.

In the same way that a mere boy is thus enabled to swing a heavy bell, the tremours of light and mobile air suffice to set in motion the heavy and solid mass of steel contained in a tuning-fork, provided that the tone which is excited in the air is exactly in unison with that of the fork, because in this case also every impact of a wave of air against the fork increases the motions excited by the like previous blows.

This experiment is most conveniently performed on a fork, Fig. 7, which is fastened to a sounding-board, the air being excited by a similar fork of precisely the same pitch. If one is struck, the other will be found after a few seconds to be sounding also. Then damp the first fork, by touching it for a moment with a finger, and the second will continue the tone. The second will then bring the first into vibration, and so on.

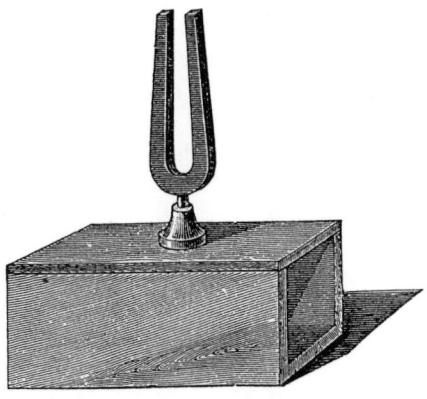

Fig. 7.

But if a very small piece of wax be attached to the ends of one of the forks, whereby its pitch will be rendered scarcely perceptibly lower than the other, the sympathetic vibration of the second fork ceases, because the times of oscillation are no longer the same in each. The blows which the waves of air excited by the first inflict upon the sounding board of the second fork, are indeed for a time in the same direction as the motions of the second fork, and consequently increase the latter, but after a very short time they cease to be so, and consequently destroy the slight motion which they had previously excited.

Lighter and more mobile elastic bodies, as for example strings, can be set in motion by a much smaller number of aerial impulses. Hence they can be set in sympathetic motion much more easily than tuning forks, and by means of a musical tone which is far less accurately in unison with themselves.

Now, then, if several tones are sounded in the neighbourhood of a pianoforte, no string can be set in sympathetic vibration unless it is in unison with one of those tones. For example, depress the forte pedal (thus raising the dampers), and put paper riders on all the strings. They will of course leap off when their strings are put in vibration. Then let several voices or instruments sound tones in the neighbourhood. All those riders, and *only* those, will leap off which are placed upon strings that correspond to tones of the same pitch as those sounded. You perceive that a pianoforte is also capable of analysing the wave confusion of the air into its elementary constituents.

The process which actually goes on in our ear is probably very like that just described. Deep in the petrous bone out of which the internal ear is hollowed, lies a peculiar organ, the cochlea or snail shell—a cavity filled with water, and so called from its resemblance to the shell of a common garden snail. This spiral passage is divided throughout its length into three sections, upper, middle, and lower, by two membranes stretched in the middle of its height. The Marchese Corti discovered some very remarkable formations in the middle section. They consist of innumerable plates, microscopically small, and arranged orderly side by side, like the keys of a piano. They are connected at one end with the fibres of the auditory nerve, and at the other with the stretched membrane.

Fig. 8 shows this extraordinarily complicated arrangement for a small part of the partition of the cochlea. The arches which leave the membrane at d and are re-inserted at e, reaching their greatest height between m and o, are probably the parts which are suited for vibration. They are spun round with innumerable fibrils, among which some nerve fibres can be recognised, coming to them through the holes near c. The transverse fibres g, h, i, k, and the cells o, also appear to belong to the nervous system. There are about three thousand arches similar to d e, lying orderly beside each other, like the keys of a piano in the whole length of the partition of the cochlea.

In the so-called vestibulum, also, where the nerves expand upon little membranous bags swimming in water, elastic appendages, similar to stiff hairs, have been lately discovered at the ends of the nerves. The anatomical arrangement of these appendages leaves scarcely any room to doubt that they are set into sympathetic vibration by the waves of sound which are conducted through the ear. Now if we venture to conjecture—it is at present only a conjecture, but after careful consideration I am led to think it very probable—that every such appendage is tuned to a certain tone like the strings of a piano, then the recent experiment with a piano shows you that when (and only when) that tone is sounded the corresponding hair-like appendage may vibrate, and the corresponding nerve-fibre experience a sensation, so that the presence of each single such tone in the midst of a whole confusion of tones must be indicated by the corresponding sensation.

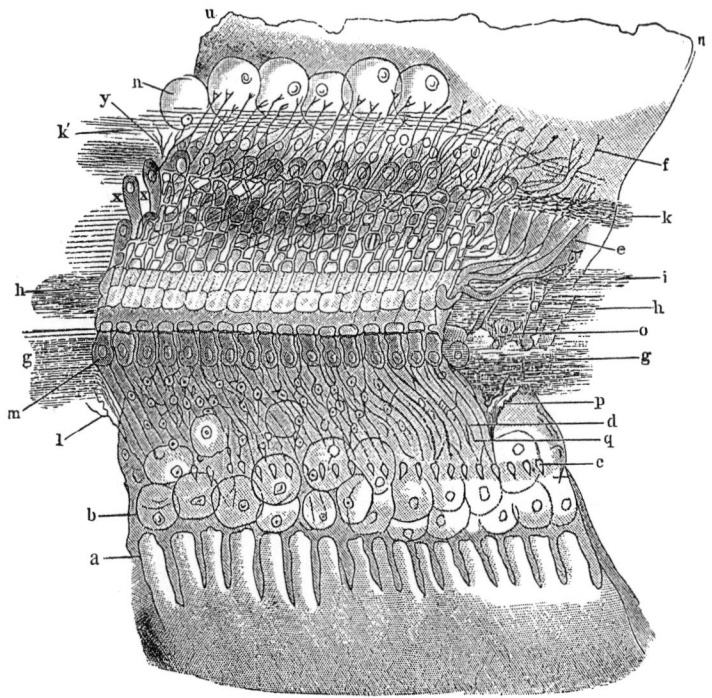

Fig. 8.

Experience then shows us that the ear really possesses the power of analysing waves of air into their elementary forms.

By compound motions of the air, we have hitherto meant such as have been caused by the simultaneous vibration of several elastic bodies. Now, since the forms of the waves of sound of different musical instruments are different, there is room to suppose that the kind of vibration excited in the passages of the ear by one such tone will be exactly the same as the kind of vibration which in another case is there excited by two or more instruments sounded together. If the ear analyses the motion into its elements in the latter case, it cannot well avoid doing so in the former, where the tone is due to a single source. And this is found to be really the case.

I have previously mentioned the form of wave with gently rounded crests and hollows, and termed it simple or pure (p. 55). In reference to this form the French mathematician Fourier has established a celebrated and important theorem which may be translated from mathematical into ordinary language thus: *Any form of wave whatever can be compounded of a number of simple waves of different lengths.* The longest of these simple waves has the same length as that of the given form of wave, the others have lengths one-half, one-third, one-fourth, &c. as great.

By the different modes of uniting the crests and hollows of these simple waves, an endless multiplicity of wave-forms may be produced.

For example, the wave-curves A and B, Fig. 9, represent waves of simple tones, B making twice as many vibrations as A in a second of time, and being consequently an octave higher in pitch. C and D, on the other hand, represent the waves which result from the superposition of B on A. The dotted curves in the first halves of C and D are repetitions of so much of the figure A. In C, the initial point e of the curve B coincides with the initial point d_0 of A. But in D, the deepest point b_2 of the first hollow in B is placed under the initial point of A. The result is two different compound-curves, the first C having steeply ascending and more gently descending crests, but so related that, by reversing the figure, the elevations would exactly fit into the depressions. But in D we have pointed crests and flattened hollows, which are, however, symmetrical with respect to right and left.

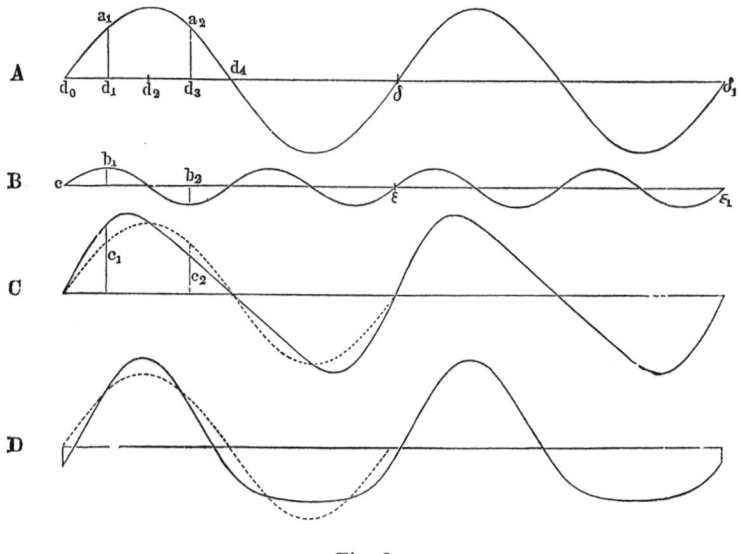

Fig. 9.

Other forms are shown in Fig 10, which are also compounded of two simple waves, A and B, of which B makes three times as many vibrations in a second as A, and consequently is the twelfth higher in pitch. The dotted curves in C and D are, as before, repetitions of A. C has flat crests and flat hollows, D has pointed crests and pointed hollows.

These extremely simple examples will suffice to give a conception of the great multiplicity of forms resulting from this method of composition. Supposing that instead of two, several simple waves were selected, with heights and initial points arbitrarily chosen, an endless variety of changes could be effected, and, in point of fact, any given form of wave could be reproduced.[3]

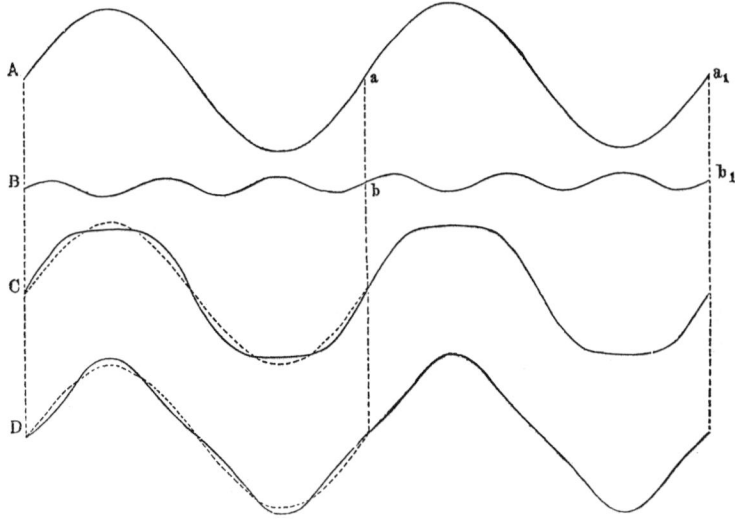

Fig. 10.

When various simple waves concur on the surface of water, the compound wave-form has only a momentary existence, because the longer waves move faster than the shorter, and consequently the two kinds of wave immediately separate, giving the eye an opportunity of recognising the presence of several systems of waves. But when waves of sound are similarly compounded, they never separate again, because long and short waves traverse air with the same velocity. Hence the compound wave is permanent, and continues its course unchanged, so that when it strikes the ear, there is nothing to indicate whether it originally left a musical instrument in this form, or whether it had been compounded on the way, out of two or more undulations.

Now what does the ear do? Does it analyse this compound wave? Or does it grasp it as a whole? The answer to these questions depends upon the sense in which we take them. We must distinguish two different points—the

3. Of course the waves could not overhang, but waves of such a form would have no possible analogue in waves of sound [which the reader will recollect are not actually in the forms here drawn, but have only condensations and rarefactions, conveniently replaced by these forms, p. 53].

audible *sensation*, as it is developed without any intellectual interference, and the *conception*, which we form in consequence of that sensation. We have, as it were, to distinguish between the material ear of the body and the spiritual ear of the mind. The material ear does precisely what the mathematician effects by means of Fourier's theorem, and what the pianoforte accomplishes when a confused mass of tones is presented to it. It analyses those wave-forms which were not originally due to simple undulations, such as those furnished by tuning forks, into a sum of simple tones, and feels the tone due to each separate simple wave separately, whether the compound wave originally proceeded from a source capable of generating it, or became compounded on the way.

For example, on striking a string, it will give a tone corresponding, as we have seen, to a wave-form widely different from that of a simple tone. When the ear analyses this wave-form into a sum of simple waves, it hears at the same time a series of simple tones corresponding to these waves.

Strings are peculiarly favourable for such an investigation, because they are themselves capable of assuming extremely different forms in the course of their vibration, and these forms may also be considered, like those of aerial undulations, as compounded of simple waves. Fig. 4, p. 56, shows the consecutive forms of a string struck by a simple rod. Fig. 11, p. 65, gives a number of other forms of vibration of a string, corresponding to simple tones. The continuous line shows the extreme displacement of the string in one direction, and the dotted line in the other. At a the string produces its fundamental tone, the deepest simple tone it can produce, vibrating in its whole length, first on one side and then on the other. At b it falls into two vibrating sections, separated by a single stationary point β, called a *node* (knot). The tone is an octave higher, the same as each of the two sections would separately produce, and it performs twice as many vibrations in a second as the fundamental tone. At c we have two nodes, γ_1 and γ_2, and three vibrating sections, each vibrating three times as fast as the fundamental tone and hence giving its twelfth. At d_1 there are three nodes, δ_1, δ_2, δ_3, and four vibrating sections, each vibrating four times as quickly as the fundamental tone, and giving the second octave above it.

In the same way forms of vibration may occur with 5, 6, 7, &c., vibrating sections, each performing respectively, 5, 6, 7, &c. times as many vibrations in a second as the fundamental tone, and all other vibrational forms of the string may be conceived as compounded of a sum of such simple vibrational forms.

The vibrational forms with stationary points or nodes may be produced, by gently touching the strings at one of these points, either with the finger or a rod, and rubbing the string with a violin bow, plucking it with the finger, or striking it with a pianoforte hammer. The bell-like harmonics or flageolet-tones of strings, so much used in violin playing, are thus produced.

Now suppose that a string has been excited, and after its tone has been allowed to continue for a moment, it is touched gently at its middle point β, Fig. 11 b, or δ_2, Fig. 11 d. The vibrational forms a and c, for which this point is in motion, will be immediately checked and destroyed; but the vibrational forms b and d, for which this point is at rest, will not be disturbed, and the tones due to them will continue to be heard. In this way we can readily discover whether certain members of the series of simple tones are contained in the compound tone of a string when excited in any given way, and the ear can be rendered sensible of their existence.

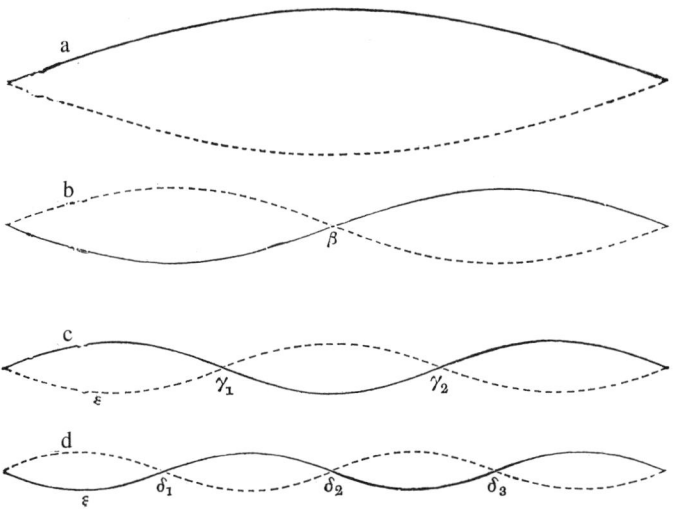

Fig. 11.

When once these simple tones in the sound of a string have been thus rendered audible, the ear will readily be able to observe them in the untouched string, after a little accurate attention.

The series of tones which are thus made to combine with a given fundamental tone, is perfectly determinate. They are tones which perform twice, thrice, four times, &c., as many vibrations in a second as the fundamental tone. They are called the *upper partials*, or harmonic overtones, of the fundamental tone. If this last be *c*, the series may be written as follows in musical notation, [it being understood that, on account of the temperament of a piano, these are not precisely the fundamental tones of the corresponding strings on that instrument, and that in particular the upper partial, b″, is necessarily much flatter than the fundamental tone of the corresponding note on the piano].

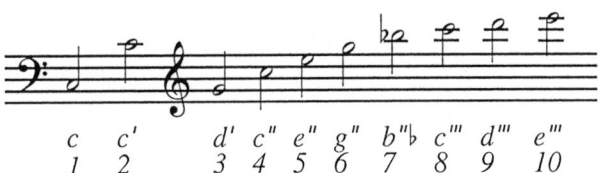

c	*c'*		*d'*	*c"*	*e"*	*g"*	*b"♭*	*c'''*	*d'''*	*e'''*
1	*2*		*3*	*4*	*5*	*6*	*7*	*8*	*9*	*10*

Not only strings, but almost all kinds of musical instruments, produce waves of sound which are more or less different from those of simple tones, and are therefore capable of being compounded out of a greater or less number of simple waves. The ear analyses them all by means of Fourier's theorem better than the best mathematician, and on paying sufficient attention can distinguish the separate simple tones due to the corresponding simple waves. This corresponds precisely to our theory of the sympathetic vibration of the organs described by Corti. Experiments with the piano, as well as the mathematical theory of sympathetic vibrations, show that any upper partials which may be present will also produce sympathetic vibrations. It follows, therefore, that in the cochlea of the ear, every external tone will set in sympathetic vibration, not merely the little plates with their accompanying nerve-fibres, corresponding to its fundamental tone, but also those corresponding to all the upper partials, and that consequently the latter must be heard as well as the former.

Hence a simple tone is one excited by a succession of simple wave-forms. All other wave-forms, such as those produced by the greater number of musical instruments, excite sensations of a variety of simple tones.

Consequently, all the tones of musical instruments must in strict language, so far as the sensation of musical tone is concerned, be regarded as chords with a predominant fundamental tone.

The whole of this theory of upper partials or harmonic overtones will perhaps seem new and singular. Probably few or none of those present, however frequently they may have heard or performed music, and however fine may be their musical ear, have hitherto perceived the existence of any such tones, although, according to my representations, they must be always and continuously present. In fact, a peculiar act of attention is requisite in order to hear them, and unless we know how to perform this act, the tones remain concealed. As you are aware, no perceptions obtained by the senses are merely sensations impressed on our nervous systems. A peculiar intellectual activity is required to pass from a nervous sensation to the conception of an external object, which the sensation has aroused. The sensations of our nerves of sense are mere symbols indicating certain external objects, and it is usually only after considerable practice that we acquire the power of drawing correct conclusions from our sensations respecting the corresponding objects. Now it is a universal law of the perceptions obtained through the senses, that we pay only so much attention to the sensations actually experienced, as is sufficient for us to recognise external objects. In this respect we are very onesided and inconsiderate partisans of practical utility; far more so indeed than we suspect. All sensations which have no direct reference to external objects, we are accustomed, as a matter of course, entirely to ignore, and we do not become aware of them till we make a scientific investigation of the action of the senses, or have our attention directed by illness to the phenomena of our own bodies. Thus we often find patients, when suffering under a slight inflammation of the eyes, become for the first time aware of those beads and fibres known as *mouches volantes* swimming about within the vitreous humour of the eye, and then they often hypochondriacally imagine all sorts of coming evils, because they fancy that these appearances are new, whereas they have generally existed all their lives.

Who can easily discover that there is an absolutely blind point, the so-called *punctum coecum*, within the retina of every healthy eye? How many people know that the only objects they see single are those at which they are looking, and that all other objects, behind or before these, appear double? I could adduce a long list of similar examples, which have not been brought to light till the actions of the senses were scientifically investigated, and which remain obstinately concealed, till attention has been drawn to them by appropriate means—often an extremely difficult task to accomplish.

To this class of phenomena belong the upper partial tones. It is not enough for the auditory nerve to have a sensation. The intellect must reflect upon it. Hence my former distinction of a material and a spiritual ear.

We always hear the tone of a string accompanied by a certain combination of upper partial tones. A different combination of such tones belongs to the tone of a flute, or of the human voice, or of a dog's howl. Whether a violin or a flute, a man or a dog is close by us is a matter of interest for us to know, and our ear takes care to distinguish the peculiarities of their tones with accuracy. The *means* by which we can distinguish them, however, is a matter of perfect indifference.

Whether the cry of the dog contains the higher octave or the twelfth of the fundamental tone, has no practical interest for us, and never occupies our attention. The upper partials are consequently thrown into that unanalysed mass of peculiarities of a tone which we call its *quality*. Now as the existence of upper partial tones depends on the *wave form*, we see, as I was able to state previously (p. 55), that the *quality of tone* corresponds to the *form of wave*.

The upper partial tones are most easily heard when they are not in harmony with the fundamental tone, as in the case of bells. The art of the bell-founder consists precisely in giving bells such a form that the deeper and stronger partial tones shall be in harmony with the fundamental tone, as otherwise the bell would be unmusical, tinkling like a kettle. But the higher partials are always out of harmony, and hence bells are unfitted for artistic music.

On the other hand, it follows, from what has been said, that the upper partial tones are all the more difficult to hear, the more accustomed we are to the compound tones of which they form a part. This is especially the case with the human voice, and many skilful observers have consequently failed to discover them there.

The preceding theory was wonderfully corroborated by leading to a method by which not only I myself, but other persons, were enabled to hear the upper partial tones of the human voice.

No particularly fine musical ear is required for this purpose, as was formerly supposed, but only proper means for directing the attention of the observer.

Let a powerful male voice sing the note e♭ ⸻ to the vowel *o* in *ore*, close to a good piano. Then lightly touch on the piano the note *b'* ♭ ⸻ in the next octave above, and listen attentively to the sound of the piano as it dies away. If this is *b'* ♭ a real upper partial in the compound tone uttered by the singer, the sound of the piano will apparently not die away at all, but the corresponding upper partial of the voice will be heard as if the note of the piano continued.[4] By properly varying the experiment, it will be found possible to distinguish the vowels from one another by their upper partial tones.

The investigation is rendered much easier by arming the ear with small globes of glass or metal, as in Fig. 12. The larger opening a is directed to

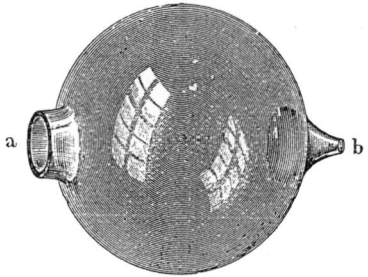

Fig. 12.

the source of sound, and the smaller funnel-shaped end is applied to the drum of the ear. The enclosed mass of air, which is almost entirely separated from that without, has its own proper tone or key-note, which will be heard, for example, on blowing across the edge of the opening a. If then this proper tone of the globe is excited in the external air, either as a fundamental or upper partial tone, the included mass of air is brought into

4. In repeating this experiment the observer must remember that the e♭ of the piano is not a true twelfth below the *b'* ♭. Hence the singer should first be given *b'* ♭ from the piano, which he will naturally sing as *b* ♭, an octave lower, and then take a true fifth below it. A skilful singer will thus hit the true twelfth and produce the required upper partial *b'* ♭. On the other hand, if he sings e♭ from the piano, his upper partial *b'* ♭ will probably beat with that of the piano.—TR.

violent sympathetic vibration, and the ear thus connected with it hears the corresponding tone with much increased intensity. By this means it is extremely easy to determine whether the proper tone of the globe is or is not contained in a compound tone or mass of tones.

On examining the vowels of the human voice, it is easy to recognise with the help of such resonators as have just been described, that the upper partial tones of each are peculiarly strong in certain parts of the scale: thus O in *ore* has its upper partials in the neighbourhood of $b' ♭$, A in *father* in the neighbourhood of $b'' ♭$ (an octave higher). The following gives a general view of those portions of the scale where the upper partials of the vowels, as pronounced in the north of Germany, are particularly strong. [5]

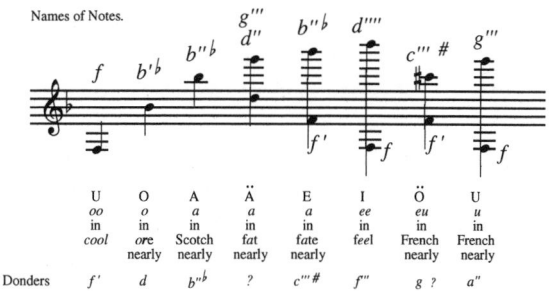

Names of Notes.								
	U	O	A	Ä	E	I	Ö	U
	oo	*o*	*a*	*a*	*a*	*ee*	*eu*	*u*
	in	in	in	in	in	in	in	in
	cool	ore	Scotch	fat	fate	feel	French	French
		nearly	nearly	nearly	nearly		nearly	nearly
Donders	f'	d	$b''^♭$	?	$c'''\#$	f''	g ?	a''

5. The corresponding English vowel sounds are probably none of them precisely the same as those pronounced by the author. It is necessary to note this, for a very slight variation in pronunciation would produce a change in the fundamental tone, and consequently a more considerable change in the position of the upper partials. The tones given by Donders, which are written below the English equivalents, are cited on the authority of Helmholtz's *Tonempfindungen*, 3rd edition, 1870, p. 171, where Helmholtz says: 'Donders's results differ somewhat from mine, partly because his refer to a Dutch, and mine to a North German pronunciation, and partly because Donders, not having had the assistance of tuning forks, could not always correctly determine the octave to which the sounds belong.' Also (*ib*. p. 167) the author remarks that $b''♭$ answers only to the deep German *a* (which is the broad Scotch *a '*, or *aw* without labialisation), and that if the brighter Italian *a* (English *a* in *father*) be used, the resonance rises a third, to d'''. Dr. C. L. Merkel, of Leipzig, in his *Physiologie der menschlishen Sprache*, 1866, p. 109, after citing Helmholtz's experiments as detailed in his *Tonempfindungen*, gives the following as 'the pitches of the vowels according to his most recent examination of his own habits of speech, as accurately as he is able to note them.'

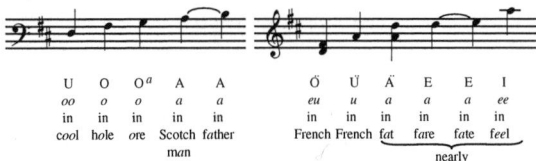

	U	O	O^a	A	A	Ö	Ü	Ä	E	E	I
	oo	*o*	*o*	*a*	*a*	*eu*	*u*	*a*	*a*	*a*	*ee*
	in	in	in	in	in	in	in	in	in	in	in
	cool	hole	ore	Scotch	father	French	French	fat	fare	fate	feel
				man						nearly	

'Here the note *a* applies to the *timbre obscur* of A with low larynx, and *b* to the *timbre clair* of A with high larynx, and similarly the vowel E may pass from d'' to c'' by narrowing the channel in the mouth. The intermediate vowels Ö, Ä, have also two different timbres and hence their pitch is not fixed; the most frequent are consequently written over one another; the lower note is for the obscure, and the higher for the bright timbre. But the vowel Ü seems to be tolerably fixed as a', just as its parents U and I are upon d and a'', and it has consequently the pitch of the ordinary a' tuning fork.'—TR.

The following easy experiment clearly shows that it is indifferent whether the several simple tones contained in a compound tone like a vowel uttered by the human voice come from one source or several. If the dampers of a pianoforte are raised, not only do the sympathetic vibrations of the strings furnish tones of the same *pitch* as those uttered beside it; but if we sing A (*a* in *father*) to any note of the piano, we hear an A quite clearly returned from the strings; and if E (*a* in *fare* or *fate*), O (*o* in *hole* or *ore*), and U (*oo* in *cool*), be similarly sung to the note, E, O, and U will also be echoed back. It is only necessary to hit the note of the piano with great exactness.[6] Now the sound of the vowel is produced solely by the sympathetic vibration of the higher strings, which correspond with the upper partial tones of the tone sung.

In this experiment the tones of numerous strings are excited by a tone proceeding from a single source, the human voice, which produces a motion of the air, equivalent in form, and therefore in quality, to that of this single tone itself.

We have hitherto spoken only of compositions of waves of different lengths. We will now compound waves of the same length which are moving in the same direction. The result will be entirely different, according as the elevations of one coincide with those of the other (in which case elevations of double the height and depressions of double the depth are produced), or the elevations of one fall on the *depressions* of the other. If both waves have the same height, so that the elevations of one exactly fit into the depressions of the other, both elevations and depressions will vanish in the second case, and the two waves will mutually destroy each other. Similarly two waves of sound, as well as two waves of water, may mutually destroy each other, when the condensations of one coincide with the rarefactions of the other. This remarkable phenomenon wherein sound is silenced by a precisely similar sound, is called the *interference* of sounds.

This is easily proved by means of the siren already described. On placing the upper box so that the puffs of air may proceed simultaneously from the rows of twelve holes in each wind chest, their effect is reinforced, and we obtain the fundamental tone of the corresponding tone of the siren very full and strong. But on arranging the boxes so that the upper puffs escape when

---

6. My own experience shows that if any vowel at any pitch be loudly and sharply sung, or called out, beside a piano, of which the dampers have been raised, that vowel will be echoed back. There is generally a sensible pause before the echo is heard. Before repeating the experiment with a new vowel, whether at the same or a different pitch, damp all the strings and then again raise the dampers. The result can easily be made audible to a hundred persons at once, and it is extremely interesting and instructive. It is peculiarly so, if different vowels be sung to the same pitch, so that they have all the same fundamental tone, and the upper partials only differ in intensity. For female voices the pitches ♯♯♯ *a'* to *c"* are favourable for all vowels. This is a fundamental experiment for the theory of vowel sounds, and should be repeated by all who are interested in speech.—Tr.

the lower series of holes is covered, and conversely, the fundamental tone vanishes, and we only hear a faint sound of the first upper partial, which is an octave higher, and which is not destroyed by interference under these circumstances.

Interference leads us to the so-called musical beats. If two tones of *exactly* the same pitch are produced simultaneously, and their elevations coincide at first, they will never cease to coincide, and if they did not coincide at first they never will coincide.

The two tones will either perpetually reinforce, or perpetually destroy each other. But if the two tones have only *approximatively* equal pitches, and their elevations at first coincide, so that they mutually reinforce each other, the elevations of one will gradually outstrip the elevations of the other. Times will come when the elevations of the one fall upon the depressions of the other, and then other times when the more rapidly advancing elevations of the one will have again reached the elevations of the other. These alternations become sensible by that alternate increase and decrease of loudness, which we call a *beat*. These beats may often be heard when two instruments which are not exactly in unison play a note of the same name. When the two or three strings which are struck by the same hammer on a piano are out of tune, the beats may be distinctly heard. Very slow and regular beats often produce a fine effect in sostenuto passages, as in sacred part-songs, by pealing through the lofty aisles like majestic waves, or by a gentle tremour giving the tone a character of enthusiasm and emotion. The greater the difference of the pitches, the quicker the beats. As long as no more than four to six beats occur in a second, the ear readily distinguishes the alternate reinforcements of the tone. If the beats are more rapid the tone grates on the ear, or, if it is high, becomes cutting. A grating tone is one interrupted by rapid breaks, like that of the letter R, which is produced by interrupting the tone of the voice by a tremour of the tongue or uvula.[7]

When the beats become more rapid, the ear finds a continuallyincreasing difficulty when attempting to hear them separately, even though there is a sensible roughness of the tone. At last they become entirely undistinguishable, and, like the separate puffs which compose a tone, dissolve as it were into a continuous sensation of tone.[8]

Hence, while every separate musical tone excites in the auditory nerve a uniform sustained sensation, two tones of different pitches mutually disturb one another, and split up into separable beats, which excite a feeling

---

7. The trill of the uvula is called the Northumbrian burr, and is not known out of Northumberland, in England. In France it is called the *r grasseyé* or *provençal*, and is the commonest Parisian sound of *r*. The uvula trill is also very common in Germany, but it is quite unknown in Italy.—TR.

8. The transition of beats into a harsh dissonance was displayed by means of two organ pipes, of which one was gradually put more and more out of tune with the other.

of discontinuity as disagreeable to the ear as similar intermittent but rapidly repeated sources of excitement are unpleasant to the other organs of sense; for example, flickering and glittering light to the eye, scratching with a brush to the skin. This roughness of tone is the essential character of dissonance. It is most unpleasant to the ear when the two tones differ by about a semitone, in which case, in the middle portions of the scale, from twenty to forty beats ensue in a second. When the difference is a whole tone, the roughness is less; and when it reaches a *third* it usually disappears, at least in the higher parts of the scale. The (minor or major) *third* may in consequence pass as a consonance. Even when the fundamental tones have such widely-different pitches that they cannot produce audible beats, the upper partial tones may beat and make the tone rough. Thus, if two tones form a *fifth* (that is, one makes two vibrations in the same time as the other makes three), there is one upper partial in both tones which makes six vibrations in the same time. Now, if the ratio of the pitches of the fundamental tones is exactly as 2 to 3, the two upper partial tones of six vibrations are precisely alike, and do not destroy the harmony of the fundamental tones. But if this ratio is only approximatively as 2 to 3, then these two upper partials are not exactly alike, and hence will beat and roughen the tone.

It is very easy to hear the beats of such imperfect fifths, because, as our pianos and organs are now tuned, all the fifths are impure, although the beats are very slow. By properly directed attention, or still better with the help of a properly tuned resonator, it is easy to hear that it is the particular upper partials here spoken of, that are beating together. The beats are necessarily weaker than those of the fundamental tones, because the beating upper partials are themselves weaker. Although we are not usually clearly conscious of these beating upper partials, the ear feels their effect as a want of uniformity or a roughness in the mass of tone, whereas a perfectly pure fifth, the pitches being precisely in the ratio of 2 to 3, continues to sound with perfect smoothness, without any alterations, reinforcements, diminutions, or roughnesses of tone. As has already been mentioned, the siren proves in the simplest manner that the most perfect consonance of the fifth precisely corresponds to this ratio between the pitches. We have now learned the reason of the roughness experienced when any deviation from that ratio has been produced.

In the same way two tones, which have their pitches exactly in the ratios of 3 to 4, or 4 to 5, and consequently form a perfect fourth or a perfect major third, sound much better when sounded together, than two others of which the pitches slightly deviate from this exact ratio. In this manner, then, any given tone being assumed as fundamental, there is a precisely determinate number of other degrees of tone which can be sounded at the same time with it, without producing any want of uniformity or any roughness of tone, or which will at least produce less roughness than any slightly greater or smaller intervals of tone under the same circumstances.

This is the reason why modern music, which is essentially based on the harmonious consonance of tones, has been compelled to limit its scale to certain determinate degrees. But even in ancient music, which allowed only one part to be sung at a time, and hence had no harmony in the modern sense of the word, it can be shown that the upper partial tones contained in all musical tones sufficed to determine a preference in favour of progressions through certain determinate intervals. When an upper partial tone is common to two successive tones in a melody, the ear recognises a certain relationship between them, serving as an artistic bond of union. Time is, however, too short for me to enlarge on this topic, as we should be obliged to go far back into the history of music.

I will but mention that there exists another kind of secondary tones, which are only heard when two or more loudish tones of different pitch are sounded together, and are hence termed *combinational*.[9] These secondary tones are likewise capable of beating, and hence producing roughness in the chords. Suppose a perfectly just major third c' e' ▦ (ratio of pitches, 4 to 5) is sounded on the siren, or with properly-tuned organ pipes, or on a violin;[10] then a faint C ▦ two octaves deeper than the c' will be heard as a combinational tone. The same C is also heard when the tones e' g' ▦ (ratio of pitches 5 to 6) are sounded together.[11]

If the three tones c', e', g', having their pitches precisely in the ratios 4, 5, and 6, are struck together, the combinational tone C is produced twice[12] in perfect unison, and without beats. But if the three notes are not exactly thus tuned,[13] the two C combinational tones will have different pitches, and produce faint beats.

The combinational tones are usually much weaker than the upper partial tones, and hence their beats are much less rough and sensible than those of the latter. They are consequently but little observable, except in tones which have scarcely any upper partials, as those produced by flutes or the closed pipes of organs. But it is indisputable that on such instruments part-music

---

9. These are of two kinds, *differential* and *summational*, according as their pitch is the difference or sum of the pitches of the two generating tones. The former are the only combinational tones here spoken of. The discovery of the latter was entirely due to the theoretical investigations of the author.—TR.

10. In the ordinary tuning of the English concertina this major third is just, and generally this instrument shows the differential tones very well. The major third is very false on the harmonium and piano.—TR.

11. This minor third is very false on the English concertina, harmonium, or piano, and the combinational tone heard is consequently very different from the true C.—TR.

12. The combinational tone c, an octave higher, is also produced once from the fifth c' g'. —TR.

13. As on the English concertina or harmonium, on both of which the consequent effect may be well heard.—TR.

scarcely presents any line of demarcation between harmony and dys-harmony, and is consequently deficient both in strength and character. On the contrary, all good musical qualities of tone are comparatively rich in upper partials, possessing the five first, which form the octaves, fifths, and major thirds of the fundamental tone. Hence, in the mixture stops of the organ, additional pipes are used, giving the series of upper partial tones corresponding to the pipe producing the fundamental tone, in order to generate a penetrating, powerful quality of tone to accompany congregational singing. The important part played by the upper partial tones in all artistic musical effects is here also indisputable.

We have now reached the heart of the theory of harmony. Harmony and dysharmony are distinguished by the undisturbed current of the tones in the former, which are as flowing as when produced separately, and by the disturbances created in the latter, in which the tones split up into separate beats. All that we have considered, tends to this end. In the first place the phenomenon of beats depends on the interference of waves. Hence they could only occur, if sound were due to undulations. Next, the determination of consonant intervals necessitated a capability in the ear of feeling the upper partial tones, and analysing the compound systems of waves into simple undulations, according to Fourier's theorem. It is entirely due to this theorem that the pitches of the upper partial tones of all serviceable musical tones must stand to the pitch of their fundamental tones in the ratios of the whole numbers to 1, and that consequently the ratios of the pitches of concordant intervals must correspond with the smallest possible whole numbers. How essential is the physiological constitution of the ear which we have just considered, becomes clear by comparing it with that of the eye. Light is also an undulation of a peculiar medium, the luminous ether, diffused through the universe, and light, as well as sound, exhibits phenomena of interference. Light, too, has waves of various periodic times of vibration, which produce in the eye the sensation of colour, red having the greatest periodic time, then orange, yellow, green, blue, violet; the periodic time of violet being about half that of the outermost red. But the eye is unable to decompose compound systems of luminous waves, that is, to distinguish compound colours from one another. It experiences from them a single, unanalysable, simple sensation, that of a mixed colour. It is indifferent to the eye whether this mixed colour results from a union of fundamental colours with simple, or with non-simple ratios of periodic times. The eye has no sense of harmony in the same meaning as the ear. There is no music to the eye.

Aesthetics endeavour to find the principle of artistic beauty in its uncon-scious conformity to law. To-day I have endeavoured to lay bare the hidden law, on which depends the agreeableness of consonant combinations. It is in the truest sense of the word unconsciously obeyed, so far as it depends on the upper partial tones, which, though felt by the nerves, are not usually consciously present to the mind. Their compatibility or incompatibility however is felt, without the hearer knowing the cause of the feeling he experiences.

These phenomena of agreeableness of tone, as determined solely by the senses, are of course merely the first step towards the beautiful in music. For the attainment of that higher beauty which appeals to the intellect, harmony and dysharmony are only means, although essential and powerful means. In dysharmony the auditory nerve feels hurt by the beats of incompatible tones. It longs for the pure efflux of the tones into harmony. It hastens towards that harmony for satisfaction and rest. Thus both harmony and dysharmony alternately urge and moderate the flow of tones, while the mind sees in their immaterial motion an image of its own perpetually streaming thoughts and moods. Just as in the rolling ocean, this movement, rhythmically repeated, and yet ever varying, rivets our attention and hurries us along. But whereas in the sea, blind physical forces alone are at work, and hence the final impression on the spectator's mind is nothing but solitude—in a musical work of art the movement follows the outflow of the artist's own emotions. Now gently gliding, now gracefully leaping, now violently stirred, penetrated or laboriously contending with the natural expression of passion, the stream of sound, in primitive vivacity, bears over into the hearer's soul unimagined moods which the artist has overheard from his own, and finally raises him up to that repose of everlasting beauty, of which God has allowed but few of his elect favourites to be the heralds.

But I have reached the confines of physical science, and must close.

A Lecture Delivered in Bonn during the Winter of 1857.

4

On the Relation of Natural Science[1]
to Science in General

TO-DAY we are met, according to annual custom, in grateful commemoration of an enlightened sovereign of this kingdom, Charles Frederick, who, in an age when the ancient fabric of European society seemed tottering to its fall, strove, with lofty purpose and untiring zeal, to promote the welfare of his subjects, and, above all, their moral and intellectual development. Rightly did he judge that by no means could he more effectually realise this beneficent intention than by the revival and the encouragement of this University. Speaking, as I do, on such an occasion, at once in the name and in the presence of the whole University, I have thought it well to try and take, as far as is permitted by the narrow standpoint of a single student, a general view of the connection of the several sciences, and of their study.

It may, indeed, be thought that, at the present day, those relations between the different sciences which have led us to combine them under the name *Universitas Literarum*, have become looser than ever. We see scholars and scientific men absorbed in specialties of such vast extent, that the most universal genius cannot hope to master more than a small section of our present range of knowledge. For instance, the philologists of the last three centuries found ample occupation in the study of Greek and Latin; at best they added to it the knowledge of two or three European languages, acquired for practical purposes. But now comparative philology aims at nothing less than an acquaintance with all the languages of all branches of the human family, in order to deduce from them the laws by which language itself has been formed, and to this gigantic task it has already applied itself with superhuman industry. Even classical philology is no longer restricted to the study of those works which, by their artistic perfection and precision of thought, or because of the importance of their

---

1. The German word *Naturwissenschaft* has no exact equivalent in modern English, including, as it does, both the Physical and the Natural Sciences. Curiously enough, in the original charter of the Royal Society, the phrase *Natural Knowledge* covers the same ground, but is there used in opposition to supernatural knowledge. (Note in Buckle's *Civilisation*, vol. ii. p. 341.)—TR.

contents, have become models of prose and poetry to all ages. On the contrary, we have learnt that every lost fragment of an ancient author, every gloss of a pedantic grammarian, every allusion of a Byzantine court-poet, every broken tombstone found in the wilds of Hungary or Spain or Africa, may contribute a fresh fact, or fresh evidence, and thus serve to increase our knowledge of the past. And so another group of scholars are busy with the vast scheme of collecting and cataloguing, for the use of their successors, every available relic of classical antiquity. Add to this, in history, the study of original documents, the critical examination of parchments and papers accumulated in the archives of states and of towns; the combination of details scattered up and down in memoirs, in correspondence, and in biographies; the deciphering of hieroglyphics and cuneiform inscriptions; in natural history the more and more comprehensive classification of minerals, plants, and animals, as well living as extinct; and there opens out before us an expanse of knowledge the contemplation of which may well bewilder us. In all these sciences the range of investigation widens as fast as the means of observation improve. The zoologists of past times were content to have described the teeth, the hair, the feet, and other external characteristics of an animal. The anatomist, on the other hand, confined himself to human anatomy, so far as he could make it out by the help of the knife, the saw, and the scalpel, with the occasional aid of injections of the vessels. Human anatomy then passed for an unusually extensive and difficult study. Now we are no longer satisfied with the comparatively rough science which bore the name of human anatomy, and which, though without reason, was thought to be almost exhausted. We have added to it comparative anatomy—that is, the anatomy of all animals—and microscopic anatomy, both of them sciences of infinitely wider range, which now absorb the interest of students.

The four elements of the ancients and of mediaeval alchemy have been increased to sixty-four, the last four of which are due to a method invented in our own University, which promises still further discoveries.[2] But not merely is the number of the elements far greater, the methods of producing complicated combinations of them have been so vastly improved, that what is called organic chemistry, which embraces only compounds of carbon with oxygen, hydrogen, nitrogen, and a few other elements, has already taken rank as an independent science.

'As the stars of heaven for multitude' was in ancient times the natural expression for a number beyond our comprehension, Pliny even thinks it almost presumption ('rem etiam Deo improbam') on the part of Hipparchus to have undertaken to count the stars and to determine their relative

---

2. That is the method of spectrum analysis, due to Bunsen and Kirchhoff, both of Heidelberg. The elements alluded to are caesium, rubidium, thallium, and iridium.

positions. And yet none of the catalogues up to the seventeenth century, constructed without the aid of telescopes, give more than from 1,000 to 1,500 stars of magnitudes from the first to the fifth. At present several observatories are engaged in continuing these catalogues down to stars of the tenth magnitude. So that upwards of 200,000 fixed stars are to be catalogued and their places accurately determined. The immediate result of these observations has been the discovery of a great number of new planets; so that, instead of the six known in 1781, there are now seventy-five.[3]

The contemplation of this astounding activity in all branches of science may well make us stand aghast at the audacity of man, and exclaim with the Chorus in the 'Antigone': 'Who can survey the whole field of knowledge? Who can grasp the clues, and then thread the labyrinth?' One obvious consequence of this vast extension of the limits of science is, that every student is forced to choose a narrower and narrower field for his own studies, and can only keep up an imperfect acquaintance even with allied fields of research. It almost raises a smile to hear that in the seventeenth century Kepler was invited to Gratz as professor of mathematics and moral philosophy; and that at Leyden, in the beginning of the eighteenth, Boerhaave occupied at the same time the chairs of botany, chemistry, and clinical medicine, and therefore practically that of pharmacy as well. At present we require at least four professors, or, in an university with its full complement of teachers, seven or eight, to represent all these branches of science. And the same is true of other faculties.

One of my strongest motives for discussing to-day the connection of the different sciences is that I am myself a student of natural philosophy; and that it has been made of late a reproach against natural philosophy that it has struck out a path of its own, and has separated itself more and more widely from the other sciences which are united by common philological and historical studies. This opposition has, in fact, been long apparent, and seems to me to have grown up mainly under the influence of the Hegelian philosophy, or, at any rate, to have been brought out into more distinct relief by that philosophy. Certainly, at the end of the last century, when the Kantian philosophy reigned supreme, such a schism had never been proclaimed; on the contrary, Kant's philosophy rested on exactly the same ground as the physical sciences, as is evident from his own scientific works, especially from his 'Cosmogony,' based upon Newton's Law of Gravitation, which afterwards, under the name of Laplace's Nebular Hypothesis, came to be universally recognised. The sole object of Kant's 'Critical Philosophy' was to test the sources and the authority of our knowledge, and to fix a definite scope and standard for the researches of philosophy, as

---

3. At the end of November 1864, the 82nd of the small planets, Alcmene, was discovered. There are now 109.

compared with other sciences. According to his teaching, a principle discovered *a priori* by pure thought was a rule applicable to the method of pure thought, and nothing further; it could contain no real, positive knowledge. The 'Philosophy of Identity'[4] was bolder. It started with the hypothesis that not only spiritual phenomena, but even the actual world—nature, that is, and man—were the result of an act of thought on the part of a creative mind, similar, it was supposed, in kind to the human mind. On this hypothesis it seemed competent for the human mind, even without the guidance of external experience, to think over again the thoughts of the Creator, and to rediscover them by its own inner activity. Such was the view with which the 'Philosophy of Identity' set to work to construct *a priori*, the results of other sciences. The process might be more or less successful in matters of theology, law, politics, language, art, history, in short, in all sciences, the subject-matter of which really grows out of our moral nature, and which are therefore properly classed together under the name of moral sciences. The state, the church, art, and language, exist in order to satisfy certain moral needs of man. Accordingly, whatever obstacles nature, or chance, or the rivalry of other men may interpose, the efforts of the human mind to satisfy its needs, being systematically directed to one end, must eventually triumph over all such fortuitous hindrances. Under these circumstances, it would not be a downright impossibility for a philosopher, starting from an exact knowledge of the mind, to predict the general course of human development under the above-named conditions, especially if he has before his eyes a basis of observed facts, on which to build his abstractions. Moreover, Hegel was materially assisted, in his attempt to solve this problem, by the profound and philosophical views on historical and scientific subjects, with which the writings of his immediate predecessors, both poets and philosophers, abound. He had, for the most part, only to collect and combine them, in order to produce a system calculated to impress people by a number of acute and original observations. He thus succeeded in gaining the enthusiastic approval of most of the educated men of his time, and in raising extravagantly sanguine hopes of solving the deepest enigma of human life; all the more sanguine doubtless, as the connection of his system was disguised under a strangely abstract phraseology, and was perhaps really understood by but few of his worshippers.

But even granting that Hegel was more or less successful in constructing, *a priori*, the leading results of the moral sciences, still it was no proof of the correctness of the hypothesis of Identity, with which he started. The facts of nature would have been the crucial test. That in the moral sciences traces

---

4. So called because it proclaimed the identity not only of subject and object, but of contradictories, such as existence and non-existence.—TR.

of the activity of the human intellect and of the several stages of its development should present themselves, was a matter of course; but surely, if nature really reflected the result of the thought of a creative mind, the system ought, without difficulty, to find a place for her comparatively simple phenomena and processes. It was at this point that Hegel's philosophy, we venture to say, utterly broke down. His system of nature seemed, at least to natural philosophers, absolutely crazy. Of all the distinguished scientific men who were his contemporaries, not one was found to stand up for his ideas. Accordingly, Hegel himself, convinced of the importance of winning for his philosophy in the field of physical science that recognition which had been so freely accorded to it elsewhere, launched out, with unusual vehemence and acrimony, against the natural philosophers, and especially against Sir Isaac Newton, as the first and greatest representative of physical investigation. The philosophers accused the scientific men of narrowness; the scientific men retorted that the philosophers were crazy. And so it came about that men of science began to lay some stress on the banishment of all philosophic influences from their work; while some of them, including men of the greatest acuteness, went so far as to condemn philosophy altogether, not merely as useless, but as mischievous dreaming. Thus, it must be confessed, not only were the illegitimate pretensions of the Hegelian system to subordinate to itself all other studies rejected, but no regard was paid to the rightful claims of philosophy, that is, the criticism of the sources of cognition, and the definition of the functions of the intellect.

In the moral sciences the course of things was different, though it ultimately led to almost the same result. In all branches of those studies, in theology, politics, jurisprudence, aesthetics, philology, there started up enthusiastic Hegelians, who tried to reform their several departments in accordance with the doctrines of their master, and, by the royal road of speculation, to reach at once the promised land and gather in the harvest, which had hitherto only been approached by long and laborious study. And so, for some time, a hard and fast line was drawn between the moral and the physical sciences; in fact, the very name of science was often denied to the latter.

The feud did not long subsist in its original intensity. The physical sciences proved conspicuously, by a brilliant series of discoveries and practical applications, that they contained a healthy germ of extraordinary fertility; it was impossible any longer to withhold from them recognition and respect. And even in other departments of science, conscientious investigators of facts soon protested against the over-bold flights of speculation. Still, it cannot be overlooked that the philosophy of Hegel and Schelling did exercise a beneficial influence; since their time the attention of investigators in the moral sciences had been constantly and more keenly directed to the scope of those sciences, and to their intellectual contents, and therefore the great amount of labour bestowed on those systems has not been entirely thrown away.

We see, then, that in proportion as the experimental investigation of facts has recovered its importance in the moral sciences, the opposition between them and the physical sciences has become less and less marked. Yet we must not forget that, though this opposition was brought out in an unnecessarily exaggerated form by the Hegelian philosophy, it has its foundation in the nature of things, and must, sooner or later, make itself felt. It depends partly on the nature of the intellectual processes the two groups of sciences involve, partly, as their very names imply, on the subjects of which they treat. It is not easy for a scientific man to convey to a scholar or a jurist a clear idea of a complicated process of nature; he must demand of them a certain power of abstraction from the phenomena, as well as a certain skill in the use of geometrical and mechanical conceptions, in which it is difficult for them to follow him. On the other hand an artist or a theologian will perhaps find the natural philosopher too much inclined to mechanical and material explanations, which seem to them commonplace, and chilling to their feeling and enthusiasm. Nor will the scholar or the historian, who have some common ground with the theologian and the jurist, fare better with the natural philosopher. They will find him shockingly indifferent to literary treasures, perhaps even more indifferent than he ought to be to the history of his own science. In short, there is no denying that, while the moral sciences deal directly with the nearest and dearest interests of the human mind, and with the institutions it has brought into being, the natural sciences are concerned with dead, indifferent matter, obviously indispensable for the sake of its practical utility, but apparently without any immediate bearing on the cultivation of the intellect.

It has been shown, then, that the sciences have branched out into countless ramifications, that there has grown up between different groups of them a real and deeply-felt opposition, that finally no single intellect can embrace the whole range, or even a considerable portion of it. Is it still reasonable to keep them together in one place of education? Is the union of the four Faculties to form one University a mere relic of the Middle Ages? Many valid arguments have been adduced for separating them. Why not dismiss the medical faculty to the hospitals of our great towns, the scientific men to the Polytechnic Schools, and form special seminaries for the theologians and jurists? Long may the German universities be preserved from such a fate! Then, indeed, would the connection between the different sciences be finally broken. How essential that connection is, not only from an university point of view, as tending to keep alive the intellectual energy of the country, but also on material grounds, to secure the successful application of that energy, will be evident from a few considerations.

First, then, I would say that union of the different Faculties is necessary to maintain a healthy equilibrium among the intellectual energies of students. Each study tries certain of our intellectual faculties more than the rest, and strengthens them accordingly by constant exercise. But any sort of one-sided development is attended with danger; it disqualifies us for using those faculties that are less exercised, and so renders us less capable of a general view; above all it leads us to overvalue ourselves. Anyone who has found himself much more successful than others in some one department of intellectual labour, is apt to forget that there are many other things which they can do better than he can: a mistake—I would have every student remember—which is the worst enemy of all intellectual activity.

How many men of ability have forgotten to practise that criticism of themselves which is so essential to the student, and so hard to exercise, or have been completely crippled in their progress, because they have thought dry, laborious drudgery beneath them, and have devoted all their energies to the quest of brilliant theories and wonder-working discoveries! How many such men have become bitter misanthropes, and put an end to a melancholy existence, because they have failed to obtain among their fellows that recognition which must be won by labour and results, but which is ever withheld from mere self-conscious genius! And the more isolated a man is, the more liable is he to this danger; while, on the other hand, nothing is more inspiriting than to feel yourself forced to strain every nerve to win the admiration of men whom you, in your turn, must admire.

In comparing the intellectual processes involved in the pursuit of the several branches of science, we are struck by certain generic differences, dividing one group of sciences from another. At the same time it must not be forgotten that every man of conspicuous ability has his own special mental constitution, which fits him for one line of thought rather than another. Compare the work of two contemporary investigators even in closely-allied branches of science, and you will generally be able to convince yourself that the more distinguished the men are, the more clearly does their individuality come out, and the less qualified would either of them be to carry on the other's researches. To-day I can, of course, do nothing more than characterise some of the most general of these differences.

I have already noticed the enormous mass of the materials accumulated by science. It is obvious that the organisation and arrangement of them must be proportionately perfect, if we are not to be hopelessly lost in the maze of erudition. One of the reasons why we can so far surpass our predecessors in each individual study is that they have shown us how to organise our knowledge.

This organisation consists, in the first place, of a mechanical arrangement of materials, such as is to be found in our catalogues, lexicons, registers, indexes, digests, scientific and literary annuals, systems of natural history, and the like. By these appliances thus much at least is gained, that such knowledge as cannot be carried about in the memory is immediately accessible to anyone who wants it. With a good lexicon a school-boy of the present day can achieve results in the interpretation of the classics, which an Erasmus, with the erudition of a lifetime, could hardly attain. Works of this kind form, so to speak, our intellectual principal, with the interest of which we trade; it is, so to speak, like capital invested in land. The learning buried in catalogues, lexicons, and indexes looks as bare and uninviting as the soil of a farm; the uninitiated cannot see or appreciate the labour and capital already invested there; to them the work of the ploughman seems infinitely dull, weary, and monotonous. But though the compiler of a lexicon or of a system of natural history must be prepared to encounter labour as weary and as obstinate as the ploughman's, yet it need not be supposed that his work is of a low type, or that it is by any means as dry and mechanical as it looks when we have it before us in black and white. In this, as in any other sort of scientific work, it is necessary to discover every fact by careful observation, then to verify and collate them, and to separate what is important from what is not. All this requires a man with a thorough grasp, both of the object of the compilation, and of the matter and methods of the science; and for such a man every detail has its bearing on the whole, and its special interest. Otherwise dictionary-making would be the vilest drudgery imaginable.[5] That the influence of the progressive development of scientific ideas extends to these works is obvious from the constant demand for new lexicons, new natural histories, new digests, new catalogues of stars, all denoting advancement in the art of methodising and organising science.

But our knowledge is not to lie dormant in the shape of catalogues. The very fact that we must carry it about in black and white shows that our intellectual mastery of it is incomplete. It is not enough to be acquainted with the facts; scientific knowledge begins only when their laws and their causes are unveiled. Our materials must be worked up by a logical process; and the first step is to connect like with like, and to elaborate a general conception embracing them all. Such a conception, as the name implies, takes a number of single facts together, and stands as their representative in our mind. We call it a general conception, or the conception of a genus, when it embraces a number of existing objects; we call it a law when it embraces a series of incidents or occurrences. When, for example, I have made out that all mammals—that is, all warm-blooded, viviparous animals—breathe through lungs, have two chambers in the heart and at least three tympanal bones, I need no longer remember these anatomical peculiarities in the individual cases of the monkey, the dog, the horse, and

---

5. Condendaque lexica mandat damnatis.—TR.

the whale; the general rule includes a vast number of single instances, and represents them in my memory. When I enunciate the law of refraction, not only does this law embrace all cases of rays falling at all possible angles on a plane surface of water, and inform me of the result, but it includes all cases of rays of any colour incident on transparent surfaces of any form and any constitution whatsoever. This law, therefore, includes an infinite number of cases, which it would have been absolutely impossible to carry in one's memory. Moreover, it should be noticed that not only does this law include the cases which we ourselves or other men have already observed, but that we shall not hesitate to apply it to new cases, not yet observed, with absolute confidence in the reliability of our results. In the same way, if we were to find a new species of mammal, not yet dissected, we are entitled to assume, with a confidence bordering on a certainty, that it has lungs, two chambers in the heart, and three or more tympanal bones.

Thus, when we combine the results of experience by a process of thought, and form conceptions, whether general conceptions or laws, we not only bring our knowledge into a form in which it can be easily used and easily retained, but we actually enlarge it, inasmuch as we feel ourselves entitled to extend the rules and the laws we have discovered to all similar cases that may be hereafter presented to us.

The above-mentioned examples are of a class in which the mental process of combining a number of single cases so as to form conceptions is unattended by farther difficulties, and can be distinctly followed in all its stages. But in complicated cases it is not so easy completely to separate like facts from unlike, and to combine them into a clear, well-defined conception. Assume that we know a man to be ambitious; we shall perhaps be able to predict with tolerable certainty that if he has to act under certain conditions, he will follow the dictates of his ambition, and decide on a certain line of action. But, in the first place, we cannot define with absolute precision what constitutes an ambitious man, or by what standard the intensity of his ambition is to be measured; nor, again, can we say precisely what degree of ambition must operate in order to impress the given direction on the actions of the man under those particular circumstances. Accordingly, we institute comparisons between the actions of the man in question, as far as we have hitherto observed them, and those of other men who in similar cases have acted as he has done, and we draw our inference respecting his future actions without being able to express either the major or the minor premiss in a clear, sharply-defined form—perhaps even without having convinced ourselves that our anticipation rests on such an analogy as I have described. In such cases our decision proceeds only from a certain psychological instinct, not from conscious reasoning, though in reality we have gone through an intellectual process identical with that which leads us to assume that a newly-discovered mammal has lungs.

This latter kind of induction, which can never be perfectly assimilated to forms of logical reasoning, nor pressed so far as to establish universal laws, plays a most important part in human life. The whole of the process by which we translate our sensations into perceptions depends upon it, as appears especially from the investigation of what are called illusions. For instance, when the retina of the eye is irritated by a blow, we imagine we see a light in our field of vision, because we have, throughout our lives, felt irritation in the optic nerves only when there was light in the field of vision, and have become accustomed to identify the sensations of those nerves with the presence of light in the field of vision. Moreover, such is the complexity of the influences affecting the formation both of character in general and of the mental condition at any given moment, that this same kind of induction necessarily plays a leading part in the investigation of psychological processes. In fact, in ascribing to ourselves free-will, that is, full power to act as we please, without being subject to a stern inevitable law of causality, we deny *in toto* the possibility of referring at least one of the ways in which our mental activity expresses itself to a rigorous law.

We might possibly, in opposition to *logical induction* which reduces a question to clearly-defined universal propositions, call this kind of reasoning *aesthetic induction*, because it is most conspicuous in the higher class of works of art. It is an essential part of an artist's talent to reproduce by words, by form, by colour, or by music, the external indications of a character or a state of mind, and by a kind of instinctive intuition, uncontrolled by any definable rule, to seize the necessary steps by which we pass from one mood to another. If we do find that the artist has consciously worked after general rules and abstractions, we think his work poor and commonplace, and cease to admire. On the contrary, the works of great artists bring before us characters and moods with such a lifelikeness, with such a wealth of individual traits and such an overwhelming conviction of truth, that they almost seem to be more real than the reality itself, because all disturbing influences are eliminated.

Now if, after these reflections, we proceed to review the different sciences, and to classify them according to the method by which they must arrive at their results, we are brought face to face with a generic difference between the natural and the moral sciences. The natural sciences are for the most part in a position to reduce their inductions to sharply-defined general rules and principles; the moral sciences, on the other hand, have, in by far the most numerous cases, to do with conclusions arrived at by psychological instinct. Philology, in so far as it is concerned with the interpretation and emendation of the texts handed down to us, must seek to feel out, as it were, the meaning which the author intended to express, and the accessory notions which he wished his words to suggest; and for that purpose it is necessary to start with a correct insight, both into the personality of the author, and into the genius of the language in which he wrote. All this affords scope for aesthetic, but not for strictly logical induction. It is only possible to pass judgment, if you have ready in your memory a great number of similar facts, to be instantaneously confronted with the question you are trying to solve. Accordingly, one of the first requisites for studies of this class is an accurate and ready memory. Many celebrated historians and philologists have, in fact, astounded their contemporaries by their extraordinary strength of memory. Of course memory alone is insufficient without a knack of everywhere discovering real resemblance, and without a delicately and fully trained insight into the springs of human action; while this again is unattainable without a certain warmth of sympathy and an interest in observing the working of other men's minds. Intercourse with our fellow-men in daily life must lay the foundation of this insight, but the study of history and art serves to make it richer and completer, for there we see men acting under comparatively unusual conditions, and thus come to appreciate the full scope of the energies which lie hidden in our breasts.

None of this group of sciences, except grammar, lead us, as a rule, to frame and enunciate general laws, valid under all circumstances. The laws of grammar are a product of the human will, though they can hardly be said to have been framed deliberately, but rather to have grown up gradually, as they were wanted. Accordingly, they present themselves to a learner rather in the form of commands, that is, of laws imposed by external authority.

With these sciences theology and jurisprudence are naturally connected. In fact, certain branches of history and philology serve both as stepping-stones and as handmaids to them. The general laws of theology and jurisprudence are likewise commands, laws imposed by external authority to regulate, from a moral or juridical point of view, the actions of mankind; not laws which, like those of nature, contain generalisations from a vast multitude of facts. At the same time the application of a grammatical, legal, moral, or theological rule is couched, like the application of a law of nature to a particular case, in the forms of logical inference. The rule forms the major premiss of the syllogism, while the minor must settle whether the case in question satisfies the conditions to which the rule is intended to apply. The solution of this latter problem, whether in grammatical analysis, where the meaning of a sentence is to be evolved, or in the legal criticism of the credibility of the facts alleged, of the intentions of the parties, or of the meaning of the documents they have put into court, will, in most cases, be again a matter of psychological insight. On the other hand, it should not be forgotten that both the syntax of fully-developed languages and a system of jurisprudence gradually elaborated, as ours has been, by the practice of more than 2,000 years,[6] have reached a high pitch of logical completeness and consistency; so that, speaking generally, the cases which do not obviously fall under some one or other of the laws actually laid down are quite exceptional. Such exceptions there will always be, for the legislation of man can never have the absolute consistency and perfection of the laws of nature. In such cases there is no course open but to try and guess the intention of the legislator; or, if needs be, to supplement it after the analogy of his decisions in similar cases.

Grammar and jurisprudence have a certain advantage as means of training the intellect, inasmuch as they tax pretty equally all the intellectual powers. On this account secondary education among modern European nations is based mainly upon the grammatical study of foreign languages. The mother-tongue and modern foreign languages, when acquired solely by practice, do not call for any conscious logical exercise of thought, though we may cultivate by means of them an appreciation for artistic beauty of expression. The two classical languages, Latin and Greek, have, besides their exquisite logical subtlety and aesthetic beauty, an additional advantage, which they seem to possess in common with most ancient and original languages—they indicate accurately the relations of words and sentences to each other by numerous and distinct inflections. Languages are, as it were, abraded by long use; grammatical distinctions are cut down to a minimum for the sake of brevity and rapidity of expression, and are thus made less and less definite, as is obvious from the comparison of any modern European language with Latin; in English the process has gone further than in any other. This seems to me to be really the reason why the modern languages are far less fitted than the ancient for instruments of education.[7]

As grammar is the staple of school education, legal studies are used, and rightly, as a means of training persons of maturer age, even when not specially required for professional purposes.

We now come to those sciences which, in respect of the kind of intellectual labour they require, stand at the opposite end of the series to philology and history; namely, the natural and physical sciences. I do not mean to say that in many branches even of these sciences an instinctive appreciation of analogies and a certain artistic sense have no part to play. On the contrary, in natural history the decision which characteristics are to be looked upon as important for classification, and which as unimportant, what divisions of the animal and vegetable kingdoms are more natural than

---

6. It should be remembered that the Roman law, which has only partially and indirectly influenced English practice, is the recognised basis of German jurisprudence.—Tr.

7. Those to whom German is not a foreign tongue may, perhaps, be permitted to hold different views on the efficacy of modern languages in education.—Tr.

others, is really left to an instinct of this kind, acting without any strictly definable rule. And it is a very suggestive fact that it was an artist, Goethe, who gave the first impulse to the researches of comparative anatomy into the analogy of corresponding organs in different animals, and to the parallel theory of the metamorphosis of leaves in the vegetable kingdom; and thus, in fact, really pointed out the direction which the science has followed ever since. But even in those departments of science where we have to do with the least understood vital processes it is, speaking generally, far easier to make out general and comprehensive ideas and principles, and to express them in definite language, than in cases where we must base our judgment on the analysis of the human mind. It is only when we come to the experimental sciences to which mathematics are applied, and especially when we come to pure mathematics, that we see the peculiar characteristics of the natural and physical sciences fully brought out.

The essential *differentia* of these sciences seems to me to consist in the comparative ease with which the individual results of observation and experiment are combined under general laws of unexceptionable validity and of an extraordinarily comprehensive character. In the moral sciences, on the other hand, this is just the point where insuperable difficulties are encountered. In mathematics the general propositions which, under the name of axioms, stand at the head of the reasoning, are so few in number, so comprehensive, and so immediately obvious, that no proof whatever is needed for them. Let me remind you that the whole of algebra and arithmetic is developed out of the three axioms:

'Things which are equal to the same things are equal to one another.'
'If equals be added to equals, the wholes are equal.'
'If unequals be added to equals, the wholes are unequal.'

And the axioms of geometry and mechanics are not more numerous. The sciences we have named are developed out of these few axioms by a continual process of deduction from them in more and more complicated cases. Algebra, however, does not confine itself to finding the sum of the most heterogeneous combinations of a finite number of magnitudes, but in the higher analysis it teaches us to sum even infinite series, the terms of which increase or diminish according to the most various laws; to solve, in fact, problems which could never be completed by direct addition. An instance of this kind shows us the conscious logical activity of the mind in its purest and most perfect form. On the one hand we see the laborious nature of the process, the extreme caution with which it is necessary to advance, the accuracy required to determine exactly the scope of such universal principles as have been attained, the difficulty of forming and understanding abstract conceptions. On the other hand, we gain confidence in the certainty, the range, and the fertility of this kind of intellectual work.

The fertility of the method comes out more strikingly in applied mathematics, especially in mathematical physics, including, of course, physical astronomy. From the time when Newton discovered, by analysing the motions of the planets on mechanical principles, that every particle of ponderable matter in the universe attracts every other particle with a force varying inversely as the square of the distance, astronomers have been able, in virtue of that one law of gravitation, to calculate with the greatest accuracy the movements of the planets to the remotest past and the most distant future, given only the position, velocity, and mass of each body of our system at any one time. More than that, we recognise the operation of this law in the movements of double stars, whose distances from us are so great that their light takes years to reach us; in some cases, indeed, so great that all attempts to measure them have failed.

This discovery of the law of gravitation and its consequences is the most imposing achievement that the logical power of the human mind has hitherto performed. I do not mean to say that there have not been men who in power of abstraction have equalled or even surpassed Newton and the other astronomers, who either paved the way for his discovery, or have carried it out to its legitimate consequences; but there has never been presented to the human mind such an admirable subject as those involved and complex movements of the planets, which hitherto had served merely as food for the astrological superstitions of ignorant star-gazers, and were now reduced to a single law, capable of rendering the most exact account of the minutest detail of their motions.

The principles of this magnificent discovery have been successfully applied to several other physical sciences, among which physical optics and the theory of electricity and magnetism are especially worthy of notice. The experimental sciences have one great advantage over the natural sciences in the investigation of general laws of nature: they can change at pleasure the conditions under which a given result takes place, and can thus confine themselves to a small number of characteristic instances, in order to discover the law. Of course its validity must then stand the test of application to more complex cases. Accordingly the physical sciences, when once the right methods have been discovered, have made proportionately rapid progress. Not only have they allowed us to look back into primeval chaos, where nebulous masses were forming themselves into suns and planets, and becoming heated by the energy of their contraction; not only have they permitted us to investigate the chemical constituents of the solar atmosphere and of the remotest fixed stars, but they have enabled us to turn the forces of surrounding nature to our own uses and to make them the ministers of our will.

Enough has been said to show how widely the intellectual processes involved in this group of sciences differ, for the most part, from those required by the moral sciences. The mathematician need have no memory whatever for detached facts, the physicist hardly any. Hypotheses based on the recollection of similar cases may, indeed, be useful to guide one into the right track, but they have no real value till they have led to a precise and strictly defined law. Nature does not allow us for a moment to doubt that we have to do with a rigid chain of cause and effect, admitting of no exceptions. Therefore to us, as her students, goes forth the mandate to labour on till we have discovered unvarying laws; till then we dare not rest satisfied, for then only can our knowledge grapple victoriously with time and space and the forces of the universe.

The iron labour of conscious logical reasoning demands great perseverance and great caution; it moves on but slowly, and is rarely illuminated by brilliant flashes of genius. It knows little of that facility with which the most varied instances come thronging into the memory of the philologist or the historian. Rather is it an essential condition of the methodical progress of mathematical reasoning that the mind should remain concentrated on a single point, undisturbed alike by collateral ideas on the one hand, and by wishes and hopes on the other, and moving on steadily in the direction it has deliberately chosen. A celebrated logician, Mr. John Stuart Mill, expresses his conviction that the inductive sciences have of late done more for the advance of logical methods than the labours of philosophers properly so called. One essential ground for such an assertion must undoubtedly be that in no department of knowledge can a fault in the chain of reasoning be so easily detected by the incorrectness of the results as in those sciences in which the results of reasoning can be most directly compared with the facts of nature.

Though I have maintained that it is in the physical sciences, and especially in such branches of them as are treated mathematically, that the solution of scientific problems has been most successfully achieved, you will not, I trust, imagine that I wish to depreciate other studies in comparison with them. If the natural and physical sciences have the advantage of great perfection in form, it is the privilege of the moral sciences to deal with a richer material, with questions that touch more nearly the interests and the feelings of men, with the human mind itself, in fact, in its motives and the different branches of its activity. They have, indeed, the loftier and the more difficult task, but yet they cannot afford to lose sight of the example of their rivals, which, in form at least, have, owing to the more ductile nature of their materials, made greater progress. Not only have they something to learn from them in point of method, but they may also draw encouragement from the greatness of their results. And I do think that our age has learnt many lessons from the physical sciences. The absolute, unconditional reverence for facts, and the fidelity with which they are collected, a certain distrustfulness of appearances, the effort to detect in all cases relations of cause and effect, and the tendency to assume their existence, which distinguish our century from preceding ones, seem to me to point to such an influence.

I do not intend to go deeply into the question how far mathematical studies, as the representatives of conscious logical reasoning, should take a more important place in school education. But it is, in reality, one of the questions of the day. In proportion as the range of science extends, its system and organisation must be improved, and it must inevitably come about that individual students will find themselves compelled to go through a stricter course of training than grammar is in a position to supply. What strikes me in my own experience of students who pass from our classical schools to scientific and medical studies, is first, a certain laxity in the application of strictly universal laws. The grammatical rules, in which they have been exercised, are for the most part followed by long lists of exceptions; accordingly they are not in the habit of relying implicitly on the certainty of a legitimate deduction from a strictly universal law. Secondly, I find them for the most part too much inclined to trust to authority, even in cases where they might form an independent judgment. In fact, in philological studies, inasmuch as it is seldom possible to take in the whole of the premises at a glance, and inasmuch as the decision of disputed questions often depends on an aesthetic feeling for beauty of expression, and for the genius of the language, attainable only by long training, it must often happen that the student is referred to authorities even by the best teachers. Both faults are traceable to a certain indolence and vagueness of thought, the sad effects of which are not confined to subsequent scientific studies. But certainly the best remedy for both is to be found in mathematics, where there is absolute certainty in the reasoning, and no authority is recognised but that of one's own intelligence.

So much for the several branches of science considered as exercises for the intellect, and as supplementing each other in that respect. But knowledge is not the sole object of man upon earth. Though the sciences arouse and educate the subtlest powers of the mind, yet a man who should study simply for the sake of knowing, would assuredly not fulfil the purpose of his existence. We often see men of considerable endowments, to whom their good or bad fortune has secured a comfortable livelihood or good social position, without giving them, at the same time, ambition or energy enough to make them work, dragging out a weary, unsatisfied existence, while all the time they fancy they are following the noblest aim of life by constantly devoting themselves to the increase of their knowledge, and the cultivation of their minds. Action alone gives a man a life worth living; and therefore he must aim either at the practical application of his knowledge, or at the extension of the limits of science itself. For to extend the limits of science is really to work for the progress of humanity. Thus we pass to the second link, uniting the different sciences, the connection, namely, between the subjects of which they treat.

Knowledge is power. Our age, more than any other, is in a position to demonstrate the truth of this maxim. We have taught the forces of inanimate nature to minister to the wants of human life and the designs of the human intellect. The application of steam has multiplied our physical strength a million-fold; weaving and spinning machines have relieved us of labours, the only merit of which consisted in a deadening monotony. The intercourse between men, with its far-reaching influence on material and intellectual progress, has increased to an extent of which no one could have even dreamed within the lifetime of the older among us. But it is not merely on the machines by which our powers are multiplied; not merely on rifled cannon, and armour-plated ships; not merely on accumulated stores of money and the necessaries of life, that the power of a nation rests; though these things have exercised so unmistakable an influence, that even the proudest and most obstinate despotisms of our times have been forced to think of removing restrictions on industry, and of conceding to the industrious middle classes a due voice in their counsels. But political organisation, the administration of justice, and the moral discipline of individual citizens are no less important conditions of the preponderance of civilised nations; and so surely as a nation remains inaccessible to the influences of civilisation in these respects, so surely is it on the high road to destruction. The several conditions of national prosperity act and react on each other; where the administration of justice is uncertain, where the interests of the majority cannot be asserted by legitimate means, the development of the national resources, and of the power depending upon them, is impossible; nor again, is it possible to make good soldiers except out of men who have learnt under just laws to educate the sense of honour that characterises an independent man, certainly not out of those who have lived the submissive slaves of a capricious tyrant.

Accordingly every nation is interested in the progress of knowledge on the simple ground of self-preservation, even were there no higher wants of an ideal character to be satisfied; and not merely in the development of the physical sciences, and their technical application, but also in the progress of legal, political, and moral sciences, and of the accessory historical and philological studies. No nation which would be independent and influential can afford to be left behind in the race. Nor has this escaped the notice of the cultivated peoples of Europe. Never before was so large a part of the public resources devoted to universities, schools, and scientific institutions. We in Heidelberg have this year occasion to congratulate ourselves on another rich endowment granted by our government and our parliament.

I was speaking, at the beginning of my address, of the increasing division of labour and the improved organisation among scientific workers. In fact, men of science form, as it were, an organised army, labouring on behalf of the whole nation, and generally under its direction and at its expense, to augment the stock of such knowledge as may serve to promote industrial enterprise, to increase wealth, to adorn life, to improve political and social relations, and to further the moral development of individual citizens. After the immediate practical results of their work we forbear to inquire; that we leave to the uninstructed. We are convinced that whatever contributes to the knowledge of the forces of nature or the powers of the human mind is worth cherishing, and may, in its own due time, bear practical fruit, very often where we should least have expected it. Who, when Galvani touched the muscles of a frog with different metals, and noticed their contraction, could have dreamt that eighty years afterwards, in virtue of the self-same process, whose earliest manifestations attracted his attention in his anatomical researches, all Europe would be traversed with wires, flashing intelligence from Madrid to St. Petersburg with the speed of lightning? In the hands of Galvani, and at first even in Volta's, electrical currents were phenomena capable of exerting only the feeblest forces, and could not be detected except by the most delicate apparatus. Had they been neglected, on the ground that the investigation of them promised no immediate practical result, we should now be ignorant of the most important and most interesting of the links between the various forces of nature. When young Galileo, then a student at Pisa, noticed one day during divine service a chandelier swinging backwards and forwards, and convinced himself, by counting his pulse, that the duration of the oscillations was independent of the arc through which it moved, who could know that this discovery would eventually put it in our power, by means of the pendulum, to attain an accuracy in the measurement of time till then deemed impossible, and would enable the storm-tossed seaman in the most distant oceans to determine in what degree of longitude he was sailing?

Whoever, in the pursuit of science, seeks after immediate practical utility, may generally rest assured that he will seek in vain. All that science can achieve is a perfect knowledge and a perfect understanding of the action of natural and moral forces. Each individual student must be content to find his reward in rejoicing over new discoveries, as over new victories of mind over reluctant matter, or in enjoying the aesthetic beauty of a well-ordered field of knowledge, where the connection and the filiation of every detail is clear to the mind, and where all denotes the presence of a ruling intellect; he must rest satisfied with the consciousness that he too has contributed something to the increasing fund of knowledge on which the dominion of man over all the forces hostile to intelligence reposes. He will, indeed, not always be permitted to expect from his fellow-men appreciation and reward adequate to the value of his work. It is only too true, that many a man to whom a monument has been erected after his death, would have been delighted to receive during his lifetime a tenth part of the money spent in doing honour to his memory. At the same time, we must acknowledge that the value of scientific discoveries is now far more fully recognised than formerly by public opinion, and that instances of the authors of great advances in science starving in obscurity have become rarer and rarer. On the contrary, the governments and peoples of Europe have, as a rule, admitted it to be their duty to recompense distinguished achievements in science by appropriate appointments or special rewards.

The sciences have then, in this respect, all one common aim, to establish the supremacy of intelligence over the world: while the moral sciences aim directly at making the resources of intellectual life more abundant and more interesting, and seek to separate the pure gold of Truth from alloy, the physical sciences are striving indirectly towards the same goal, inasmuch as they labour to make mankind more and more independent of the material restraints that fetter their activity. Each student works in his own department, he chooses for himself those tasks for which he is best fitted by his abilities and his training. But each one must be convinced that it is only in connection with others that he can further the great work, and that therefore he is bound, not only to investigate, but to do his utmost to make the results of his investigation completely and easily accessible. If he does this, he will derive assistance from others, and will in his turn be able to render them his aid. The annals of science abound in evidence how such mutual services have been exchanged, even between departments of science apparently most remote. Historical chronology is essentially based on astronomical calculations of eclipses, accounts of which are preserved in ancient histories. conversely, many of the important data of astronomy—for instance, the invariability of the length of the day, and the periods of several comets—rest upon ancient historical notices. Of late years, physiologists, especially Brücke, have actually undertaken to draw up a complete system of all the vocables that can be produced by the organs of speech, and to base upon it propositions for an universal alphabet, adapted to all human languages. Thus physiology has entered the service of comparative philology, and has already succeeded in accounting for many apparently anomalous substitutions, on the ground that they are governed, not as hitherto supposed, by the laws of euphony, but by similarity between the movements of the mouth that produce them. Again, comparative philology gives us information about the relationships, the separations and the migrations of tribes in prehistoric times, and of the degree of civilisation which they had reached at the time when they parted. For the names of objects to which they had already learnt to give distinctive appellations reappear as words common to their later languages. So that the study of languages actually gives us historical data for periods respecting which no other historical evidence exists.[8] Yet again I may notice the help which not only the sculptor, but the archaeologist, concerned with the investigation of ancient statues, derives from anatomy. And if I may be permitted to refer to my own most recent studies, I would mention that it is possible, by reference to physical acoustics and to the physiological theory of the sensation of hearing, to account for the elementary principles on which our musical system is constructed, a problem essentially within the sphere of aesthetics. In fact, it is a general principle that the physiology of the organs of sense is most intimately connected with psychology, inasmuch as physiology traces in our sensations the results of mental processes which do not fall within the sphere of consciousness, and must therefore have remained inaccessible to us.

I have been able to quote only some of the most striking instances of this interdependence of different sciences, and such as could be explained in a few words. Naturally, too, I have tried to choose them from the most widely-separated sciences. But far wider is of course the influence which allied sciences exert upon each other. Of that I need not speak, for each of you knows it from his own experience.

In conclusion, I would say, let each of us think of himself, not as a man seeking to gratify his own thirst for knowledge, or to promote his own private advantage, or to shine by his own abilities, but rather as a fellow-labourer in one great common work bearing upon the highest interests of humanity. Then assuredly we shall not fail of our reward in the approval of our own conscience and the esteem of our fellow-citizens. To keep up these relations between all searchers after truth and all branches of knowledge, to animate them all to vigorous co-operation towards their common end, is the great office of the Universities. Therefore is it necessary that the four Faculties should ever go hand in hand, and in this conviction will we strive, so far as in us lies, to press onward to the fulfilment of our great mission.

Academical Discourse Delivered at Heidelberg, November 22, 1862.

---

8. See, for example, Mommsen's *Rome*, Book I.ch.ii.—TR.

5

On the Conservation of Force

As I have undertaken to deliver here a series of lectures, I think the best way in which I can discharge that duty will be to bring before you, by means of a suitable example, some view of the special character of those sciences to the study of which I have devoted myself. The natural sciences, partly in consequence of their practical applications, and partly from their intellectual influence on the last four centuries, have so profoundly, and with such increasing rapidity, transformed all the relations of the life of civilised nations; they have given these nations such increase of riches, of enjoyment of life, of the preservation of health, of means of industrial and of social intercourse, and even such increase of political power, that every educated man who tries to understand the forces at work in the world in which he is living, even if he does not wish to enter upon the study of a special science, must have some interest in that peculiar kind of mental labour which works and acts in the sciences in question.

On a former occasion I have already discussed the characteristic differences which exist between the natural and the mental sciences as regards the kind of scientific work. I then endeavoured to show that it is more especially in the thorough conformity with law which natural phenomena and natural products exhibit, and in the comparative ease with which laws can be stated, that this difference exists. Not that I wish by any means to deny, that the mental life of individuals and peoples is also in conformity with law, as is the object of philosophical, philological, historical, moral, and social sciences to establish. But in mental life, the influences are so interwoven, that any definite sequence can but seldom be demonstrated. In Nature the converse is the case. It has been possible to discover the law of the origin and progress of many enormously extended series of natural phenomena with such accuracy and completeness that we can predict their future occurrence with the greatest certainty; or in cases in which we have power over the conditions under which they occur, we can direct them just according to our will. The greatest of all instances of what the human mind can effect by means of a well-recognized law of natural phenomena is that afforded by modern astronomy. The one simple law of gravitation regulates the motions of the heavenly bodies not only of our own planetary system, but also of the far more distant double stars; from which, even the ray of light, the quickest of all messengers, needs years to reach our eye; and just on account of this simple conformity with law, the motions of the bodies in question, can be accurately predicted and determined both for the past and for future years and centuries to a fraction of a minute.

On this exact conformity with law depends also the certainty with which we know how to tame the impetuous force of steam, and to make it the obedient servant of our wants. On this conformity depends, moreover, the intellectual fascination which chains the physicist to his subjects. It is an interest of quite a different kind to that which mental and moral sciences afford. In the latter it is man in the various phases of his intellectual activity who chains us. Every great deed of which history tells us, every mighty passion which art can represent, every picture of manners, of civic arrangements, of the culture of peoples of distant lands, or of remote times, seizes and interests us, even if there is no exact scientific connection among them. We continually find points of contact and comparison in our own conceptions and feelings; we get to know the hidden capacities and desires of the mind, which in the ordinary peaceful course of civilised life remain unawakened.

It is not to be denied that, in the natural sciences, this kind of interest is wanting. Each individual fact, taken of itself, can indeed arouse our curiosity or our astonishment, or be useful to us in its practical applications. But intellectual satisfaction we obtain only from a connection of the whole, just from its conformity with law. *Reason* we call that faculty innate in us of discovering laws and applying them with thought. For the unfolding of the peculiar forces of pure reason in their entire certainty and in their entire bearing, there is no more suitable arena than inquiry into nature in the wider sense, the mathematics included. And it is not only the pleasure at the successful activity of one of our most essential mental powers; and the victorious subjections to the power of our thought and will of an external world, partly unfamiliar, and partly hostile, which is the reward of this labour; but there is a kind, I might almost say, of artistic satisfaction, when we are able to survey the enormous wealth of Nature as a regularly-ordered whole—a kosmos, an image of the logical thought of our own mind.

The last decades of scientific development have led us to the recognition of a new universal law of all natural phenomena, which, from its extraordinarily extended range, and from the connection which it constitutes between natural phenomena of all kinds, even of the remotest times and the most distant places, is especially fitted to give us an idea of what I have described as the character of the natural sciences, which I have chosen as the subject of this lecture.

This law is *the Law of the Conservation of Force*, a term the meaning of which I must first explain. It is not absolutely new; for individual domains of natural phenomena it was enunciated by Newton and Daniel Bernoulli; and Rumford and Humphry Davy have recognised distinct features of its presence in the laws of heat.

The possibility that it was of universal application was first stated by Dr. Julius Robert Mayer, a Schwabian physician (now living in Heilbronn) in the year 1842, while almost simultaneously with, and independently of him, James Prescot Joule, an English manufacturer, made a series of important and difficult experiments on the relation of heat to mechanical force, which supplied the chief points in which the comparison of the new theory with experience was still wanting.

The law in question asserts, that the *quantity of force which can be brought into action in the whole of Nature is unchangeable*, and can neither be increased nor diminished. My first object will be to explain to you what is understood by *quantity of force*; or as the same idea is more popularly expressed with reference to its technical application, what we call *amount of work* in the mechanical sense of the word.

The idea of work for machines, or natural processes, is taken from comparison with the working power of man; and we can therefore best illustrate from human labour, the most important features of the question with which we are concerned. In speaking of the work of machines, and of natural forces, we must, of course, in this comparison eliminate anything in which activity of intelligence comes into play. The latter is also capable of the hard and intense work of thinking, which tries a man just as muscular exertion does. But whatever of the actions of intelligence is met with in the work of machines, of course is due to the mind of the constructor and cannot be assigned to the instrument at work.

Now, the external work of man is of the most varied kind as regards the force or ease, the form and rapidity, of the motions used on it, and the kind of work produced. But both the arm of the blacksmith who delivers his powerful blows with the heavy hammer, and that of the violinist who produces the most delicate variations in sound, and the hand of the lace-maker who works with threads so fine that they are on the verge of the invisible, all these acquire the force which moves them in the same manner and by the same organs, namely, the muscles of the arm. An arm the muscles of which are lamed is incapable of doing any work; the moving force of the muscle must be at work in it, and these must obey the nerves, which bring to them orders from the brain. That member is then capable of the greatest variety of motions; it can compel the most varied instruments to execute the most diverse tasks.

Just so is it with machines: they are used for the most diversified arrangements. We produce by their agency an infinite variety of movements, with the most various degrees of force and rapidity, from powerful steam-hammers and rolling-mills, where gigantic masses of iron are cut and shaped like butter, to spinning and weaving-frames, the work of which rivals that of the spider. Modern mechanism has the richest choice of means of transferring the motion of one set of rolling wheels to another with greater or less velocity; of changing the rotating motion of wheels into the up-and-down motion of the piston-rod, of the shuttle, of falling hammers and stamps; or, conversely, of changing the latter into the former; or it can, on the other hand, change movements of uniform into those of varying velocity, and so forth. Hence this extraordinarily rich utility of machines for so extremely varied branches of industry. But one thing is common to all these differences; they all need a *moving force*, which sets and keeps them in motion, just as the works of the human hand all need the moving force of the muscles.

Now, the work of the smith requires a far greater and more intense exertion of the muscles than that of the violin-player; and there are in machines corresponding differences in the power and duration of the moving force required. These differences, which correspond to the different degree of exertion of the muscles in human labour, are alone what we have to think of when we speak of the *amount of work* of a machine. We have nothing to do here with the manifold character of the actions and arrangements which the machines produce; we are only concerned with an expenditure of force.

This very expression which we use so fluently, 'expenditure of force,' which indicates that the force applied has been expended and lost, leads us to a further characteristic analogy between the effects of the human arm and those of machines. The greater the exertion, and the longer it lasts, the more is the arm *tired*, and the more *is the store of its moving force for the time exhausted*. We shall see that this peculiarity of becoming exhausted by work is also met with in the moving forces of inorganic nature; indeed, that this capacity of the human arm of being tired is only one of the consequences of the law with which we are now concerned. When fatigue sets in, recovery is needed, and this can only be effected by rest and nourishment. We shall find that also in the inorganic moving forces, when their capacity for work is spent, there is a possibility of reproduction, although in general other means must be used to this end than in the case of the human arm.

From the feeling of exertion and fatigue in our muscles, we can form a general idea of what we understand by amount of work; but we must endeavour, instead of the indefinite estimate afforded by this comparison, to form a clear and precise idea of the standard by which we have to measure the amount of work. This we can do better by the simplest inorganic moving forces than by the actions of our muscles, which are a very complicated apparatus, acting in an extremely intricate manner.

Let us now consider that moving force which we know best, and which is simplest—gravity. It acts, for example as such, in those clocks which are driven by a weight. This weight fastened to a string, which is wound round a pulley connected with the first toothed wheel of the clock, cannot obey the pull of gravity without setting the whole clockwork in motion. Now I must beg you to pay special attention to the following points: the weight cannot put the clock in motion without itself sinking; did the weight not move, it could not move the clock, and its motion can only be such a one as obeys the action of gravity. Hence, if the clock is to go, the weight must continually sink lower and lower, and must at length sink so far that the string which supports it is run out. The clock then stops. The useful effect of its weight is for the present exhausted. Its gravity is not lost or diminished; it is attracted by the earth as before, but the capacity of this gravity to produce the motion of the clockwork is lost. It can only keep the weight at rest in the lowest point of its path, it cannot farther put it in motion.

But we can wind up the clock by the power of the arm, by which the weight is again raised. When this has been done, it has regained its former capacity, and can again set the clock in motion.

We learn from this that a raised weight possesses a *moving force*, but that it must necessarily sink if this force is to act; that by sinking, this moving force is exhausted, but by using another extraneous moving force—that of the arm—its activity can be restored.

The work which the weight has to perform in driving the clock is not indeed great. It has continually to overcome the small resistances which the friction of the axles and teeth, as well as the resistance of the air, oppose to the motion of the wheels, and it has to furnish the force for the small impulses and sounds which the pendulum produces at each oscillation. If the weight is detached from the clock, the pendulum swings for a while before coming to rest, but its motion becomes each moment feebler, and ultimately ceases entirely, being gradually used up by the small hindrances I have mentioned. Hence, to keep the clock going, there must be a moving force, which, though small, must be continually at work. Such a one is the weight.

We get, moreover, from this example, a measure for the amount of work. Let us assume that a clock is driven by a weight of a pound, which falls five feet in twenty-four hours. If we fix ten such clocks, each with a weight of one pound, then ten clocks will be driven twenty-four hours; hence, as each has to overcome the same resistances in the same time as the others, ten times as much work is performed for ten pounds fall through five feet. Hence, we conclude that the height of the fall being the same, the work increases directly as the weight.

Now, if we increase the length of the string so that the weight runs down ten feet, the clock will go two days instead of one; and, with double the height of fall, the weight will overcome on the second day the same resistances as on the first, and will therefore do twice as much work as when it can only run down five feet. The weight being the same, the work increases as the height of fall. Hence, we may take the product of the weight into the height of fall as a measure of work, at any rate, in the present case. The application of this measure is, in fact, not limited to the individual case, but the universal standard adopted in manufactures for measuring magnitude of work is a *foot pound*—that is, the amount of work which a pound raised through a foot can produce.[1]

We may apply this measure of work to all kinds of machines, for we should be able to set them all in motion by means of a weight sufficient to turn a pulley. We could thus always express the magnitude of any driving force, for any given machine, by the magnitude and height of fall of such a weight as would be necessary to keep the machine going with its arrangements until it had performed a certain work. Hence it is that the measurement of work by foot pounds is universally applicable. The use of such a weight as a driving force would not indeed be practically advantageous in those cases in which we were compelled to raise it by the power of our own arm; it would in that case be simpler to work the machine by the direct action of the arm. In the clock we use a weight so that we need not stand the whole day at the clockwork, as we should have to do to move it directly. By winding up the clock we accumulate a store of working capacity in it, which is sufficient for the expenditure of the next twenty-four hours.

The case is somewhat different when Nature herself raises the weight, which then works for us. She does not do this with solid bodies, at least not with such regularity as to be utilised; but she does it abundantly with water, which, being raised to the tops of mountains by meteorological processes, returns in streams from them. The gravity of water we use as moving force, the most direct application being in what are called *overshot* wheels, one of which is represented in Fig. 1. Along the circumference of such a wheel are a series of buckets, which act as receptacles for the water, and, on the side turned to the observer, have the tops uppermost; on the opposite side the tops of the buckets are upside-down. The water flows at M into the buckets of the front of the wheel, and at F, where the mouth begins to incline downwards, it flows out. The buckets on the circumference are filled on the side turned to the observer, and empty on the other side. Thus the former are weighted by the water contained in them, the latter not; the weight of the water acts continuously on only one side of the wheel, draws this down, and thereby turns the wheel; the other side of the wheel offers no resistance, for it contains no water. It is thus the weight of the falling water which turns the

---

1. This is the *technical* measure of work; to convert it into scientific measure it must be multiplied by the intensity of gravity.

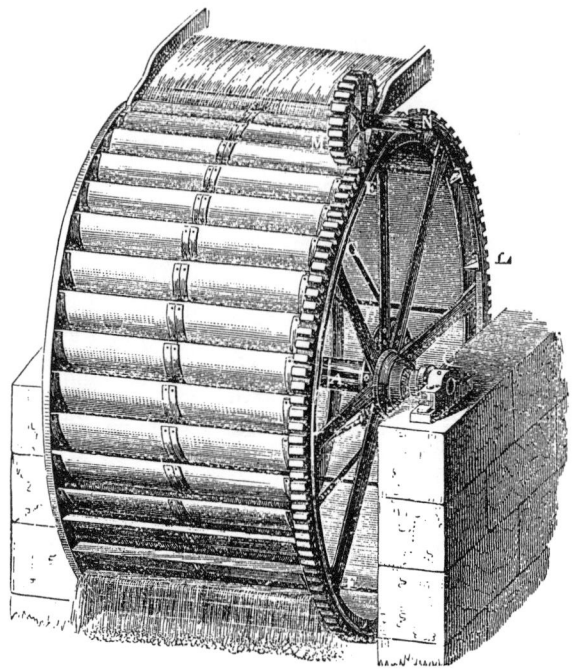

Fig. 1.

wheel, and furnishes the motive power. But you will at once see that the mass of water which turns the wheel must necessarily fall in order to do so, and that though, when it has reached the bottom, it has lost none of its gravity, it is no longer in a position to drive the wheel, if it is not restored to its original position, either by the power of the human arm or by means of some other natural force. If it can flow from the mill-stream to still lower levels, it may be used to work other wheels. But when it has reached its lowest level, the sea, the last remainder of the moving force is used up, which is due to gravity—that is, to the attraction of the earth, and it cannot act by its weight until it has been again raised to a high level. As this is actually effected by meteorological processes, you will at once observe that these are to be considered as sources of moving force.

Water-power was the first inorganic force which man learnt to use instead of his own labour or of that of domestic animals. According to Strabo, it was known to King Mithridates, of Pontus, who was also otherwise celebrated for his knowledge of nature; near his palace there was a water-wheel. Its use was first introduced among the Romans in the time of the first Emperors. Even now we find water-mills in all mountains, valleys, or wherever there are rapidly-flowing, regularly-filled, brooks and streams. We find water-power used for all purposes which can possibly be effected by machines. It drives mills which grind corn, saw-mills, hammers and oil-presses, spinning-frames and looms, and so forth. It is the cheapest of all motive powers, it flows spontaneously from the inexhaustible stores of nature; but it is restricted to a particular place, and only in mountainous countries is it present in any quantity; in level countries extensive reservoirs are necessary for damming the rivers to produce any amount of water-power.

Before passing to the discussion of other motive forces, I must answer an objection which may readily suggest itself. We all know that there are numerous machines, systems of pulleys, levers and cranes, by the aid of which heavy burdens may be lifted by a comparatively small expenditure of force. We have all of us often seen one or two workmen hoist heavy masses of stones to great heights, which they would be quite unable to do directly; in like manner, one or two men, by means of a crane, can transfer the largest and heaviest chests from a ship to the quay. Now it may be asked, If a large, heavy weight had been used for driving a machine, would it not be very easy, by means of a crane or a system of pulleys, to raise it anew, so that it could again be used as a motor, and thus acquire motive power, without being compelled to use a corresponding exertion in raising the weight?

The answer to this is, that all these machines, in that degree in which for the moment they facilitate the exertion, also prolong it, so that by their help no motive power is ultimately gained. Let us assume that four labourers have to raise a load of four hundredweight, by means of a rope passing over a single pulley. Every time the rope is pulled down through four feet, the load is also raised through four feet. But now, for the sake of comparison, let us suppose the same load hung to a block of four pulleys, as represented in Fig. 2. A single labourer would now be able to raise the load by the same exertion of force as each one of the four put forth. But when he pulls the rope through four feet, the load only rises one foot, for the length through which he pulls the rope, at *a*, is uniformly distributed in the block over four ropes, so that each of these is only shortened by a foot. To raise the load, therefore, to the same height, the one man must necessarily work four times as long as the four together did. But the total expenditure of work is the same, whether four labourers work for a quarter of an hour or one works for an hour.

If, instead of human labour, we introduce the work of a weight, and hang to the block a load of 400, and at *a*, where otherwise the labourer works, a weight of 100 pounds, the block is then in equilibrium, and, without any appreciable exertion of the arm, may be set in motion. The weight of 100 pounds sinks, that of 400 rises. Without any measurable expenditure of force, the heavy weight has been raised by the sinking of the smaller one. But observe that the smaller weight will have sunk through four times the distance that the greater one has risen. But a fall of 100 pounds through four feet is just as much 400 foot pounds as a fall of 400 pounds through one foot.

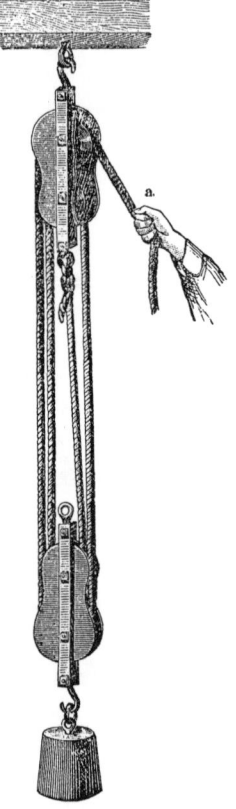

Fig. 2.

The action of levers in all their various modifications is precisely similar. Let *ab*, Fig. 3, be a simple lever, supported at *c*, the arm *cb* being four times as long as the other arm *ac*. Let a weight of one pound be hung at *b*, and a weight of four pounds at *a*, the lever is then in equilibrium, and the least pressure of the finger is sufficient, without any appreciable exertion of force, to place it in the position *a'b'*, in which the heavy weight of four pounds has been raised, while the one-pound weight has sunk. But here, also, you will observe no work has been gained, for while the heavy weight has been raised through one inch, the lighter one has fallen through four inches; and four pounds through one inch is, as work, equivalent to the product of one pound through four inches.

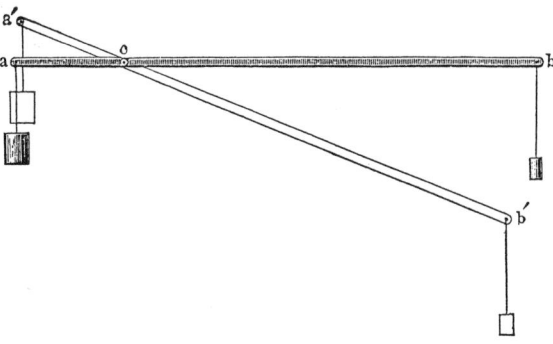

Fig. 3.

Most other fixed parts of machines may be regarded as modified and compound levers; a toothed-wheel, for instance as a series of levers, the ends of which are represented by the individual teeth, and one after the other of which is put in activity, in the degree in which the tooth in question seizes, or is seized by the adjacent pinion. Take, for instance, the crabwinch, represented in Fig. 4. Suppose the pinion on the axis of the barrel of the winch has twelve teeth, and the toothed-wheel, HH, seventy-two teeth, that is six times as many as the former. The winch must now be turned round six times before the toothed-wheel, H, and the barrel, D, have made one turn, and before the rope which raises the load has been lifted by a length equal to the circumference of the barrel. The workman thus requires six times the

Fig. 4.

time, though to be sure only one-sixth of the exertion, which he would have to use if the handle were directly applied to the barrel, D. In all these machines, and parts of machines, we find it confirmed that in proportion as the velocity of the motion increases its power diminishes, and that when the power increases the velocity diminishes, but that the amount of work is never thereby increased.

In the overshot mill-wheel, described above, water acts by its weight. But there is another form of mill-wheels, what is called the *undershot wheel*, in which it only acts by its impact, as represented in Fig. 5. These are used where the height from which the water comes is not great enough to flow on the upper part of the wheel. The lower part of undershot wheels dips in

Fig. 5.

the flowing water which strikes against their float-boards and carries them along. Such wheels are used in swift-flowing streams which have a scarcely perceptible fall, as, for instance, on the Rhine. In the immediate neighbour-hood of such a wheel, the water need not necessarily have a great fall if it only strikes with considerable velocity. It is the velocity of the water, exerting an impact against the float-boards, which acts in this case, and which produces the motive power.

Windmills, which are used in the great plains of Holland and North Germany to supply the want of falling water, afford another instance of the action of velocity. The sails are driven by air in motion—by wind. Air at rest could just as little drive a windmill as water at rest a water-wheel. The driving force depends here on the velocity of moving masses.

A bullet resting in the hand is the most harmless thing in the world; by its gravity it can exert no great effect; but when fired and endowed with great velocity it drives through all obstacles with the most tremendous force.

If I lay the head of a hammer gently on a nail, neither its small weight nor the pressure of my arm is quite sufficient to drive the nail into wood; but if I swing the hammer and allow it to fall with great velocity, it acquires a new force, which can overcome far greater hindrances. These examples teach us that the velocity of a moving mass can act as motive force. In mechanics, velocity in so far as it is motive force, and can produce work, is called *vis viva*. The name is not well chosen; it is too apt to suggest to us the force of living beings. Also in this case you will see, from the instances of the hammer and of the bullet, that velocity is lost as such, when it produces working power. In the case of the water-mill, or of the windmill, a more careful investigation of the moving masses of water and air is necessary to prove that part of their velocity has been lost by the work which they have performed.

The relation of velocity to working power is most simply and clearly seen in a simple pendulum, such as can he constructed by any weight which we suspend to a cord. Let M, Fig. 6, be such a weight, of a spherical form;

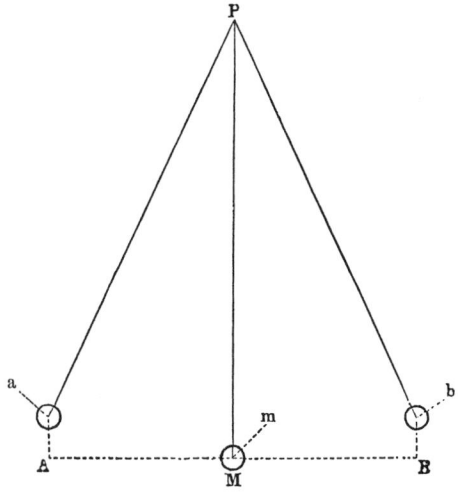

Fig. 6.

AB, a horizontal line drawn through the centre of the sphere; P the point at which the cord is fastened. If now I draw the weight M on one side towards A, it moves in the arc M*a*, the end of which, *a*, is somewhat higher than the point A in the horizontal line. The weight is thereby raised to the height A*a*. Hence my arm must exert a certain force to bring the weight to *a*. Gravity resists this motion and endeavours to bring back the weight to M, the lowest point which it can reach.

Now, if after I have brought the weight to *a* I let it go, it obeys this force of gravity and returns to M, arrives there with a certain velocity, and no longer remains quietly hanging at M as it did before, but swings beyond M towards *b*, where its motion stops as soon as it has traversed on the side of B an arc equal in length to that on the side of A, and after it has risen to a distance B*b* above the horizontal line, which is equal to the height A*a*, to which my arm had previously raised it. In *b* the pendulum returns, swings the same way back through M towards *a*, and so on, until its oscillations are gradually diminished, and ultimately annulled by the resistance of the air and by friction.

You see here that the reason why the weight, when it comes from *a* to M, and does not stop there, but ascends to *b*, in opposition to the action of gravity, is only to be sought in its velocity. The velocity which it has acquired in moving from the height A*a* is capable of again raising it to an equal height, B*b*. The velocity of the moving mass, M, is thus capable of raising this mass; that is to say, in the language of mechanics, of performing work. This would also be the case if we had imparted such a velocity to the suspended weight by a blow.

From this we learn further how to measure the working power of velocity—or, what is the same thing, the *vis viva* of the *moving mass*. It is equal to the work, expressed in foot pounds, which the same mass can exert after its velocity has been used to raise it, under the most favourable circumstances, to as great a height as possible.[2] This does not depend on the direction of the velocity; for if we swing a weight attached to a thread in a circle, we can even change a downward motion into an upward one.

The motion of the pendulum shows us very distinctly how the forms of working power hitherto considered—that of a raised weight and that of a moving mass—may merge into one another. In the points *a* and *b*, Fig. 6, the mass has no velocity; at the point M it has fallen as far as possible, but possesses velocity. As the weight goes from *a* to *m* the work of the raised weight is changed into *vis viva*; as the weight goes further from *m* to *b* the *vis viva* is changed into the work of a raised weight. Thus the work which the arm originally imparted to the pendulum is not lost in these oscillations, provided we may leave out of consideration the influence of the resistance of the air and of friction. Neither does it increase, but it continually changes the form of its manifestation.

Let us now pass to other mechanical forces, those of elastic bodies. Instead of the weights which drive our clocks, we find in time-pieces and in watches, steel springs which are coiled in winding up the clock, and are

---

2. The measure of *vis viva* in theoretical mechanics is half the product of the weight into the square of the velocity. To reduce it to the technical measure of the work we must divide it by the intensity of gravity; that is, by the velocity at the end of the first second of a freely falling body.

uncoiled by the working of the clock. To coil up the spring we consume the force of the arm; this has to overcome the resisting elastic force of the spring as we wind it up, just as in the clock we have to overcome the force of gravity which the weight exerts. The coiled spring can, however, perform work; it gradually expends this acquired capability in driving the clock-work.

If I stretch a crossbow and afterwards let it go, the stretched string moves the arrow; it imparts to it force in the form of velocity. To stretch the cord my arm must work for a few seconds; this work is imparted to the arrow at the moment it is shot off. Thus the crossbow concentrates into an extremely short time the entire work which the arm had communicated in the operation of stretching; the clock, on the contrary, spreads it over one or several days. In both cases no work is produced which my arm did not originally impart to the instrument, it is only expended more conveniently.

The case is somewhat different if by any other natural process I can place an elastic body in a state of tension without having to exert my arm. This is possible and is most easily observed in the case of gases.

If, for instance, I discharge a fire-arm loaded with gunpowder, the greater part of the mass of the powder is converted into gases at a very high temperature, which have a powerful tendency to expand, and can only be retained in the narrow space in which they are formed, by the exercise of the most powerful pressure. In expanding with enormous force they propel the bullet, and impart to it a great velocity, which we have already seen is a form of work.

In this case, then, I have gained work which my arm has not performed. Something, however, has been lost; the gunpowder, that is to say, whose constituents have changed into other chemical compounds, from which they cannot, without further ado, be restored to their original condition. Here, then, a chemical change has taken place, under the influence of which work has been gained.

Elastic forces are produced in gases by the aid of heat, on a far greater scale.

Let us take, as the most simple instance, atmospheric air. In Fig. 7 an apparatus is represented such as Regnault used for measuring the expansive force of heated gases. If no great accuracy is required in the measurement, the apparatus may be arranged more simply. At C is a glass globe filled with dry air, which is placed in a metal vessel, in which it can be heated by steam. It is connected with the U-shaped tube, ss, which contains a liquid, and the limbs of which communicate with each other when the stop-cock R is closed. If the liquid is in equilibrium in the tube ss when the globe is cold, it rises in the leg s, and ultimately overflows when the globe is heated. If, on the contrary, when the globe is heated, equilibrium be restored by allowing some of the liquid to flow out at R, as the globe cools it will be drawn up towards *n*. In both cases liquid is raised, and work thereby produced.

The same experiment is continuously repeated on the largest scale in steam engines, though in order to keep up a continual disengagement of

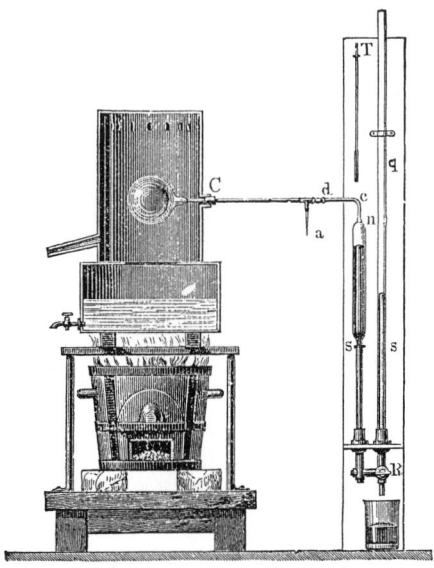

Fig. 7.

compressed gases from the boiler, the air in the globe in Fig. 7, which would soon reach the maximum of its expansion, is replaced by water, which is gradually changed into steam by the application of heat. But steam, so long as it remains as such, is an elastic gas which endeavours to expand exactly like atmospheric air. And instead of the column of liquid which was raised in our last experiment, the machine is caused to drive a solid piston which imparts its motion to other parts of the machine. Fig. 8 represents a front view of the working parts of a high pressure engine, and Fig. 9 a section. The boiler in which steam is generated is not represented; the steam passes through the tube zz, Fig. 9, to the cylinder AA in which moves a tightly fitting piston C. The parts between the tube zz and the cylinder AA, that is the slide valve in the valve-chest KK, and the two tubes d and e allow the steam to pass first below and then above the piston, while at the same time the steam has free exit from the other half of the cylinder. When the steam passes under the piston, it forces it upward; when the piston has reached the top of its course the position of the valve in KK changes, and the steam passes above the piston and forces it down again. The piston-rod acts by means of the connecting-rod P, on the crank Q of the fly-wheel X and sets this in motion. By means of the rod s, the motion of the rod regulates the opening and closing of the valve. But we need not here enter into those mechanical arrangements, however ingeniously they have been devised. We are only interested in the manner in which heat produces elastic vapour, and how this vapour, in its endeavour to expand, is compelled to move the solid parts of the machine, and furnish work.

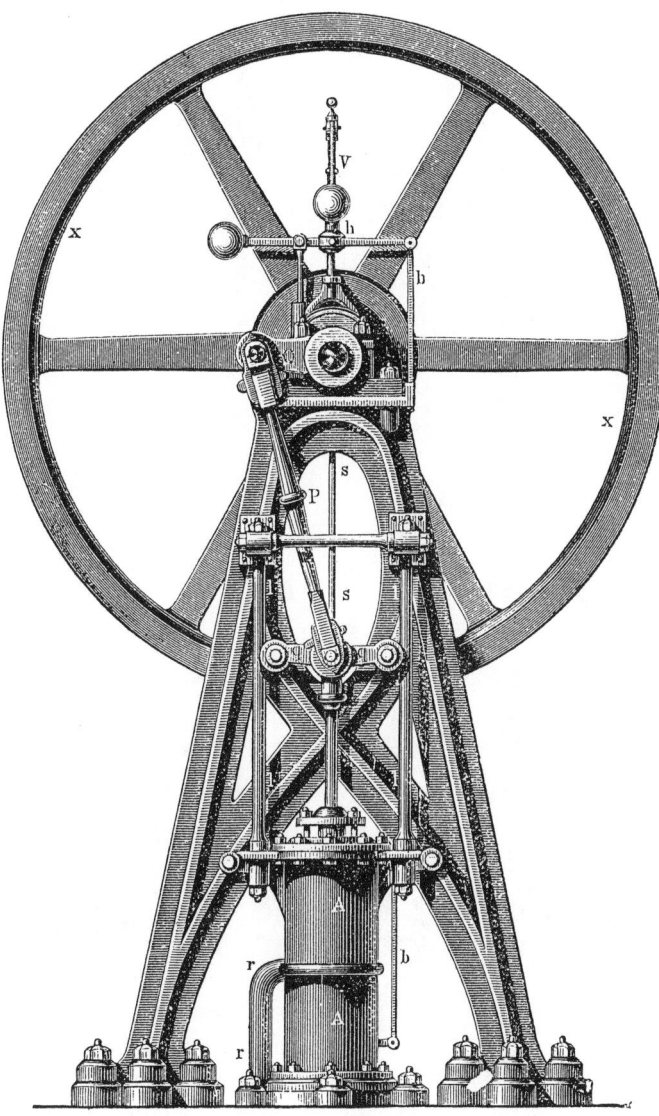

Fig. 8.

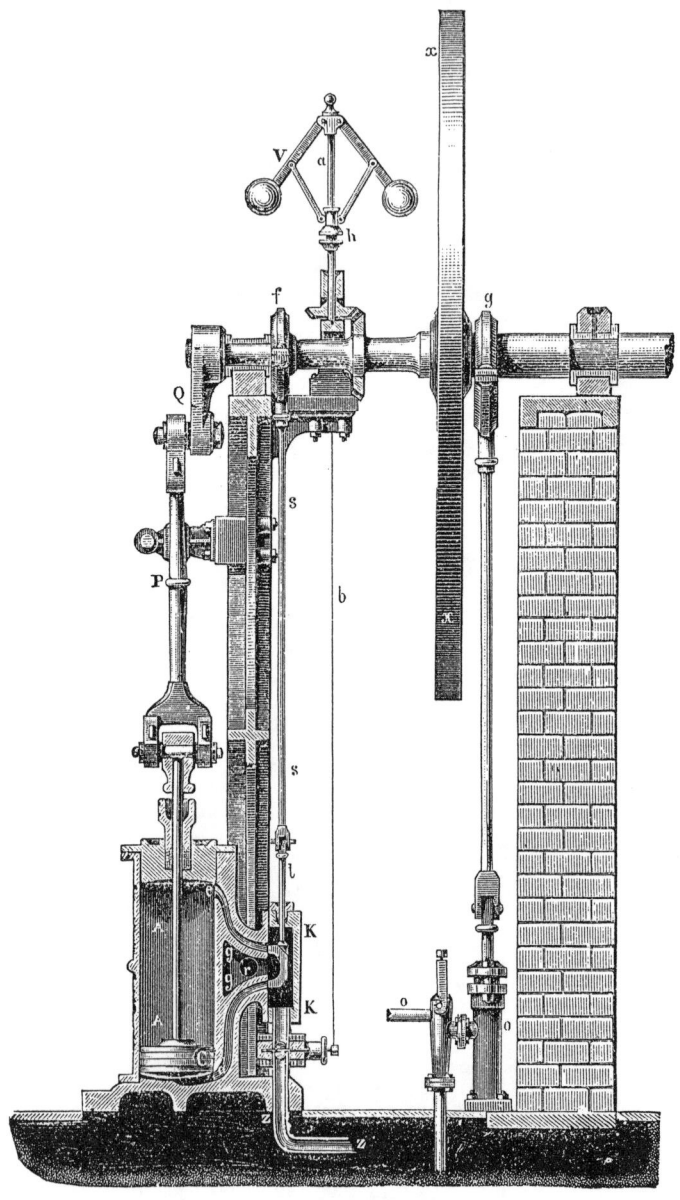

Fig. 9.

You all know how powerful and varied are the effects of which steam engines are capable; with them has really begun the great development of industry which has characterised our century before all others. Its most essential superiority over motive powers formerly known, is that it is not restricted to a particular place. The store of coal and the small quantity of water which are the sources of its power can be brought everywhere, and steam engines can even be made movable, as is the case with steam-ships and locomotives. By means of these machines we can develop motive power to almost an indefinite extent at any place on the earth's surface, in deep mines and even on the middle of the ocean; while water and wind-mills are bound to special parts of the surface of the land. The locomotive transports travellers and goods over the land in numbers and with a speed which must have seemed an incredible fable to our forefathers, who looked upon the mail-coach with its six passengers in the inside and its ten miles an hour, as an enormous progress. Steam-engines traverse the ocean independently of the direction of the wind, and, successfully resisting storms which would drive sailing-vessels far away, reach their goal at the appointed time. The advantages which the concourse of numerous, and variously skilled workmen in all branches offers in large towns where wind and water power are wanting, can be utilised, for steam-engines find place everywhere, and supply the necessary crude force; thus the more intelligent human force may be spared for better purposes; and, indeed, wherever the nature of the ground or the neighbourhood of suitable lines of communication present a favourable opportunity for the development of industry, the motive power is also present in the form of steam-engines.

We see, then, that heat can produce mechanical power; but in the cases which we have discussed we have seen that the quantity of force which can be produced by a given measure of a physical process is always accurately defined, and that the further capacity for work of the natural forces, is either diminished or exhausted by the work which has been performed. How is it now with *Heat* in this respect?

This question was of decisive importance in the endeavour to extend the law of the Conservation of Force to all natural processes. In the answer lay the chief difference between the older and newer views in these respects. Hence it is that many physicists designate that view of Nature corresponding to the law of the conservation of force with the name of *the Mechanical Theory of Heat.*

The older view of the nature of heat was that it is a substance, very fine and imponderable indeed, but indestructible, and unchangeable in quantity, which is an essential fundamental property of all matter. And, in fact, in a large number of natural processes, the quantity of heat which can he demonstrated by the thermometer is unchangeable.

By conduction and radiation, it can indeed pass from hotter to colder bodies; but the quantity of heat which the former lose can be shown by the thermometer to have reappeared in the latter. Many processes, too, were known, especially in the passage of bodies from the solid to the liquid and gaseous states, in which heat disappeared—at any rate, as regards the thermometer. But when the gaseous body was restored to the liquid, and the liquid to the solid state, exactly the same quantity of heat reappeared which formerly seemed to have been lost. Heat was said *to have become latent.* On this view, liquid water differed from solid ice in containing a certain quantity of heat bound, which, just because it was bound, could not pass to the thermometer, and therefore was not indicated by it. Aqueous vapour contains a far greater quantity of heat thus bound. But if the vapour be precipitated, and the liquid water restored to the state of ice, exactly the same amount of heat is liberated as had become latent in the melting of the ice and in the vaporisation of the water.

Finally, heat is sometimes produced and sometimes disappears in chemical processes. But even here it might be assumed that the various chemical elements and chemical compounds contain certain constant quantities of latent heat, which, when they change their composition, are sometimes liberated and sometimes must be supplied from external sources. Accurate experiments have shown that the quantity of heat which is developed by a chemical process, for instance, in burning a pound of pure carbon into carbonic acid, is perfectly constant, whether the combustion is slow or rapid, whether it takes place all at once or by intermediate stages. This also agreed very well with the assumption, which was the basis of the theory of heat, that heat is a substance entirely unchangeable in quantity. The natural processes which have here been briefly mentioned, were the subject of extensive experimental and mathematical investigations, especially of the great French physicists in the last decade of the former, and the first decade of the present, century; and a rich and accurately-worked chapter of physics had been developed, in which everything agreed excellently with the hypothesis—that heat is a substance. On the other hand, the invariability in the quantity of heat in all these processes could at that time be explained in no other manner than that heat is a substance.

But one relation of heat—namely, that to mechanical work—had not been accurately investigated. A French engineer, Sadi Carnot, son of the celebrated War Minister of the Revolution, had indeed endeavoured to deduce the work which heat performs, by assuming that the hypothetical caloric endeavoured to expand like a gas; and from this assumption he deduced in fact a remarkable law as to the capacity of heat for work, which even now, though with an essential alteration introduced by Clausius, is among the bases of the modern mechanical theory of heat, and the practical conclusions from which, so far as they could at that time be compared with experiments, have held good.

But it was already known that whenever two bodies in motion rubbed against each other, heat was developed anew, and it could not be said whence it came.

The fact is universally recognised; the axle of a carriage which is badly greased and where the friction is great, becomes hot—so hot, indeed, that it may take fire; machine-wheels with iron axles going at a great rate may become so hot that they weld to their sockets. A powerful degree of friction is not, indeed, necessary to disengage an appreciable degree of heat; thus, a lucifer-match, which by rubbing is so heated that the phosphoric mass ignites, teaches this fact. Nay, it is enough to rub the dry hands together to feel the heat produced by friction, and which is far greater than the heating which takes place when the hands lie gently on each other. Uncivilized people use the friction of two pieces of wood to kindle a fire. With this view, a sharp spindle of hard wood is made to revolve rapidly on a base of soft wood in the manner represented in Fig. 10.

Fig. 10.

So long as it was only a question of the friction of solids, in which particles from the surface become detached and compressed, it might be supposed that some changes in structure of the bodies rubbed might here liberate latent heat, which would thus appear as heat of friction.

But heat can also be produced by the friction of liquids, in which there could be no question of changes in structure, or of the liberation of latent heat. The first decisive experiment of this kind was made by Sir Humphry Davy in the commencement of the present century. In a cooled space he made two pieces of ice rub against each other, and thereby caused them to melt. The latent heat which the newly formed water must have here assimilated could not have been conducted to it by the cold ice, or have been produced by a change of structure; it could have come from no other cause than from friction, and must have been created by friction.

Heat can also be produced by the impact of imperfectly elastic bodies as well as by friction. This is the case, for instance, when we produce fire by striking flint against steel, or when an iron bar is worked for some time by powerful blows of the hammer.

If we inquire into the mechanical effects of friction and of inelastic impact, we find at once that these are the processes by which all terrestrial movements are brought to rest. A moving body whose motion was not retarded by any resisting force would continue to move to all eternity. The motions of the planets are an instance of this. This is apparently never the case with the motion of the terrestrial bodies, for they are always in contact with other bodies which are at rest, and rub against them. We can, indeed, very much diminish their friction, but never completely annul it. A wheel which turns about a well-worked axle, once set in motion continues it for a long time; and the longer, the more truly and smoother the axle is made to turn, the better it is greased, and the less the pressure it has to support. Yet the *vis viva* of the motion which we have imparted to such a wheel when we started it, is gradually lost in consequence of friction. It disappears, and if we do not carefully consider the matter, it seems as if the *vis viva* which the wheel had possessed had been simply destroyed without any substitute.

A bullet which is rolled on a smooth horizontal surface continues to roll until its velocity is destroyed by friction on the path, caused by the very minute impacts on its little roughnesses.

A pendulum which has been put in vibration can continue to oscillate for hours if the suspension is good, without being driven by a weight; but by the friction against the surrounding air, and by that at its place of suspension, it ultimately comes to rest.

A stone which has fallen from a height has acquired a certain velocity on reaching the earth; this we know is the equivalent of a mechanical work; so long as this velocity continues as such, we can direct it upwards by means of suitable arrangements, and thus utilise it to raise the stone again. Ultimately the stone strikes against the earth and comes to rest; the impact has destroyed its velocity, and therewith apparently also the mechanical work which this velocity could have effected.

If we review the result of all these instances, which each of you could easily add to from your own daily experience, we shall see that friction and inelastic impact are processes in which mechanical work is destroyed, and heat produced in its place.

The experiments of Joule, which have been already mentioned, lead us a step further. He has measured in foot pounds the amount of work which is destroyed by the friction of solids and by the friction of liquids; and, on the other hand, he has determined the quantity of heat which is thereby produced, and has established a definite relation between the two. His experiments show that when heat is produced by the consumption of work, a definite quantity of work is required to produce that amount of heat which is known to physicists as the *unit of heat*; the heat, that is to say, which is necessary to raise one gramme of water through one degree centigrade. The quantity of work necessary for this is, according to Joule's best experiments, equal to the work which a gramme would perform in falling through a height of 425 metres.

In order to show how closely concordant are his numbers, I will adduce the results of a few series of experiments which he obtained after introducing the latest improvements in his methods.

1. A series of experiments in which water was heated by friction in a brass vessel. In the interior of this vessel a vertical axle provided with sixteen paddles was rotated, the eddies thus produced being broken by a series of projecting barriers, in which parts were cut out large enough for the paddles to pass through. The value of the equivalent was 424·9 metres.

2. Two similar experiments, in which mercury in an iron vessel was substituted for water in a brass one, gave 425 and 426·3 metres.

3. Two series of experiments, in which a conical ring rubbed against another, both surrounded by mercury, gave 426·7 and 425·6 metres.

Exactly the same relations between heat and work were also found in the reverse process—that is, when work was produced by heat. In order to execute this process under physical conditions that could be controlled as perfectly as possible, permanent gases and not vapours were used, although the latter are, in practice, more convenient for producing large quantities of work, as in the case of the steam-engine. A gas which is allowed to expand with moderate velocity becomes cooled. Joule was the first to show the reason of this cooling. For the gas has, in expanding, to overcome the resistance, which the pressure of the atmosphere and the slowly yielding side of the vessel oppose to it; or, if it cannot of itself overcome this resistance, it supports the arm of the observer which does it. Gas thus performs work, and this work is produced at the cost of its heat. Hence the cooling. If, on the contrary, the gas is suddenly allowed to issue into a perfectly exhausted space where it finds no resistance, it does not become cool as Joule has shown; or if individual parts of it become cool, others become warm; and, after the temperature has become equalised, this is exactly as much as before the sudden expansion of the gaseous mass.

How much heat the various gases disengage when they are compressed, and how much work is necessary for their compression; or, conversely, how much heat disappears when they expand under a pressure equal to their own counterpressure, and how much work they thereby effect in overcoming this counterpressure, was partly known from the older physical experiments, and has partly been determined by the recent experiments of Regnault by extremely perfect methods. Calculations with the best data of this kind give us the value of the thermal equivalent from experiments:—

With atmospheric air . . . . 426·0 metres.
" oxygen . . . . 425·7   "
" nitrogen . . . . 431·3   "
" hydrogen . . . . 425·3   "

Comparing these numbers with those which determine the equivalence of heat and mechanical work in friction, as close an agreement is seen as can at all be expected from numbers which have been obtained by such varied investigations of different observers.

Thus then: a certain quantity of heat may be changed into a definite quantity of work; this quantity of work can also be retransformed into heat, and, indeed, into exactly the same quantity of heat as that from which it originated; in a mechanical point of view, they are exactly equivalent. Heat is a new form in which a quantity of work may appear.

These facts no longer permit us to regard heat as a substance, for its quantity is not unchangeable. It can be produced anew from the *vis viva* of motion destroyed; it can be destroyed, and then produces motion. We must rather conclude from this that heat itself is a motion, an internal invisible motion of the smallest elementary particles of bodies. If, therefore, motion seems lost in friction and impact, it is not actually lost, but only passes from the great visible masses to their smallest particles; while in steam-engines the internal motion of the heated gaseous particles is transferred to the piston of the machine, accumulated in it, and combined in a resultant whole.

But what is the nature of this internal motion, can only be asserted with any degree of probability in the case of gases. Their particles probably cross one another in rectilinear paths in all directions, until, striking another particle, or against the side of the vessel, they are reflected in another direction. A gas would thus be analogous to a swarm of gnats, consisting, however, of particles infinitely small and infinitely more closely packed. This hypothesis, which has been developed by Krönig, Clausius, and Maxwell, very well accounts for all the phenomena of gases.

What appeared to the earlier physicists to be the constant quantity of heat is nothing more than the whole motive power of the motion of heat, which remains constant so long as it is not transformed into other forms of work, or results afresh from them.

We turn now to another kind of natural forces which can produce work—I mean the chemical. We have to-day already come across them. They are the ultimate cause of the work which gunpowder and the steam-engine produce; for the heat which is consumed in the latter, for example, originates in the combustion of carbon—that is to say, in a chemical process. The burning of coal is the chemical union of carbon with the oxygen of the air, taking place under the influence of the chemical affinity of the two substances.

We may regard this force as an attractive force between the two, which, however, only acts through them with extraordinary power, if the smallest particles of the two substances are in closest proximity to each other. In combustion this force acts; the carbon and oxygen atoms strike against each other and adhere firmly, inasmuch as they form a new compound—carbonic acid—a gas known to all of you as that which ascends from all fermenting and fermented liquids—from beer and champagne. Now this attraction between the atoms of carbon and of oxygen performs work just as much as that which the earth in the form of gravity exerts upon a raised weight. When the weight falls to the ground, it produces an agitation, which is partly transmitted to the vicinity as sound waves, and partly remains as the motion of heat. The same result we must expect from chemical action. When carbon and oxygen atoms have rushed against each other, the newly-formed particles of carbonic acid must be in the most violent molecular motion—that is, in the motion of heat. And this is so. A pound of carbon burned with oxygen to form carbonic acid, gives as much heat as is necessary to raise 80.9 pounds of water from the freezing to the boiling point; and just as the same amount of work is produced when a weight falls, whether it falls slowly or fast, so also the same quantity of heat is produced by the combustion of carbon, whether this is slow or rapid, whether it takes place all at once, or by successive stages.

When the carbon is burned, we obtain in its stead, and in that of the oxygen, the gaseous product of combustion—carbonic acid. Immediately after combustion it is incandescent. When it has afterwards imparted heat to the vicinity, we have in the carbonic acid the entire quantity of carbon and the entire quantity of oxygen, and also the force of affinity quite as strong as before. But the action of the latter is now limited to holding the atoms of carbon and oxygen firmly united; they can no longer produce either heat or work any more than a fallen weight can do work if it has not been again raised by some extraneous force. When the carbon has been burnt we take no further trouble to retain the carbonic acid; it can do no more service, we endeavour to get it out of the chimneys of our houses as fast as we can.

Is it possible, then, to tear asunder the particles of carbonic acid, and give to them once more the capacity of work which they had before they were combined, just as we can restore the potentiality of a weight by raising it from the ground? It is indeed possible. We shall afterwards see how it occurs in the life of plants; it can also be effected by inorganic processes, though in roundabout ways, the explanation of which would lead us too far from our present course.

This can, however, be easily and directly shown for another element, hydrogen, which can be burnt just like carbon. Hydrogen with carbon is a constituent of all combustible vegetable substances, among others, it is also an essential constituent of the gas which is used for lighting our streets and rooms; in the free state it is also a gas, the lightest of all, and burns when ignited with a feebly luminous blue flame. In this combustion—that is, in the chemical combination of hydrogen with oxygen, a very considerable quantity of heat is produced; for a given weight of hydrogen, four times as much heat as in the combustion of the same weight of carbon. The product of combustion is water, which, therefore, is not of itself further combustible, for the hydrogen in it is completely saturated with oxygen. The force of affinity, therefore, of hydrogen for oxygen, like that of carbon for oxygen, performs work in combustion, which appears in the form of heat. In the water which has been formed during combustion, the force of affinity is exerted between the elements as before, but its capacity for work is lost. Hence the two elements must be again separated, their atoms torn apart, if new effects are to be produced from them.

This we can do by the aid of currents of electricity. In the apparatus depicted in Fig. 11, we have two glass vessels filled with acidulated water, a and a¹, which are separated in the middle by a porous plate moistened with water. In both sides are fitted platinum wires, k, which are attached to

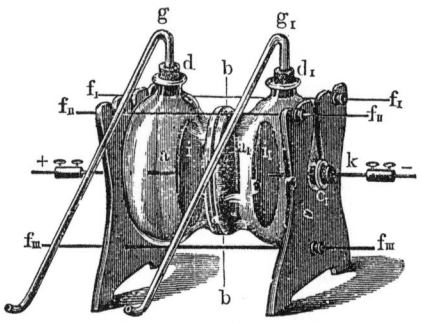

Fig. 11.

platinum plates, i and i¹. As soon as a galvanic current is transmitted through the water by the platinum wires, k, you see bubbles of gas ascend from the plates i and i¹. These bubbles are the two elements of water, hydrogen on the one hand, and oxygen on the other. The gases emerge through the tubes g and g¹. If we wait until the upper part of the vessels and the tubes have been filled with it, we can inflame hydrogen at one side; it burns with a blue flame. If I bring a glimmering spill near the mouth of the other tube it bursts into flame, just as happens with oxygen gas, in which the processes of combustion are far more intense than in atmospheric air, where the oxygen mixed with nitrogen is only one-fifth of the whole volume.

If I hold a glass flask filled with water over the hydrogen flame, the water, newly formed in combustion, condenses upon it.

If a platinum wire be held in the almost non-luminous flame, you see how intensely it is ignited; in a plentiful current of a mixture of the gases, hydrogen and oxygen, which have been liberated in the above experiment, the almost infusible platinum might even be melted. The hydrogen which has here been liberated from the water by the electrical current has regained the capacity of producing large quantities of heat by a fresh combination with oxygen; its affinity for oxygen has regained for it its capacity for work.

We here become acquainted with a new source of work, the electric current which decomposes water. This current is itself produced by a galvanic battery, Fig. 12. Each of the four vessels contains nitric acid, in

Fig. 12.

which there is a hollow cylinder of very compact carbon. In the middle of the carbon cylinder is a cylindrical porous vessel of white clay, which contains dilute sulphuric acid; in this dips a zinc cylinder. Each zinc cylinder is connected by a metal ring with the carbon cylinder of the next vessel, the last zinc cylinder n is connected with one platinum plate, and the first carbon cylinder, p, with the other platinum plate of the apparatus for the decomposition of water.

If now the conducting circuit of this galvanic apparatus is completed, and the decomposition of water begins, a chemical process takes place simultaneously in the cells of the voltaic battery. Zinc takes oxygen from the surrounding water and undergoes a slow combustion. The product of combustion thereby produced, oxide of zinc, unites further with sulphuric acid, for which it has a powerful affinity, and sulphate of zinc, a saline kind of substance, dissolves in the liquid. The oxygen, moreover, which is withdrawn from it is taken by the water from the nitric acid surrounding the cylinder of carbon, which is very rich in it, and readily gives it up. Thus, in the galvanic battery zinc burns to sulphate of zinc at the cost of the oxygen of nitric acid.

Thus, while one product of combustion, water, is again separated, a new combustion is taking place—that of zinc. While we there reproduce chemical affinity which is capable of work, it is here lost. The electrical current is, as it were, only the carrier which transfers the chemical force of the zinc uniting with oxygen and acid to water in the decomposing cell, and uses it for overcoming the chemical force of hydrogen and oxygen.

In this case, we can restore work which has been lost, but only by using another force, that of oxidizing zinc.

Here we have overcome chemical forces by chemical forces, through the instrumentality of the electrical current. But we can attain the same object by mechanical forces, if we produce the electrical current by a magneto-electrical machine, Fig. 13. If we turn the handle, the anker RR[1], on which is coiled copper-wire, rotates in front of the poles of the horse-shoe magnet, and in these coils electrical currents are produced, which can be led from the points a and b. If the ends of these wires are connected with the apparatus for decomposing water we obtain hydrogen and oxygen, though in far smaller quantity than by the aid of the battery which we used before. But this process is interesting, for the mechanical force of the arm which turns the wheel produces the work which is required for separating the combined chemical elements. Just as the steam-engine changes chemical into mechanical force, the magneto-electrical machine transforms mechanical force into chemical.

The application of electrical currents opens out a large number of relations between the various natural forces. We have decomposed water into its elements by such currents, and should be able to decompose a large number of other chemical compounds. On the other hand, in ordinary galvanic batteries electrical currents are produced by chemical forces.

In all conductors through which electrical currents pass they produce heat; I stretch a thin platinum wire between the ends n and p of the galvanic battery, Fig. 12; it becomes ignited and melts. On the other hand, electrical currents are produced by heat in what are called thermo-electric elements.

Iron which is brought near a spiral of copper wire, traversed by an electrical current, becomes magnetic, and then attracts other pieces of iron, or a suitably placed steel magnet. We thus obtain mechanical actions which meet with extended applications in the electrical telegraph, for instance. Fig. 14 represents a Morse's telegraph in one-third of the natural size. The essential part is a horse-shoe shaped iron core, which stands in the copper spirals bb. Just over the top of this is a small steel magnet cc, which is attracted the moment an electrical current, arriving by the telegraph wire, traverses the spirals bb. The magnet cc is rigidly fixed in the lever dd, at the other end of which is a style; this makes a mark on a paper band, drawn by

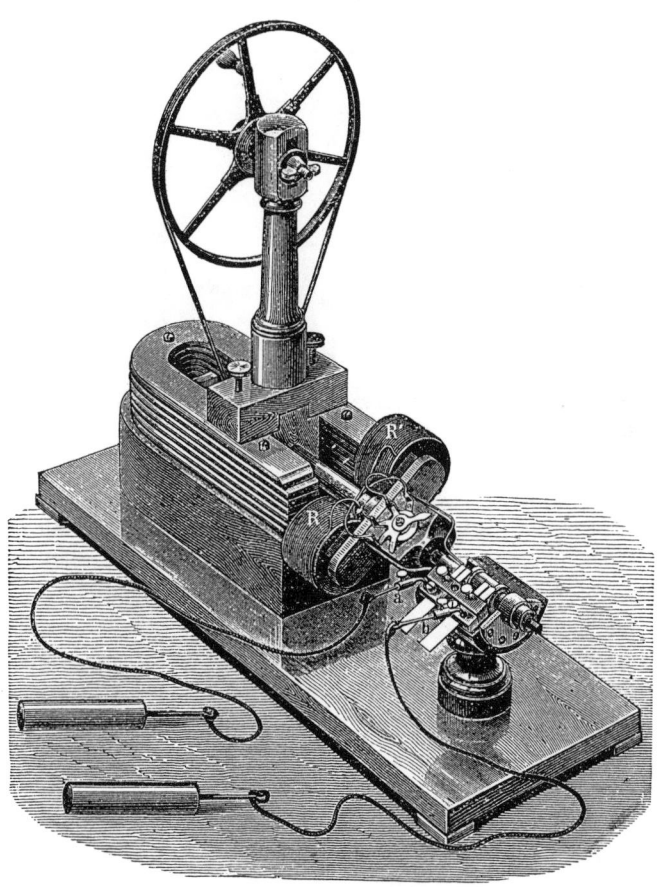

Fig. 13.

a clock-work, as often and as long as cc is attracted by the magnetic action of the electrical current. Conversely, by reversing the magnetism in the iron core of the spirals bb, we should obtain in them an electrical current just as we have obtained such currents in the magneto-electrical machine, Fig. 13; in the spirals of that machine there is an iron core which, by being approached to the poles of the large horse-shoe magnet, is sometimes magnetised in one and sometimes in the other direction.

I will not accumulate examples of such relations; in subsequent lectures we shall come across them. Let us review these examples once more, and recognise in them the law which is common to all.

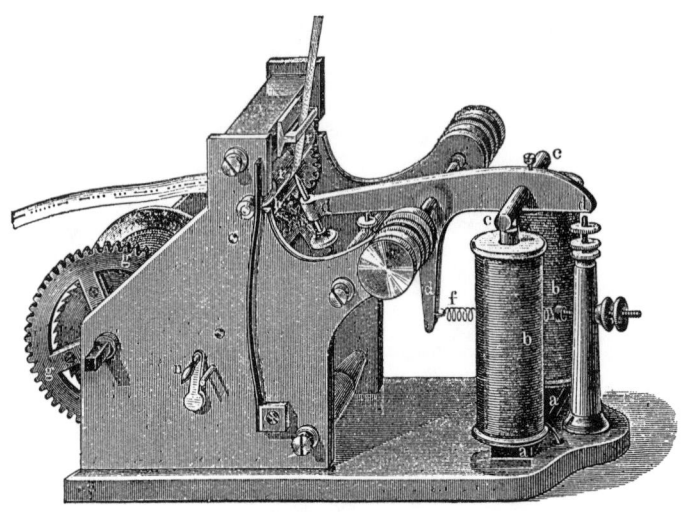

Fig. 14.

A raised weight can produce work, but in doing so it must necessarily sink from its height, and, when it has fallen as deep as it can fall, its gravity remains as before, but it can no longer do work.

A stretched spring can do work, but in so doing it becomes loose. The velocity of a moving mass can do work, but in doing so it comes to rest. Heat can perform work; it is destroyed in the operation. Chemical forces can perform work, but they exhaust themselves in the effort.

Electrical currents can perform work, but to keep them up we must consume either chemical or mechanical forces, or heat.

We may express this generally. *It is a universal character of all known natural forces that their capacity for work is exhausted in the degree in which they actually perform work.*

We have seen, further, that when a weight fell without performing any work, it *either* acquired velocity or produced heat. We might also drive a magneto-electrical machine by a falling weight; it would then furnish electrical currents.

We have seen that chemical forces, when they come into play, produce either heat or electrical currents or mechanical work.

We have seen that heat may be changed into work; there are apparatus (thermo-electric batteries) in which electrical currents are produced by it. Heat can directly separate chemical compounds; thus, when we burn limestone, it separates carbonic acid from lime.

Thus, whenever the capacity for work of one natural force is destroyed, it is transformed into another kind of activity. Even within the circuit of inorganic natural forces, we can transform each of them into an active condition by the aid of any other natural force which is capable of work. The connections between the various natural forces which modern physics has revealed, are so extraordinarily numerous that several entirely different methods may be discovered for each of these problems. I have stated how we are accustomed to measure mechanical work, and how the equivalent in work of heat may be found. The equivalent in work of chemical processes is again measured by the heat which they produce. By similar relations, the equivalent in work of the other natural forces may be expressed in terms of mechanical work.

If, now, a certain quantity of mechanical work is lost, there is obtained, as experiments made with the object of determining this point show, an equivalent quantity of heat, or, instead of this, of chemical force; and, conversely, when heat is lost, we gain an equivalent quantity of chemical or mechanical force; and, again, when chemical force disappears, an equivalent of heat or work; so that in all these interchanges between various inorganic natural forces working force may indeed disappear in one form, but then it reappears in exactly equivalent quantity in some other form; it is thus neither increased nor diminished, but always remains in exactly the same quantity. We shall subsequently see that the same law holds good also for processes in organic nature, so far as the facts have been tested.

It follows thence *that the total quantity of all the forces* capable of work *in the whole universe remains eternal and unchanged throughout all their changes.* All change in nature amounts to this, that force can change its form and locality without its quantity being changed. The universe possesses, once for all, a store of force which is not altered by any change of phenomena, can neither be increased nor diminished, and which maintains any change which takes place on it.

You see how, starting from considerations based on the immediate practical interests of technical work, we have been led up to a universal natural law, which, as far as all previous experience extends, rules and embraces all natural processes; which is no longer restricted to the practical objects of human utility, but expresses a perfectly general and particularly characteristic property of all natural forces, and which, as regards generality, is to be placed by the side of the laws of the unalterability of mass, and the unalterability of the chemical elements.

At the same time, it also decides a great practical question which has been much discussed in the last two centuries, to the decision of which an infinity of experiments have been made and an infinity of apparatus constructed—that is, the question of the possibility of a perpetual motion. By this was understood a machine which was to work continuously without the aid of any external driving force. The solution of this problem promised enormous gains. Such a machine would have had all the advantages of steam without requiring the expenditure of fuel. Work is wealth. A machine which could produce work from nothing was as good as one which made gold. This problem had thus for a long time occupied the place of gold making, and had confused many a pondering brain. That a perpetual motion could not be produced by the aid of the then known mechanical forces could be demonstrated in the last century by the aid of the mathematical mechanics which had at that time been developed. But to show also that it is not possible even if heat, chemical forces, electricity, and magnetism were made to co-operate, could not be done without a knowledge of our law in all its generality. The possibility of a perpetual motion was first finally negatived by the law of the conservation of force, and this law might also be expressed in the practical form that no perpetual motion is possible, that force cannot be produced from nothing, something must be consumed.

You will only be ultimately able to estimate the importance and the scope of our law when you have before your eyes a series of its applications to individual processes on nature.

What I have to-day mentioned as to the origin of the moving forces which are at our disposal, directs us to something beyond the narrow confines of our laboratories and our manufactories, to the great operations at work in the life of the earth and of the universe. The force of falling water can only flow down from the hills when rain and snow bring it to them. To furnish these, we must have aqueous vapour in the atmosphere, which can only be effected by the aid of heat, and this heat comes from the sun. The steam-engine needs the fuel which the vegetable life yields, whether it be the still active life of the surrounding vegetation, or the extinct life which has produced the immense coal deposits in the depths of the earth. The forces of man and animals must be restored by nourishment; all nourishment comes ultimately from the vegetable kingdom, and leads us back to the same source.

You see then that when we inquire into the origin of the moving forces which we take into our service, we are thrown back upon the meteorological processes in the earth's atmosphere, on the life of plants in general, and on the sun.

Introduction to a Series of Lectures Delivered at Carlsruhe in the Winter of 1862–1863.

6

The Recent Progress of the Theory of Vision

I. THE EYE AS AN OPTICAL INSTRUMENT.

THE physiology of the senses is a border land in which the two great divisions of human knowledge, natural and mental science, encroach on one another's domain; in which problems arise which are important for both, and which only the combined labour of both can solve.

No doubt the first concern of physiology is only with material changes in material organs, and that of the special physiology of the senses is with the nerves and their sensations, so far as these are excitations of the nerves. But, in the course of investigation into the functions of the organs of the senses, science cannot avoid also considering the apprehension of external objects, which is the result of these excitations of the nerves, and for the simple reason that the fact of a particular state of mental apprehension often reveals to us a nervous excitation which would otherwise have escaped our notice. on the other hand, apprehension of external objects must always be an act of our power of realization, and must therefore be accompanied by consciousness, for it is a mental function. Indeed the further exact investigation of this process has been pushed, the more it has revealed to us an ever-widening field of such mental functions, the results of which are involved in those acts of apprehension by the senses which at first sight appear to be most simple and immediate. These concealed functions have been but little discussed, because we are so accustomed to regard the apprehension of any external object as a complete and direct whole, which does not admit of analysis.

It is scarcely necessary for me to remind my present readers of the fundamental importance of this field of inquiry to almost every other department of science. For apprehension by the senses supplies after all, directly or indirectly, the material of all human knowledge, or at least the stimulus necessary to develop every inborn faculty of the mind. It supplies the basis for the whole action of man upon the outer world; and if this stage of mental processes is admitted to be the simplest and lowest of its kind, it is none the less important and interesting. For there is little hope that he who does not begin at the beginning of knowledge will ever arrive at its end.

It is by this path that the art of experiment, which has become so important in natural science, found entrance into the hitherto inaccessible field of mental processes. At first this will be only so far as we are able by experiment to determine the particular sensible impressions which call up one or another conception in our consciousness. But from this first step will follow numerous deductions as to the nature of the mental processes which contribute to the result. I will therefore endeavour to give some account of the results of physiological inquiries so far as they bear on the questions above mentioned.

I am the more desirous of doing so because I have lately completed[1] a complete survey of the field of physiological optics, and am happy to have an opportunity of putting together in a compendious form the views and deductions on the present subject which might escape notice among the numerous details of a book devoted to the special objects of natural science. I may state that in that work I took great pains to convince myself of the truth of every fact of the slightest importance by personal observation and experiment. There is no longer much controversy on the more important facts of observation, the chief difference of opinion being as to the extent of certain individual differences of apprehension by the senses. During the last few years a great number of distinguished investigators have, under the influence of the rapid progress of ophthalmic medicine, worked at the physiology of vision; and in proportion as the number of observed facts has increased, they have also become more capable of scientific arrangement and explanation. I need not remind those of my readers who are conversant with the subject how much labor must be expended to establish many facts which appear comparatively simple and almost self-evident.

To render what follows understood in all its bearings, I shall first describe the *physical* characters of the eye as an optical instrument; next the *physiological* processes of excitation and conduction in the parts of the nervous system which belong to it; and lastly I shall take up the *psychological* question, how mental apprehensions are produced by the changes which take place in the optic nerve.

The first part of our inquiry, which cannot be passed over because it is the foundation of what follows, will be in great part a repetition of what is already generally known, in order to bring in what is new in its proper place. But it is just this part of the subject which excites so much interest, as the real starting point of that remarkable progress which ophthalmic

1. Prof. Helmholtz's *Handbook of Physiological Optics* was published at Leipzig in 1867.

medicine has made during the last twenty years—a progress which for its rapidity and scientific character is perhaps without parallel in the history of the healing art.

Every lover of his kind must rejoice in these achievements which ward off or remove so much misery that formerly we were powerless to help, but a man of science has peculiar reason to look on them with pride. For this wonderful advance has not been achieved by groping and lucky finding but by deduction rigidly followed out, and thus carries with it the pledge of still future successes. As once astronomy was the pattern from which the other sciences learned how the right method will lead to success, so does ophthalmic medicine now display how much may be accomplished in the treatment of disease by extended application of well-understood methods of investigation and accurate insight into the causal connection of phenomena. It is no wonder that the right sort of men were drawn to an arena which offered a prospect of new and noble victories over the opposing powers of nature to the true scientific spirit—the spirit of patient and cheerful work. It was because there were so many of them that the success was so brilliant. Let me be permitted to name out of the whole number a representative of each of the three nations of common origin which have contributed most to the result: Von Graefe in Germany, Donders in Holland, and Bowman in England.

There is another point of view from which this advance in ophthalmology may be regarded, and that with equal satisfaction. Schiller says of science:—

Wer um die Göttin freit, suche in ihr nicht das Weib.[2]
Who woos the goddess must not hope the wife.

And history teaches us, what we shall have opportunity of seeing in the present inquiry, that the most important practical results have sprung unexpectedly out of investigations which might seem to the ignorant mere busy trifling, and which even those better able to judge could only regard with the intellectual interest which pure theoretical inquiry excites.

Of all our members the eye has always been held the choicest gift of Nature—the most marvellous product of her plastic force. Poets and orators have celebrated its praises; philosophers have extolled it as a crowning instance of perfection in an organism; and opticians have tried to imitate it as an unsurpassed model. And indeed the most enthusiastic admiration of this wonderful organ is only natural, when we consider what functions it

---

2. From Schiller's *Sprüche*. Literally, 'Let not him who seeks the love of a goddess expect to find in her the woman.'

performs; when we dwell on its penetrating power, on the swiftness of succession of its brilliant pictures, and on the riches which it spreads before our sense. It is by the eye alone that we know the countless shining worlds that fill immeasurable space, the distant landscapes of our own earth, with all the varieties of sunlight that reveal them, the wealth of form and colour among flowers, the strong and happy life that moves in animals. Next to the loss of life itself that of eyesight is the heaviest.

But even more important, than the delight in beauty and admiration of majesty in the creation which we owe to the eye, is the security and exactness with which we can judge by sight of the position, distance, and size of the objects which surround us. For this knowledge is the necessary foundation for all our actions, from threading a needle through a tangled skein of silk to leaping from cliff to cliff when life itself depends on the right measurement of the distance. In fact, the success of the movements and actions dependent on the accuracy of the pictures that the eye gives us forms a continual test and confirmation of that accuracy. If sight were to deceive us as to the position and distance of external objects, we should at once become aware of the delusion on attempting to grasp or to approach them. This daily verification by our other senses of the impressions we receive by sight produces so firm a conviction of its absolute and complete truth that the exceptions taken by philosophy or physiology, however well grounded they may seem, have no power to shake it.

No wonder then that, according to a wide-spread conviction, the eye is looked on as an optical instrument so perfect that none formed by human hands can ever be compared with it, and that its exact and complicated construction should be regarded as the full explanation of the accuracy and variety of its functions.

Actual examination of the performances of the eye as an optical instrument carried on chiefly during the last ten years has brought about a remarkable change in these views, just as in so many other cases the test of facts has disabused our minds of similar fancies. But as again in similar cases reasonable admiration rather increases than diminishes when really important functions are more clearly understood and their object better estimated, so it may well be with our more exact knowledge of the eye. For the great performances of this little organ can never be denied; and while we might consider ourselves compelled to withdraw our admiration from one point of view, we must again experience it from another.

Regarded as an optical instrument, the eye is a camera obscura. This apparatus is well known in the form used by photographers (Fig. 1). A box constructed of two parts, of which one slides in the other, and blackened, has in front a combination of lenses fixed in the tube *h i* on the inside, which refract the incident rays of light, and unite them at the back of the instrument into an optical image of the objects which lie in front of the camera. When the photographer first arranges his instrument, he receives the image upon a plate of ground glass, *g*. It is there seen as a small and elaborate picture in its natural colours, more clear and beautiful than the most skilful painter could imitate, though indeed it is upside down. The next step is to substitute for this glass a prepared plate upon which the light exerts a permanent chemical effect, stronger on the more brightly illuminated parts, weaker on those which are darker. These chemical changes having once taken place are permanent: by their means the image is fixed upon the plate.

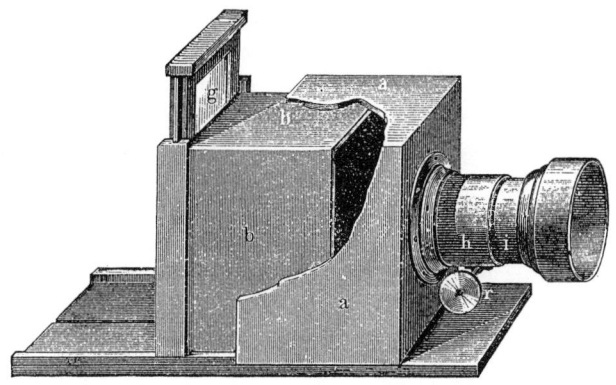

Fig. 1.

The natural camera obscura of the eye (seen in a diagrammatic section in Fig. 2) has its blackened chamber globular instead of cubical, and made not of wood, but of a thick, strong, white substance known as the *sclerotic coat*. It is this which is partly seen between the eyelids as 'the white of the eye.' This globular chamber is lined with a delicate coat of winding blood-vessels covered inside by black pigment. But the apple of the eye is not empty like the camera: it is filled with a transparent jelly as clear as water. The lens of the camera obscura is represented, first, by a convex transparent window like a pane of horn (the *cornea*), which is fixed in front of the sclerotic like a watch glass in front of its metal case. This union and its own firm texture make its position and its curvature constant. But the glass lenses of the photographer are not fixed; they are moveable by means of a sliding tube which can be adjusted by a screw (Fig. 1, *r*), so as to bring the objects in front of the camera into focus. The nearer they are, the further the lens is pushed forward; the farther off, the more it is screwed in. The eye has the same task of bringing at one time near, at another distant, objects to a focus at the back of its dark chamber. So that some power of adjustment or 'accommodation' is necessary. This is accomplished by the movements

of the *crystalline lens* (Fig. 2, L), which is placed a short distance behind
the cornea. It is covered by a curtain of varying colour, the *iris* (J), which
is perforated in the centre by a round hole, the *pupil*, the edges of which are
in contact with the front of the lens. Through this opening we see through

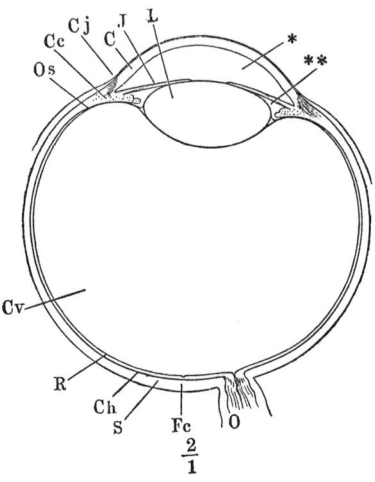

Fig. 2.

the transparent and, of course, invisible lens the black chamber within. The
crystalline lens is circular, biconvex, and elastic. It is attached at its edge to
the inside of the eye by means of a circular band of folded membrane which
surrounds it like a plaited ruff, and is called the *ciliary body* or Zonule of
Zinn (Fig. 2,**). The tension of this ring (and so of the lens itself) is
regulated by a series of muscular fibres known as the *ciliary muscle* (Cc).
When this muscle contracts, the tension of the lens is diminished, and its
surfaces—but chiefly the front one—become by its physical property of
elasticity more convex than when the eye is at rest; its refractive power is
thus increased, and the images of near objects are brought to a focus on the
back of the dark chamber of the eye.

Accordingly the healthy eye when at rest sees distant objects distinctly:
by the contraction of the ciliary muscle it is 'accommodated' for those
which are near. The mechanism by which this is accomplished, as above
shortly explained, was one of the greatest riddles of the physiology of the
eye since the time of Kepler; and the knowledge of its mode of action is of
the greatest practical importance from the frequency of defects in the power
of accommodation. No problem in optics has given rise to so many
contradictory theories as this. The key to its solution was found when the

French surgeon Sanson first observed very faint reflexions of light through the pupil from the two surfaces of the crystalline lens, and thus acquired the character of an unusually careful observer. For this phenomenon was anything but obvious; it can only be seen by strong side illumination, in darkness otherwise complete, only when the observer takes a certain position, and then all he sees is a faint misty reflexion. But this faint reflexion was destined to become a shining light in a dark corner of science. It was in fact the first appearance observed in the living eye which came directly from the lens. Sanson immediately applied his discovery to ascertain whether the lens was in its place in cases of impaired vision. Max Langenbeck made the next step by observing that the reflexions from the lens alter during accommodation. These alterations were employed by Cramer of Utrecht, and also independently by the present writer, to arrive at an exact knowledge of all the changes which the lens undergoes during the process of accommodation. I succeeded in applying to the moveable eye in a modified form the principle of the heliometer, an instrument by which astronomers are able so accurately to measure small distances between stars in spite of their constant apparent motion in the heavens, that they can thus sound the depths of the region of the fixed stars. An instrument constructed for the purpose, the *ophthalmometer*, enables us to measure in the living eye the curvature of the cornea, and of the two surfaces of the lens, the distance of these from each other, &c., with greater precision than could before be done even after death. By this means we can ascertain the entire range of the changes of the optical apparatus of the eye so far as it affects accommodation.

The physiological problem was therefore solved. Oculists, and especially Donders, next investigated the individual defects of accommodation which give rise to the conditions known as *long sight* and *short sight*. It was necessary to devise trustworthy methods in order to ascertain the precise limits of the power of accommodation even with inexperienced and uninstructed patients. It became apparent that very different conditions had been confounded as short sight and long sight, and this confusion had made the choice of suitable glasses uncertain. It was also discovered that some of the most obstinate and obscure affections of the sight, formerly reputed to be 'nervous,' simply depended on certain defects of accommodation, and could be readily removed by using suitable glasses. Moreover Donders[3] proved that the same defects of accommodation are the most frequent cause of squinting, and Von Graefe[4] had already shown that neglected and progressive shortsightedness tends to produce the most dangerous expansion and deformity of the back of the globe of the eye.

3. Professor of Physiology in the University of Utrecht.
4. This great ophthalmic surgeon died in Berlin at the early age of forty-two.

Thus connections were discovered, where least expected between the optical discovery and important diseases, and the result was no less beneficial to the patient than interesting to the physiologist.

We must now speak of the curtain which receives the optical image when brought to a focus in the eye. This is the *retina*, a thin membranous expansion of the optic nerve which forms the innermost of the coats of the eye. The optic nerve (Fig. 2,O) is a cylindrical cord which contains a multitude of minute fibres protected by a strong tendinous sheath. The nerve enters the apple of the eye from behind, rather to the inner (nasal) side of the middle of its posterior hemisphere. Its fibres then spread out in all directions over the front of the retina. They end by becoming connected, first, with *ganglion cells* and *nuclei*, like those found in the brain; and, secondly, with structures not elsewhere found, called *rods* and *cones*. The rods are slender cylinders; the cones, or bulbs, somewhat thicker, flask-shaped structures. All are ranged perpendicular to the surface of the retina, closely packed together, so as to form a regular mosaic layer behind it. Each rod is connected with one of the minutest nerve fibres, each cone with one somewhat thicker. This layer of rods and bulbs (also known as *membrana Jacobi*) has been proved by direct experiments to be the really sensitive layer of the retina, the structure in which alone the action of light is capable of producing a nervous excitation.

There is in the retina a remarkable spot which is placed near its centre, a little to the outer (temporal) side, and which from its colour is called *the yellow spot*. The retina is here somewhat thickened, but in the middle of the yellow spot is found a depression, the *fovea centralis*, where the retina is reduced to those elements alone which are absolutely necessary for exact vision. Fig. 3, from Henle, shows a thin transverse section of this central depression made on a retina which had been hardened in alcohol. *Lh* (*Lamina hyalina, membrana limitans*) is an elastic membrane which divides the retina from the vitreous. The bulbs (seen at *b*) are here smaller than elsewhere, measuring only the 400th part of a millimeter in diameter, and form a close and regular mosaic. The other, more or less opaque, elements of the retina are seen to be wanting, except the corpuscles (*g*), which belong to the cones. At *f* are seen the fibres which unite these with the rest of the retina. This consists of a layer of fibres of the optic nerve (*n*) in front, and two layers of nerve cells (*gli* and *gle*), known as the internal and external ganglion layers, with a stratum of fine granules (*gri*) between them. All these parts of the retina are absent at the bottom of the *fovea centralis*, and their gradual thinning away at its borders is seen in the diagram. Nor do the blood vessels of the retina enter the *fovea*, but end in a circle of delicate capillaries around it.

This *fovea*, or pit of the retina, is of great importance for vision, since it is the spot where the most exact discrimination of distances is made. The cones are here packed most closely together, and receive light which has not been impeded by other semi-transparent parts of the retina. We may assume that a single nervous fibril runs from each of these cones through the trunk of the optic nerve to the brain, without touching its neighbours, and there produces its special impression, so that the excitation of each individual cone will produce a distinct and separate effect upon the sense.

The production of optical images in a camera obscura depends on the well-known fact that the rays of light which come off from an illuminated object are so broken or refracted in passing through the lenses of the

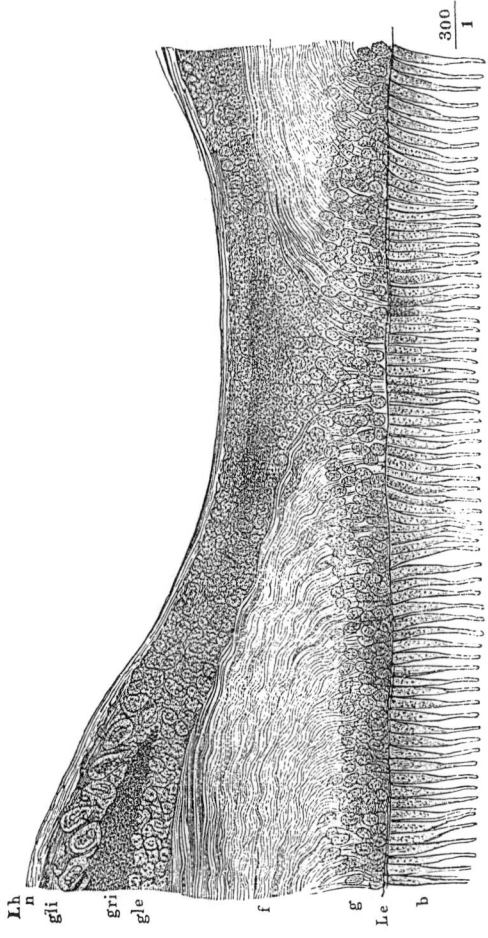

Fig. 3.

instrument, that they follow new directions which bring them all to a single point, the *focus*, at the back of the camera. A common burning glass has the same property; if we allow the rays of the sun to pass through it, and hold a sheet of white paper at the proper distance behind it, we may notice two effects. In the first place (and this is often disregarded) the burning lens, although made of transparent glass, throws a shadow like any opaque body; and next we see in the middle of this shadow a spot of dazzling brilliance, the image of the sun. The rays which, if the lens had not been there, would have illuminated the whole space occupied by the shadow, are concentrated by the refracting power of the burning glass upon the bright spot in the middle, and so both light and heat are more intense there than where the unrefracted solar rays fall. If, instead of the disc of the sun, we choose a star or any other point as the source of light, its light will be united into a point at the focus of the lens, and the image of the star will appear as such upon the white paper. If there is another fixed star near the one first chosen, its light will be collected at a second illuminated point on the paper; and if the star happen to send out red rays, its image on the paper will also appear red. The same will be true of any number of neighbouring stars, the image of each corresponding to it in brilliance, colour, and relative position. And if, instead of a multitude of separate luminous points, we have a continuous series of them in a bright line or surface, a similar line or surface will be produced upon the paper. But here also, if the piece of paper be put to the proper distance, all the light that proceeds from any one point will be brought to a focus at a point which corresponds to it in strength and colour of illumination, and (as a corollary) no point of the paper receives light from more than a single point of the object.

If now we replace our sheet of white paper by a prepared photographic plate, each point of its surface will be altered by the light which is concentrated on it. This light is all derived from the corresponding point in the object, and answers to it in intensity. Hence the changes which take place on the plate will correspond in amount to the chemical intensity of the rays which fall upon it.

This is exactly what takes place in the eye. Instead of the burning glass we have the cornea and crystalline lens; and instead of the piece of paper, the retina. Accordingly, if an optically accurate image is thrown upon the retina, each of its cones will be reached by exactly so much light as proceeds from the corresponding point in the field of vision; and also the nerve fibre which arises from each cone will be excited only by the light proceeding from the corresponding point in the field, while other nerve fibres will be excited by the light proceeding from other points of the field. Fig. 4 illustrates this effect. The rays which come from the point A in the object of vision are so broken that they all unite at *a* on the retina, while those from B unite at *b*. Thus it results that the light of each separate bright point of the field of vision excites a separate impression; that the difference of the several points of the field of vision in degree of brightness can be appreciated by the sense; and lastly, that separate impressions may each arrive separately at the seat of consciousness.

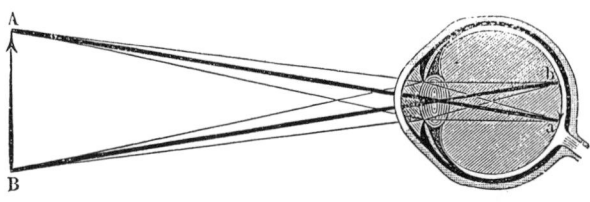

Fig. 4.

If now we compare the eye with other optical instruments, we observe the advantage it has over them in its very large field of vision. This for each eye separately is 160° (nearly two right angles) laterally, and 120° vertically, and for both together somewhat more than two right angles from right to left. The field of view of instruments made by art is usually very small, and becomes smaller with the increased size of the image.

But we must also admit, that we are accustomed to expect in these instruments complete precision of the image in its entire extent, while it is only necessary for the image on the retina to be exact over a very small surface, namely, that of the yellow spot. The diameter of the central pit corresponds in the field of vision to an angular magnitude which can be covered by the nail of one's forefinger when the hand is stretched out as far as possible. In this small part of the field our power of vision is so accurate that it can distinguish the distance between two points, of only *one minute* angular magnitude, i.e. a distance equal to the sixtieth part of the diameter of the finger-nail. This distance corresponds to the width of one of the cones of the retina. All the other parts of the retinal image are seen imperfectly, and the more so the nearer to the limit of the retina they fall. So that the image which we receive by the eye is like a picture, minutely and elaborately finished in the centre, but only roughly sketched in at the borders. But although at each instant we only see a very small part of the field of vision accurately, we see this in combination with what surrounds it, and enough of this outer and larger part of the field, to notice any striking object, and particularly any change that takes place in it. All of this is unattainable in a telescope.

But if the objects are too small, we cannot discern them at all with the greater part of the retina.

When, lost in boundless blue on high,
The lark pours forth his thrilling song,[5]

the 'ethereal minstrel' is lost until we can bring her image to a focus upon the central pit of our retina. Then only are we able to *see* her. To *look* at anything means to place the eye in such a position that the image of the object falls on the small region of perfectly clear vision. This we may call *direct* vision, applying the term *indirect* to that exercised with the lateral parts of the retina—indeed with all except the yellow spot.

The defects which result from the inexactness of vision and the smaller number of cones in the greater part of the retina are compensated by the rapidity with which we can turn the eye to one point after another of the field of vision, and it is this rapidity of movement which really constitutes the chief advantage of the eye over other optical instruments.

Indeed the peculiar way in which we are accustomed to give our attention to external objects, by turning it only to one thing at a time, and as soon as this has been taken in hastening to another, enables the sense of vision to accomplish as much as is necessary; and so we have practically the same advantage as if we enjoyed an accurate view of the whole field of vision at once. It is not in fact until we begin to examine our sensations closely that we become aware of the imperfections of indirect vision. Whatever we want to see we look at, and see it accurately; what we do not look at, we do not as a rule care for at the moment, and so do not notice how imperfectly we see it.

Indeed, it is only after long practice that we are able to turn our attention to an object in the field of indirect vision (as is necessary for some physiological observations) without looking at it, and so bringing it into direct view. And it is just as difficult to fix the eye on an object for the number of seconds required to produce the phenomenon of an after-image.[6] To get this well defined requires a good deal of practice.

A great part of the importance of the eye as an organ of expression depends on the same fact; for the movements of the eyeball—its glances—are among the most direct signs of the movement of the attention, of the movements of the mind, of the person who is looking at us.

Just as quickly as the eye turns upwards, downwards, and from side to side, does the accommodation change, so as to bring the object to which our attention is at the moment directed into focus; and thus near and distant objects pass in rapid succession into accurate view.

All these changes of direction and of accommodation take place far more

---

5. The lines in the well-known passage of Faust:—
   Wenn über uns im blauen Raum verloren
   Ihr schmetternd Lied die Lerche singt.
6. Vide infra, p. 164.

slowly in artificial instruments. A photographic camera can never show near and distant objects clearly at once, nor can the eye; but the eye shows them so rapidly one after another that most people, who have not thought how they see, do not know that there is any change at all.

Let us now examine the optical properties of the eye further. We will pass over the individual defects of accommodation which have been already mentioned as the cause of short and long sight. These defects appear to be partly the result of our artificial way of life, partly of the changes of old age. Elderly persons lose their power of accommodation, and their range of clear vision becomes confined within more or less narrow limits. To exceed these they must resort to the aid of glasses.

But there is another quality which we expect of optical instruments, namely, that they shall be free from dispersion—that they be achromatic. Dispersion of light depends on the fact that the coloured rays which united make up the white light of the sun are not refracted in exactly the same degree by any transparent substance known. Hence the size and position of the optical images thrown by these differently coloured rays are not quite the same; they do not perfectly overlap each other in the field of vision, and thus the white surface of the image appears fringed with a violet or orange, according as the red or blue rays are broader. This of course takes off so far from the sharpness of the outline.

Many of my readers know what a curious part the inquiry into the chromatic dispersion of the eye has played in the invention of achromatic telescopes. It is a celebrated instance of how a right conclusion may sometimes be drawn from two false premisses. Newton thought he had discovered a relation between the refractive and dispersive powers of various transparent materials, from which it followed that no achromatic refraction was possible. Euler,[7] on the other hand, concluded that, since the eye is achromatic, the relation discovered by Newton could not be correct. Reasoning from this assumption, he constructed theoretical rules for making achromatic instruments, and Dollond[8] carried them out. But Dollond himself observed that the eye could not be achromatic, because its construction did not answer to Euler's rules; and at last Fraunhofer[9] actually measured the degree of chromatic aberration of the eye. An eye constructed to bring red light from infinite distance to a focus on the retina can only do the same with violet rays from a distance of two feet. With ordinary light this is not noticed because these extreme colours are the least luminous of all, and so the images they produce are scarcely observed beside the more

7. Leonard Euler born at Basel, 1707; died at St. Petersburgh, 1783.
8. John Dollond, F.R.S. born 1706; died in London, 1761.
9. Joseph Fraunhofer born in Bavaria, 1787; died at Münich, 1826.

intense images of the intermediate yellow, green, and blue rays. But the effect is very striking when we isolate the extreme rays of the spectrum by means of violet glass. Glasses coloured with cobalt oxide allow the red and blue rays to pass, but stop the green and yellow ones, that is, the brightest rays of the spectrum. If those of my readers who have eyes of ordinary focal distance will look at lighted street lamps from a distance with this violet glass, they will see a red flame surrounded by a broad bluish violet halo. This is the dispersive image of the flame thrown by its blue and violet light. The phenomenon is a simple and complete proof of the fact of chromatic aberration in the eye.

Now the reason why this defect is so little noticed under ordinary circumstances, and why it is in fact somewhat less than a glass instrument of the same construction would have, is that the chief refractive medium of the eye is water, which possesses a less dispersive power than glass.[10] Hence it is that the chromatic aberration of the eye, though present, does not materially affect vision with ordinary white illumination.

A second defect which is of great importance in optical instruments of high magnifying power is what is known as spherical aberration. Spherical refracting surfaces approximately unite the rays which proceed from a luminous point into a single focus, only when each ray falls nearly perpendicularly upon the corresponding part of the refracting surface. If all those rays which form the centre of the image are to be exactly united, a lens with other than spherical surfaces must be used, and this cannot be made with sufficient mechanical perfection. Now the eye has its refracting surfaces partly elliptical; and so here again the natural prejudice in its favour led to the erroneous belief that spherical aberration was thus prevented. But this was a still greater blunder. More accurate investigation showed that much greater defects than that of spherical aberration are present in the eye, defects which are easily avoided with a little care in making optical instruments, and compared with which the amount of spherical aberration becomes very unimportant. The careful measurements of the curvature of the cornea, first made by Senff of Dorpat, next, with a better adapted instrument, the writer's ophthalmometer already referred to, and afterwards carried out in numerous cases by Donders, Knapp, and others, have proved that the cornea of most human eyes is not a perfectly symmetrical curve, but is variously bent in different directions. I have also devised a method of testing the 'centering' of an eye during life, i.e. ascertaining whether the cornea and the crystalline lens are symmetrically placed with regard to their common axis. By this means I discovered in the eyes I examined slight but

---

10. But still the diffraction in the eye is rather greater than an instrument made with water would produce under the same conditions.

distinct deviations from accurate centering. The result of these two defects of construction is the condition called *astigmatism*, which is found more or less in most human eyes, and prevents our seeing vertical and horizontal lines at the same distance perfectly clearly at once. If the degree of astigmatism is excessive, it can be obviated by the use of glasses with cylindrical surfaces, a circumstance which has lately much attracted the attention of oculists.

Nor is this all. A refracting surface which is imperfectly elliptical, an ill-centered telescope, does not give a single illuminated point as the image of a star, but, according to the surface and arrangement of the refracting media, elliptic, circular, or linear images. Now the images of an illuminated point, as the human eye brings them to focus, are even more inaccurate: they are irregularly radiated. The reason of this lies in the construction of the crystalline lens, the fibres of which are arranged around six diverging axes (shown in Fig. 5). So that the rays which we see around stars and other distant lights are images of the radiated structure of our lens; and the universality of this optical defect is proved by any figure with diverging rays being called 'star-shaped.' It is from the same cause that the moon, while her crescent is still narrow, appears to many persons double or threefold.

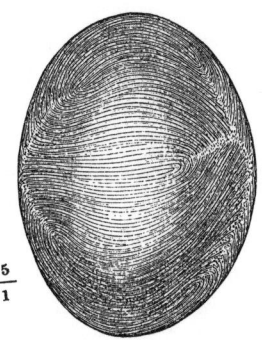

Fig. 5.

Now it is not too much to say that if an optician wanted to sell me an instrument which had all these defects, I should think myself quite justified in blaming his carelessness in the strongest terms, and giving him back his instrument. Of course, I shall not do this with my eyes, and shall be only too glad to keep them as long as I can—defects and all. Still, the fact that, however bad they may be, I can get no others, does not at all diminish their defects, so long as I maintain the narrow but indisputable position of a critic on purely optical grounds.

We have, however, not yet done with the list of the defects of the eye. We expect that the optician will use good, clear, perfectly transparent glass for his lenses. If it is not so, a bright halo will appear around each illuminated surface in the image: what should be black looks grey, what should be white is dull. But this is just what occurs in the image our eyes give us of the outer world. The obscurity of dark objects when seen near very bright ones depends essentially on this defect; and if we throw a strong light[11] through the cornea and crystalline lens, they appear of a dingy white, less transparent than the 'aqueous humour' which lies between them. This defect is most apparent in the blue and violet rays of the solar spectrum; for there comes in the phenomenon of fluorescence[12] to increase it.

In fact, although the crystalline lens looks so beautifully clear when taken out of the eye of an animal just killed, it is far from optically uniform in structure. It is possible to see the shadows and dark spots within the eye (the so-called 'entoptic objects') by looking at an extensive bright surface—the clear sky, for instance—through a very narrow opening. And these shadows are chiefly due to the fibres and spots in the lens.

There are also a number of minute fibres, corpuscles and folds of membrane, which float in the vitreous humour, and are seen when they come close in front of the retina, even under the ordinary conditions of vision. They are then called *muscoe volitantes*, because when the observer tries to look[13] at them, they naturally move with the movement of the eye. They seem continually to flit away from the point of vision, and thus look like flying insects. These objects are present in everyone's eyes, and usually float in the highest part of the globe of the eye, out of the field of vision, whence on any sudden movement of the eye they are dislodged and swim freely in the vitreous humour. They may occasionally pass in front of the central pit, and so impair sight. It is a remarkable proof of the way in which we observe, or fail to observe, the impressions made on our senses, that these *muscoe volitantes* often appear something quite new and disquieting to persons whose sight is beginning to suffer from any cause; although, of course, there must have been the same conditions long before.

A knowledge of the way in which the eye is developed in man and other vertebrates explains these irregularities in the structure of the lens and the

11. Eg. from a lamp, concentrated by a bull's-eye condenser.

12. This term is given to the property which certain substances possess of becoming for a time faintly luminous as long as they receive violet and blue light. The bluish tint of a solution of quinine, and the green colour of uranium glass, depend on this property. The fluorescence of the cornea and crystalline lens appears to depend upon the presence in their tissue of a very small quantity of a substance like quinine. For the physiologist this property is most valuable, for by its aid he can see the lens in a living eye by throwing on it a concentrated beam of blue light, and thus ascertain that it is placed close behind the iris, not separated by a large 'posterior chamber,' as was long supposed. But for seeing, the fluorescence of the cornea and lens is simply disadvantageous.

13. Vide *supra*, pp. 137–38.

vitreous body. Both are produced by an invagination of the integument of the embryo. A dimple is first formed, this deepens to a round pit, and then expands until its orifice becomes relatively minute, when it is finally closed and the pit becomes completely shut off. The cells of the scarf-skin which line this hollow form the crystalline lens, the true skin beneath them becomes its capsule, and the loose tissue which underlies the skin is developed into the vitreous humour. The mark where the neck of the fossa was sealed is still to be recognised as one of the 'entoptic images' of many adult eyes.

The last defect of the human eye which must be noticed is the existence of certain inequalities of the surface which receives the optical image. Not far from the centre of the field of vision there is a break in the retina, where the optic nerve enters. Here there is nothing but nerve fibres and blood-vessels; and, as the cones are absent, any rays of light which fall on the optic nerve itself are unperceived. This 'blind spot' will therefore produce a corresponding gap in the field of vision where nothing will be visible. Fig. 6 shows the posterior half of the globe of a right eye which has been cut across. R is the retina with its branching blood-vessels. The point from which these diverge is that at which the optic nerve enters. To the reader's left is seen the 'yellow spot.'

Now the gap caused by the presence of the optic nerve is no slight one. It is about 6° in horizontal and 8° in vertical dimension. Its inner border is about 12° horizontally distant from the 'temporal' or external side of the centre of distinct vision. The way to recognise this blind spot most readily is doubtless known to many of my readers. Take a sheet of white paper and mark on it a little cross; then to the right of this, on the same level, and about three inches off, draw a round black spot half an inch in diameter. Now, holding the paper at arm's length, shut the left eye, fix the right upon the cross, and bring the paper gradually nearer. When it is about eleven inches from the eye, the black spot will suddenly disappear, and will again come into sight as the paper is moved nearer.

This blind spot is so large that it might prevent our seeing eleven full moons if placed side by side, or a man's face at a distance of only six or seven feet. Mariotte,[14] who discovered the phenomenon, amused Charles II and his courtiers by showing them how they might see each other with their heads cut off.

There are, in addition, a number of smaller gaps in the field of vision, in which a small bright point, a fixed star for example, may be lost. These are caused by the blood-vessels of the retina. The vessels run in the front layers, and so cast their shadow on the part of the sensitive mosaic which lies

14. Edme. Mariotte born in Burgundy, died at Paris, 1684.

behind them. The larger ones shut off the light from reaching the rods and cones altogether, the more slender at least limit its amount.

These splits in the picture presented by the eye may be recognised by making a hole in a card with a fine needle, and looking through it at the sky, moving the card a little from side to side all the time. A still better experiment is to throw sunlight through a small lens upon the white of the eye at the outer angle (temporal canthus), while the globe is turned as much as possible inwards. The shadow of the blood-vessels is then thrown across on to the inner wall of the retina, and we see them as gigantic branching lines, like fig. 6 magnified. These vessels lie in the front layer of the retina itself,

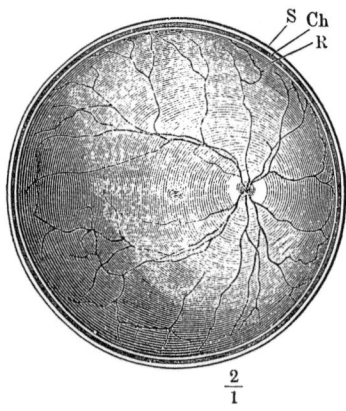

Fig. 6.

and, of course, their shadow can only be seen when it falls on the proper sensitive layer. So that this phenomenon furnishes a proof that the hindmost layer is that which is sensitive to light. And by its help it has become possible actually to measure the distance between the sensitive and the vascular layers of the retina. It is done as follows:—

If the focus of the light thrown on to the white of the eye (the sclerotic) is moved slightly backwards and forwards, the shadow of the blood-vessels and its image in the field of vision will, of course, move also. The extent of these movements can be easily measured, and from these data Heinrich Müller, of Würzburg—whose too early loss to science we still deplore—determined the distance between the two foci, and found it exactly to equal the thickness which actually separates the layer of rods and cones from the vascular layer of the retina.

The condition of the point of clearest vision (the yellow spot) is disadvantageous in another way. It is less sensitive to weak light than the other parts of the retina. It has been long known that many stars of inferior magnitude—for example, the *Coma Berenicoe* and the Pleiades—are seen more brightly if looked at somewhat obliquely than when their rays fall full upon the eye. This can be proved to depend partly on the yellow colour of the *macula*, which weakens blue more than other rays. It may also be partly the result of the absence of vessels at this yellow spot which has been noticed above, which interferes with its free communication with the life-giving blood.

All these imperfections would be exceedingly troublesome in an artificial camera obscura and in the photographic picture it produced. But they are not so in the eye—so little, indeed, that it was very difficult to discover some of them. The reason of their not interfering with our perception of external objects is not simply that we have two eyes, and so one makes up for the defects of the other. For even when we do not use both, and in the case of persons blind of one eye, the impression we receive from the field of vision is free from the defects which the irregularity of the retina would otherwise occasion. The chief reason is that we are continually moving the eye, and also that the imperfections almost always affect those parts of the field to which we are not at the moment directing our attention.

But, after all it remains a wonderful paradox, that we are so slow to observe these and other peculiarities of vision (such as the after-images of bright objects), so long as they are not strong enough to prevent our seeing external objects. It is a fact which we constantly meet, not only in optics, but in studying the perceptions produced by other senses on the consciousness. The difficulty with which we perceive the defect of the blind spot is well shown by the history of its discovery. Its existence was first demonstrated by theoretical arguments. While the long controversy whether the perception of light resided in the retina or the choroid was still undecided, Mariotte asked himself what perception there was where the choroid is deficient. He made experiments to ascertain this point, and in the course of them discovered the blind spot. Millions of men had used their eyes for ages, thousands had thought over the nature and cause of their functions, and, after all, it was only by a remarkable combination of circumstances that a simple phenomenon was noticed which would apparently have revealed itself to the slightest observation. Even now, anyone who tries for the first time to repeat the experiment which demonstrates the existence of the blind spot, finds it difficult to divert his attention from the fixed point of clear vision, without losing sight of it in the attempt. Indeed, it is only by long practice in optical experiments that even an experienced observer is able, as soon as he shuts one eye, to recognise the blank space in the field of vision which corresponds to the blind spot.

Other phenomena of this kind have only been discovered by accident, and usually by persons whose senses were peculiarly acute, and whose power of observation was unusually stimulated. Among these may be mentioned Goethe, Purkinje,[15] and Johannes Müller.[16] When a subsequent observer tries to repeat on his own eyes these experiments as he finds them described, it is of course easier for him than for the discoverer; but even now there are many of the phenomena described by Purkinje which have never been seen by anyone else, although it cannot be certainly held that they depended on individual peculiarities of this acute observer's eyes.

The phenomena of which we have spoken, and a number of others also, may be explained by the general rule that it is much easier to recognise any change in the condition of a nerve than a constant and equable impression on it. In accordance with this rule, all peculiarities in the excitation of separate nerve fibres, which are equally present during the whole of life (such as the shadow of the blood-vessels of the eye, the yellow colour of the central pit of the retina, and most of the fixed entoptic images), are never noticed at all; and if we want to observe them we must employ unusual modes of illumination and, particularly, constant change of its direction.

According to our present knowledge of the conditions of nervous excitation, it seems to me to be very unlikely that we have here to do with a simple property of sensation; it must, I think, be rather explained as a phenomenon belonging to our power of attention, and I now only refer to the question in passing, since its full discussion will come afterwards in its proper connection.

So much for the physical properties of the Eye. If I am asked why I have spent so much time in explaining its imperfection to my readers, I answer, as I said at first, that I have not done so in order to depreciate the performances of this wonderful organ or to diminish our admiration of its construction. It was my object to make the reader understand, at the first step of our inquiry, that it is not any mechanical perfection of the organs of our senses which secures for us such wonderfully true and exact impressions of the outer world. The next section of this inquiry will introduce much bolder and more paradoxical conclusions than any I have yet stated. We have now seen that the eye in itself is not by any means so complete an optical instrument as it first appears: its extraordinary value depends upon the way in which we use it: its perfection is practical, not absolute, consisting not in the avoidance of every error, but in the fact that all its defects do not prevent its rendering us the most important and varied services.

15. A distinguished embryologist, for many years professor at Breslau: he died at Prague, 1869, aet. 82.
16. A great biologist, in the full sense of the term. He was professor of physiology at Berlin, and died 1858, aet. 57. His *Manual of Physiology* was translated into English by the late Dr. Baly.—TR.

From this point of view, the study of the eye gives us a deep insight into the true character of organic adaptation generally. And this consideration becomes still more interesting when brought into relation with the great and daring conceptions which Darwin has introduced into science, as to the means by which the progressive perfection of the races of animals and plants has been carried on. Wherever we scrutinise the construction of physiological organs, we find the same character of practical adaptation to the wants of the organism; although, perhaps, there is no instance which we can follow out so minutely as that of the eye.

For the eye has every possible defect that can be found in an optical instrument, and even some which are peculiar to itself; but they are all so counteracted, that the inexactness of the image which results from their presence very little exceeds, under ordinary conditions of illumination, the limits which are set to the delicacy of sensation by the dimensions of the retinal cones. But as soon as we make our observations under somewhat changed conditions, we become aware of the chromatic aberration, the astigmatism, the blind spots, the venous shadows, the imperfect transparency of the media, and all the other defects of which I have spoken.

The adaptation of the eye to its function is, therefore, most complete, and is seen in the very limits which are set to its defects. Here the result which may be reached by innumerable generations working under the Darwinian law of inheritance, coincides with what the wisest Wisdom may have devised beforehand. A sensible man will not cut firewood with a razor, and so we may assume that each step in the elaboration of the eye must have made the organ more vulnerable and more slow in its development. We must also bear in mind that soft, watery animal textures must always be unfavorable and difficult material for an instrument of the mind.

One result of this mode of construction of the eye, of which we shall see the importance bye and bye, is that clear and complete apprehension of external objects by the sense of sight is only possible when we direct our attention to one part after another of the field of vision in the manner partly described above. Other conditions, which tend to produce the same limitation, will afterwards come under our notice.

But, apparently, we are not yet come much nearer to understanding sight. We have only made one step: we have learnt how the optical arrangement of the eye renders it possible to separate the rays of light which come in from all parts of the field of vision, and to bring together again all those that have proceeded from a single point, so that they may produce their effect upon a single fibre of the optic nerve.

Let us see, therefore, how much we know of the sensations of the eye, and how far this will bring us towards the solution of the problem.

II. THE SENSATION OF SIGHT.

IN the first section of our subject we have followed the course of the rays of light as far as the retina, and seen what is the result produced by the peculiar arrangement of the optical apparatus. The light which is reflected from the separate illuminated points of external objects is again united in the sensitive terminal structures of separate nerve fibres, and thus throws them into action without affecting their neighbours. At this point the older physiologists thought they had solved the problem, so far as it appeared to them to be capable of solution. External light fell directly upon a sensitive nervous structure in the retina, and was, as it seemed, directly felt there.

But during the last century, and still more during the first quarter of this, our knowledge of the processes which take place in the nervous system was so far developed, that Johannes Müller, as early as the year 1826,[17] when writing that great work on the 'Comparative Physiology of Vision,' which marks an epoch in science, was able to lay down the most important principles of the theory of the impressions derived from the senses. These principles have not only been confirmed in all important points by subsequent investigation, but have proved of even more extensive application than this eminent physiologist could have suspected.

The conclusions which he arrived at are generally comprehended under the name of the theory of the Specific Action of the Senses. They are no longer so novel that they can be reckoned among the latest advances of the theory of vision, which form the subject of the present essay. Moreover, they have been frequently expounded in a popular form by others as well as by myself.[18] But that part of the theory of vision with which we are now occupied is little more than a further development of the theory of the specific action of the senses. I must, therefore, beg my reader to forgive me if, in order to give him a comprehensive view of the whole subject in its proper connection, I bring before him much which he already knows, while I also introduce the more recent additions to our knowledge in their appropriate places.

All that we apprehend of the external world is brought to our consciousness by means of certain changes which are produced in our organs of sense by external impressions, and transmitted to the brain by the nerves. It is in the brain that these impressions first become conscious sensations, and are combined so as to produce our conceptions of surrounding objects. If the nerves which convey these impressions to the

17. The year in which he was appointed Extraordinary Professor of Physiology in the University of Bonn.

18. 'On the Nature of Special Sensations in Man,' *Königsberger naturwissenschaftliche Unterhaltungen*, vol. iii. 1852. 'Human Vision,' a popular Scientific Lecture by H. Helmholtz, Leipzig, 1855.

brain are cut through, the sensation, and the perception of the impression, immediately cease. In the case of the eye, the proof that visual perception is not produced directly in each retina, but only in the brain itself by means of the impressions transmitted to it from both eyes, lies in the fact (which I shall afterwards more fully explain) that the visual impression of any solid object of three dimensions is only produced by the combination of the impressions derived from both eyes.

What, therefore, we directly apprehend is not the immediate action of the external exciting cause upon the ends of our nerves, but only the changed condition of the nervous fibres which we call the state of *excitation* or functional activity.

Now all the nerves of the body, so far as we at present know, have the same structure, and the change which we call excitation is in each of them a process of precisely the same kind, whatever be the function it subserves. For while the task of some nerves is that already mentioned, of carrying sensitive impressions from the external organs to the brain, others convey voluntary impulses in the opposite direction, from the brain to the muscles, causing them to contract, and so moving the limbs. Other nerves, again, carry an impression from the brain to certain glands, and call forth their secretion, or to the heart and to the blood-vessels, and regulate the circulation. But the fibres of all these nerves are the same clear, cylindrical threads of microscopic minuteness, containing the same oily and albuminous material. It is true that there is a difference in the diameter of the fibres, but this, so far as we know, depends only upon minor causes, such as the necessity of a certain strength and of getting room for a certain number of independent conducting fibres. It appears to have no relation to their peculiarities of function.

Moreover, all nerves have the same electro-motor actions, as the researches of Du Bois Reymond[19] prove. In all of them the condition of excitation is called forth by the same mechanical, electrical, chemical, or thermometric changes. It is propagated with the same rapidity, of about one hundred feet in the second, to each end of the fibres, and produces the same changes in their electro-motor properties. Lastly, all nerves die when submitted to like conditions, and, with a slight apparent difference according to their thickness, undergo the same coagulation of their contents. In short, all that we can ascertain of nervous structure and function, apart from the action of the other organs with which they are united and in which during life we see the proofs of their activity, is precisely the same for all the different kinds of nerves. Very lately the French physiologists, Philippeaux and Vulpian, after dividing the motor and sensitive nerves of the tongue, succeeded in getting the upper half of the sensitive nerve to

---

19. Professor of Physiology in the University of Berlin.

unite with the lower half of the motor. After the wound had healed, they found that irritation of the upper half, which in normal conditions would have been felt as a sensation, now excited the motor branches below, and thus caused the muscles of the tongue to move. We conclude from these facts that all the difference which is seen in the excitation of different nerves depends only upon the difference of the organs to which the nerve is united, and to which it transmits the state of excitation.

The nerve-fibres have been often compared with telegraphic wires traversing a country, and the comparison is well fitted to illustrate this striking and important peculiarity of their mode of action. In the net-work of telegraphs we find everywhere the same copper or iron wires carrying the same kind of movement, a stream of electricity, but producing the most different results in the various stations according to the auxiliary apparatus with which they are connected. At one station the effect is the ringing of a bell, at another a signal is moved, and at a third a recording instrument is set to work. Chemical decompositions may be produced which will serve to spell out the messages, and even the human arm may be moved by electricity so as to convey telegraphic signals. When the Atlantic cable was being laid, Sir William Thomson found that the slightest signals could be recognised by the sense of taste, if the wire was laid upon the tongue. Or, again, a strong electric current may be transmitted by telegraphic wires in order to ignite gunpowder for blasting rocks. In short, everyone of the hundred different actions which electricity is capable of producing may be called forth by a telegraphic wire laid to whatever spot we please, and it is always the same process in the wire itself which leads to these diverse consequences. Nerve-fibres and telegraphic wires are equally striking examples to illustrate the doctrine that the same causes may, under different conditions, produce different results. However commonplace this may now sound, mankind had to work long and hard before it was understood, and before this doctrine replaced the belief previously held in the constant and exact correspondence between cause and effect. And we can scarcely say that the truth is even yet universally recognised, since in our present subject its consequences have been till lately disputed.

Therefore, as motor nerves, when irritated, produce movement, because they are connected with muscles, and glandular nerves secretion, because they lead to glands, so do sensitive nerves, when they are irritated, produce sensation, because they are connected with sensitive organs. But we have very different kinds of sensation. In the first place, the impressions derived from external objects fall into five groups, entirely distinct from each other. These correspond to the five senses, and their difference is so great that it is not possible to compare in quality a sensation of light with one of sound or of smell. We will name this difference, so much deeper than that between comparable qualities, a difference of the mode, or *kind*, of sensation, and will describe the differences between impressions belonging to the same sense (for example, the difference between the various sensations of colour) as a difference of *quality*.

Whether by the irritation of a nerve we produce a muscular movement, a secretion or a sensation depends upon whether we are handling a motor, a glandular, or a sensitive nerve, and not at all upon what means of irritation we may use. It may be an electrical shock, or tearing the nerve, or cutting it through, or moistening it with a solution of salt, or touching it with a hot wire. In the same way (and this great step in advance was due to Johannes Müller) the *kind* of sensation which will ensue when we irritate a sensitive nerve, whether an impression of light, or of sound, or of feeling, or of smell, or of taste, will be produced, depends entirely upon which sense the excited nerve subserves, and not at all upon the method of excitation we adopt.

Let us now apply this to the optic nerve, which is the object of our present enquiry. In the first place, we know that no kind of action upon any part of the body except the eye and the nerve which belongs to it, can ever produce the sensation of light. The stories of somnambulists, which are the only arguments that can be adduced against this belief, we may be allowed to disbelieve. But, on the other hand, it is not light alone which can produce the sensation of light upon the eye, but also any other power which can excite the optic nerve. If the weakest electrical currents are passed through the eye they produce flashes of light. A blow, or even a slight pressure made upon the side of the eyeball with the finger, makes an impression of light in the darkest room, and, under favourable circumstances, this may become intense. In these cases it is important to remember that there is no objective light produced in the retina, as some of the older physiologists assumed, for the sensation of light may be so strong that a second observer could not fail to see through the pupil the illumination of the retina which would follow, if the sensation were really produced by an actual development of light within the eye. But nothing of the sort has ever been seen. Pressure or the electric current excites the optic nerve, and therefore, according to Müller's law, a sensation of light results, but under these circumstances, at least, there is not the smallest spark of actual light.

In the same way, increased pressure of blood, its abnormal constitution in fevers, or its contamination with intoxicating or narcotic drugs, can produce sensations of light to which no actual light corresponds. Even in cases in which an eye is entirely lost by accident or by an operation, the irritation of the stump of the optic nerve while it is healing is capable of producing similar subjective effects. It follows from these facts that the peculiarity in kind which distinguishes the sensation of light from all others does not depend upon any peculiar qualities of light itself. Every action which is capable of exciting the optic nerve is capable of producing the impression of light; and the purely subjective sensation thus produced is so precisely similar to that caused by external light, that persons unacquainted with these phenomena readily suppose that the rays they see are real objective beams.

Thus we see that external light produces no other effects in the optic nerve than other agents of an entirely different nature. In one respect only does light differ from the other causes which are capable of exciting this nerve: namely, that the retina, being placed at the back of the firm globe of the eye, and further protected by the bony orbit, is almost entirely withdrawn from other exciting agents, and is thus only exceptionally affected by them, while it is continually receiving the rays of light which stream in upon it through the transparent media of the eye.

On the other hand, the optic nerve, by reason of the peculiar structures in connection with the ends of its fibres, the rods and cones of the retina, is incomparably more sensitive to rays of light than any other nervous apparatus of the body, since the rest can only be affected by rays which are concentrated enough to produce noticeable elevation of temperature.

This explains why the sensations of the optic nerve are for us the ordinary sensible sign of the presence of light in the field of vision, and why we always connect the sensation of light with light itself, even where they are really unconnected. But we must never forget that a survey of all the facts in their natural connection puts it beyond doubt that external light is only *one* of the exciting causes capable of bringing the optic nerve into functional activity, and therefore that there is no exclusive relation between the sensation of light and light itself.

Now that we have considered the action of excitants upon the optic nerve in general, we will proceed to the qualitative differences of the sensation of light, that is to say, to the various sensations of colour. We will try to ascertain how far these differences of sensation correspond to actual differences in external objects.

Light is known in Physics as a movement which is propagated by successive waves in the elastic ether distributed through the universe, a movement of the same kind as the circles which spread upon the smooth surface of a pond when a stone falls on it, or the vibration which is transmitted through our atmosphere as sound. The chief difference is, that the rate with which light spreads, and the rapidity of movement of the minute particles which form the waves of ether, are both enormously greater than that of the waves of water or of air. The waves of light sent, forth from the sun differ exceedingly in size, just as the little ripples whose summits are a few inches distant from each other differ from the waves of the ocean, between whose foaming crests lie valleys of sixty or a hundred feet. But, just as high and low, short and long waves, on the surface of water, do not differ in kind, but only in size, so the various waves of light which stream from the sun differ in their height and length, but move all in the same manner, and show (with certain differences depending upon the length of the waves) the same remarkable properties of reflection, refraction, interference, diffraction, and polarisation. Hence we conclude that the undulating movement of the ether is in all of them the same. We must particularly note that the phenomena of interference, under which light is now strengthened, and now obscured by light of the same kind, according to the distance it has traversed, prove that all the rays of light depend upon oscillations of waves; and further, that the phenomena of polarisation, which differ according to different lateral directions of the rays, show that the particles of ether vibrate at right angles to the direction in which the ray is propagated.

All the different sorts of rays which I have mentioned produce one effect in common. They raise the temperature of the objects on which they fall, and accordingly are all felt by our skin as rays of heat.

On the other hand, the eye only perceives one part of these vibrations of ether as light. It is not at all cognisant of the waves of great length, which I have compared with those of the ocean; these, therefore, are named the *dark heat-rays*. Such are those which proceed from a warm but not red-hot stove, and which we recognise as heat, but not as light.

Again, the waves of shortest length, which correspond with the very smallest ripples produced by a gentle breeze, are so slightly appreciated by the eye, that such rays are also generally regarded as invisible, and are known as the *dark chemical rays*.

Between the very long and the very short waves of ether there are waves of intermediate length, which strongly affect the eye, but do not essentially differ in any other physical property from the dark rays of heat and the dark chemical rays. The distinction between the visible and invisible rays depends only on the different length of their waves and the different physical relations which result therefrom. We call these middle rays Light, because they alone illuminate our eyes.

When we consider the heating property of these rays we also call them luminous heat; and because they produce such a very different impression on our skin and on our eyes, heat was universally considered as an entirely different kind of radiation from light, until about thirty years ago. But both kinds of radiation are inseparable from one another in the illuminating rays of the sun; indeed, the most careful recent investigations prove that they are precisely identical. To whatever optical processes they may be subjected, it is impossible to weaken their illuminating power without at the same time, and in the same degree, diminishing their heating and their chemical action. Whatever produces an undulatory movement of ether, of course produces thereby all the effects of the undulation, whether light, or heat, or fluorescence, or chemical change.

Those undulations which strongly affect our eyes, and which we call light, excite the impression of different colours, according to the length of the waves. The undulations with the longest waves appear to us red; and as the length of the waves gradually diminishes they seem to be golden-yellow, yellow, green, blue, violet, the last colour being that of the illuminating rays which have the smallest wave-length. This series of colours is universally known in the rainbow. We also see it if we look towards the light through a glass prism, and a diamond sparkles with hues which follow in the same order. In passing through transparent prisms, the primitive beam of white light, which consists of a multitude of rays of various colour and various wave-length, is decomposed by the different degree of refraction of its several parts, referred to in the last essay; and thus each of its component hues appears separately. These colours of the several primary forms of light are best seen in the spectrum produced by a narrow streak of light passing through a glass prism: they are at once the fullest and the most brilliant which the external world can show.

When several of these colours are mixed together, they give the impression of a new colour, which generally seems more or less white. If they were all mingled in precisely the same proportions in which they are combined in the sun-light, they would give the impression of perfect white. According as the rays of greatest, middle, or least wave-length predominate in such a mixture, it appears as reddish-white, greenish-white, bluish-white, and so on.

Everyone who has watched a painter at work knows that two colours mixed together give a new one. Now, although the results of the mixture of coloured light differ in many particulars from those of the mixture of pigments, yet on the whole the appearance to the eye is similar in both cases. If we allow two different coloured lights to fall at the same time upon a white screen, or upon the same part of our retina, we see only a single compound colour, more or less different from the two original ones.

The most striking difference between the mixture of pigments and that of coloured light is, that while painters make green by mixing blue and yellow pigments, the union of blue and yellow rays of light produces white. The simplest way of mixing coloured light is shown in Fig. 7. *P* is a small flat piece of glass; *b* and *g* are two coloured wafers. The observer looks at *b* through the glass plate, while *g* is seen reflected in the same; and if *g* is put in a proper position, its image exactly coincides with that of *b*. It then appears as if there was a single wafer at *b*, with a colour produced by the mixture of the two real ones. In this experiment the light from *b*, which traverses the glass pane, actually unites with that from *g*, which is reflected from it, and the two combined pass on to the retina at *o*. In general, then, light, which consists of undulations of different wave-lengths, produces different impressions upon our eye, namely, those of different colours. But the number of hues which we can recognise is much smaller than that of the various possible combinations of rays with different wave-lengths which

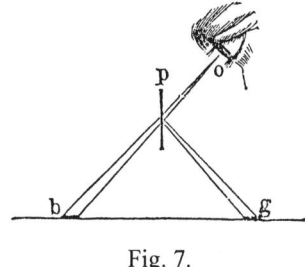

Fig. 7.

external objects can convey to our eyes. The retina cannot distinguish between the white which is produced by the union of scarlet and bluish-green light, and that which is composed of yellowish-green and violet, or of yellow and ultramarine blue, or of red, green, and violet, or of all the colours of the spectrum united. All these combinations appear identically as white; and yet, from a physical point of view, they are very different. In fact, the only resemblance between the several combinations just mentioned is, that they are indistinguishable to the human eye. For instance, a surface illuminated with red and bluish-green light would come out black in a photograph; while another lighted with yellowish green and violet would appear very bright, although both surfaces alike seem to the eye to be simply white. Again, if we successively illuminate coloured objects with white beams of light of various composition, they will appear differently coloured. And whenever we decompose two such beams by a prism, or look at them through a coloured glass, the difference between them at once becomes evident.

Other colours, also, especially when they are not strongly pronounced, may, like pure white light, be composed of very different mixtures, and yet appear indistinguishable to the eye, while in every other property, physical or chemical, they are entirely distinct.

Newton first showed how to represent the system of colours distinguishable to the eye in a simple diagrammatic form; and by the same means it is comparatively easy to demonstrate the law of the combination of colours. The primary colours of the spectrum are arranged in a series around the circumference of a circle, beginning with red, and by imperceptible degrees passing through the various hues of the rainbow to violet. The red and violet are united by shades of purple, which on the one side pass off to the indigo and blue tints, and on the other through crimson and scarlet to orange. The middle of the circle is left white, and on lines which run from the centre to the circumference are represented the various tints which can be produced by diluting the full colours of the circumference until they pass into white. A colour-disc of this kind shows all the varieties of hue which can be produced with the same amount of light.

It will now be found possible so to arrange the places of the several colours in this diagram, and the quantity of light which each reflects, that when we have ascertained the resultants of two colours of different known strength of light (in the same way as we might determine the centre of gravity of two bodies of different known weights), we shall then find their combination-colour at the 'centre of gravity' of the two amounts of light. That is to say, that in a properly constructed colour-disc, the combination-colour of any two colours will be found upon a straight line drawn from between them; and compound colours which contain more of one than of the other component hue, will be found in that proportion nearer to the former, and further from the latter.

We find, however, when we have drawn our diagram, that those colours of the spectrum which are most saturated in nature, and which must therefore be placed at the greatest distance from the central white, will not arrange themselves in the form of a circle. The circumference of the diagram presents three projections corresponding to the red, the green, and the violet, so that the colour circle is more properly a triangle, with the corners rounded off, as seen in Fig. 8. The continuous line represents the curve of the colours of the spectrum, and the small circle in the middle the white. At the corners are the three colours I have mentioned,[20] and the sides of the triangle show the transitions from red through yellow into green, from green through bluish-green and ultramarine to violet, and from violet through purple to scarlet.

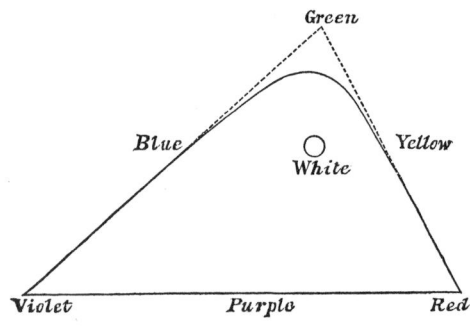

Fig. 8.

20. The author has restored violet as a primitive colour in accordance with the experiments of J. J. Müller, having in the first edition adopted the opinion of Maxwell that it is blue.

Newton used the diagram of the colours of the spectrum (in a somewhat different form from that just given) only as a convenient way of representing the facts to the eye; but recently Maxwell has succeeded in demonstrating the strict and even quantitative accuracy of the principles involved in the construction of this diagram. His method is to produce combinations of colours on swiftly rotating discs, painted of various tints in sectors. When such a disc is turned rapidly round, so that the eye can no longer follow the separate hues, they melt into a uniform combination-colour, and the quantity of light which belongs to each can be directly measured by the breadth of the sector of the circle it occupies. Now the combination-colours which are produced in this manner are exactly those which would result if the same qualities of coloured light illuminated the same surface continuously as can be experimentally proved. Thus have the relations of size and number been introduced into the apparently inaccessible region of colours, and their differences in quality have been reduced to relations of quantity.

All differences between colours may be reduced to three, which may be described as difference of tone, difference of fulness, or, as it is technically called, 'saturation,' and difference of brightness. The differences of *tone* are those which exist between the several colours of the spectrum, and to which we give the names red, yellow, blue, violet, purple. Thus, with regard to tone, colours form a series which returns upon itself; a series which we complete when we allow the terminal colours of the rainbow to pass into one another through purple and crimson. It is in fact the same which we described as arranged around the circumference of the colour-disc.

The *fulness* or saturation of colours is greatest in the pure tints of the spectrum, and becomes less in proportion as they are mixed with white light. This, at least, is true for colours produced by external light, but for our sensations it is possible to increase still further the apparent saturation of colour, as we shall presently see. Pink is a whitish-crimson, flesh-colour a whitish-scarlet, and so pale green, straw-colour, light blue, &c., are all produced by diluting the corresponding colours with white. All compound colours are, as a rule, less saturated than the simple tints of the spectrum.

Lastly, we have the difference of *brightness*, or strength of light, which is not represented in the colour-disc. As long as we observe coloured rays of light, difference in brightness appears to be only one of quantity, not of quality. Black is only darkness—that is, simple absence of light. But when we examine the colours of external objects, black corresponds just as much to a peculiarity of surface in reflection, as does white, and therefore has as good a right to be called a colour. And as a matter of fact, we find in common language a series of terms to express colours with a small amount of light. We call them *dark* (or rather in English, *deep*) when they have little light but are 'full' in tint, and *grey* when they are 'pale.' Thus dark blue conveys the idea of depth in tint, of a full blue with a small amount of light; while grey-blue is a pale blue with a small amount of light. In the same way, the colours known as maroon, brown and olive are dark, more or less saturated tints of red, yellow and green respectively.

In this way we may reduce all possible actual (objective) differences in colour, so far as they are appreciated by the eye, to three kinds; difference of hue (*tone*), difference of fulness (*saturation*), and difference of amount of illumination (*brightness*). It is in this way that we describe the system of colours in ordinary language. But we are able to express this threefold difference in another way.

I said above that a properly constructed colour-disc approaches a triangle in its outline. Let us suppose for a moment that it is an exact rectilinear triangle, as made by the dotted line in Fig. 8; how far this differs from the actual condition we shall have afterwards to point out. Let the colours red, green, and violet be placed at the corners, then we see the law which was mentioned above: namely, that all the colours in the interior and on the sides of the triangle are compounds of the three at its corners. It follows that *all differences of hue depend upon combinations in different proportions of the three primary colours*. It is best to consider the three just named as primary; the old ones red, yellow, and blue are inconvenient, and were only chosen from experience of painters' colours. It is impossible to make a green out of blue and yellow light.

We shall better understand the remarkable fact that we are able to refer all the varieties in the composition of external light to mixtures of three primitive colours, if in this respect we compare the eye with the ear.

Sound, as I mentioned before, is, like light, an undulating movement, spreading by waves. In the case of sound also, we have to distinguish waves of various length which produce upon our ear impressions of different quality. We recognise the long waves as low notes, the short as high-pitched, and the ear may receive at once many waves of sound—that is to say, many notes. But here these do not melt into compound notes, in the same way that colours, when perceived at the same time and place, melt into compound colours. The eye cannot tell the difference, if we substitute orange for red and yellow; but if we hear the notes C and E sounded at the same time, we cannot put D instead of them, without entirely changing the impression upon the ear. The most complicated harmony of a full orchestra becomes changed to our perception if we alter any one of its notes. No accord (or consonance of several tones) is, at least for the practised ear, completely like another, composed of different tones; whereas, if the ear perceived musical tones as the eye colours, every accord might be completely represented by combining only three constant notes, one very low, one very high, and one intermediate, simply changing the relative strength of these three primary notes to produce all possible musical effects.

In reality we find that an accord only remains unchanged to the ear, when the strength of each separate tone which it contains remains unchanged. Accordingly, if we wish to describe it exactly and completely, the strength of each of its component tones must be exactly stated. In the same way, the physical nature of a particular kind of light can only be fully ascertained by measuring and noting the amount of light of each of the simple colours which it contains. But in sunlight, in the light of most of the stars, and in flames, we find a continuous transition of colours into one another through numberless intermediate gradations. Accordingly, we must ascertain the amount of light of an infinite number of compound rays if we would arrive at an exact physical knowledge of sun or starlight. In the sensations of the eye we need distinguish for this purpose only the varying intensities of three components.

The practised musician is able to catch the separate notes of the various instruments among the complicated harmonies of an entire orchestra, but the optician cannot directly ascertain the composition of light by means of the eye; he must make use of the prism to decompose the light for him. As soon, however, as this is done, the composite character of light becomes apparent, and he can then distinguish the light of separate fixed stars from one another by the dark and bright lines which the spectrum shows him, and can recognise what chemical elements are contained in flames which are met with on the earth, or even in the intense heat of the sun's atmosphere, in the fixed stars, or in the nebulae. The fact that light derived from each separate source carries with it certain permanent physical peculiarities is the foundation of spectral analysis—that most brilliant discovery of recent years, which has opened the extreme limits of celestial space to chemical analysis.

There is an extremely interesting and not very uncommon defect of sight which is known as colour-blindness. In this condition the differences of colour are reduced to a still more simple system than that described above; namely, to combinations of only two primary colours. Persons so affected are called *colour-blind*, because they confound certain hues which appear very different to ordinary eyes. At the same time they distinguish other colours, and that quite as accurately, or even (as it seems) rather more accurately, than ordinary people. They are usually 'red-blind'; that is to say, there is no red in their system of colours, and accordingly they see no difference which is produced by the addition of red. All tints are for them varieties of blue and green or, as they call it, yellow. Accordingly scarlet, flesh-colour, white, and bluish-green appear to them to be identical, or at the utmost to differ in brightness. The same applies to crimson, violet, and blue, and to red, orange, yellow, and green. The scarlet flowers of the geranium have for them exactly the same colours as its leaves. They cannot distinguish between the red and the green signals of trains. They cannot see the red end of the spectrum at all. Very full scarlet appears to them almost black, so that a red-blind Scotch clergyman went to buy scarlet cloth for his gown, thinking it was black.[21]

In this particular of discrimination of colours, we find remarkable inequalities in different parts of the retina. In the first place all of us are red-blind in the outermost part of our field of vision. A geranium-blossom when moved backwards and forwards just within the field of sight, is only recognised as a moving object. Its colour is not seen, so that if it is waved in front of a mass of leaves of the same plant it cannot be distinguished from them in hue. In fact, all red colours appear much darker when viewed indirectly. This red-blind part of the retina is most extensive on the inner or nasal side of the field of vision; and according to recent researches of Woinow, there is at the furthest limit of the visible field a narrow zone in which all distinction of colours ceases and there only remain differences of brightness. In this outermost circle everything appears white, grey, or black. Probably those nervous fibres which convey impressions of green light are alone present in this part of the retina.

In the second place, as I have already mentioned, the middle of the retina, just around the central pit, is coloured yellow. This makes all blue light appear somewhat darker in the centre of the field of sight. The effect is particularly striking with mixtures of red and greenish-blue, which appear white when looked at directly, but acquire a blue tint when viewed at a slight distance from the middle of the field; and, on the other hand, when they appear white here, are red to direct vision. These inequalities of the retina, like the others mentioned in the former essay, are rectified by the constant movements of the eye. We know from the pale and indistinct colours of the external world as usually seen, what impressions of indirect vision correspond to those of direct; and we thus learn to judge of the colours of objects according to the impression which they *would* make on us if seen directly. The result is, that only unusual combinations and unusual or special direction of attention enable us to recognise the difference of which I have been speaking.

The theory of colours, with all these marvellous and complicated relations, was a riddle which Goethe in vain attempted to solve; nor were we physicists and physiologists more successful. I include myself in the number; for I long toiled at the task, without getting any nearer my object, until I at last discovered that a wonderfully simple solution had been discovered at the beginning of this century, and had been in print ever since for any one to read who chose. This solution was found and published by

---

21. A similar story is told of Dalton, the author of the 'Atomic Theory.' He was a Quaker, and went to the Friends' Meeting, at Manchester, in a pair of scarlet stockings, which some wag had put in place of his ordinary dark grey ones.—TR.

the same Thomas Young[22] who first showed the right method of arriving at the interpretation of Egyptian hieroglyphics. He was one of the most acute men who ever lived, but had the misfortune to be too far in advance of his contemporaries. They looked on him with astonishment, but could not follow his bold speculations, and thus a mass of his most important thoughts remained buried and forgotten in the 'Transactions of the Royal Society,' until a later generation by slow degrees arrived at the rediscovery of his discoveries, and came to appreciate the force of his arguments and the accuracy of his conclusions.

In proceeding to explain the theory of colours proposed by him, I beg the reader to notice that the conclusions afterwards to be drawn upon the nature of the sensations of sight are quite independent of what is hypothetical in this theory.

Dr. Young supposes that there are in the eye three kinds of nerve-fibres, the first of which, when irritated in any way, produces the sensation of red, the second the sensation of green, and the third that of violet. He further assumes that the first are excited most strongly by the waves of ether of greatest length; the second, which are sensitive to green light, by the waves of middle length; while those which convey impressions of violet are acted upon only by the shortest vibrations of ether. Accordingly, at the red end of the spectrum the excitation of those fibres which are sensitive to that colour predominates; hence the appearance of this part as red. Further on there is added an impression upon the fibres sensitive to green light, and thus results the mixed sensation of yellow. In the middle of the spectrum, the nerves sensitive to green become much more excited than the other two kinds, and accordingly green is the predominant impression. As soon as this becomes mixed with violet the result is the colour known as blue; while at the most highly refracted end of the spectrum the impression produced on the fibres which are sensitive to violet light overcomes every other.[23]

It will be seen that this hypothesis is nothing more than a further extension of Johannes Müller's law of special sensation. Just as the difference of sensation of light and warmth depends demonstrably upon whether the rays of the sun fall upon nerves of sight or nerves of feeling, so it is supposed in Young's hypothesis that the difference of sensation of colours depends simply upon whether one or the other kind of nervous fibres are more strongly affected. When all three kinds are equally excited,

22. Born at Milverton, in Somersetshire, 1773, died 1829.
23. The precise tint of the three primary colours cannot yet be precisely ascertained by experiment. The red alone, it is certain from the experience of the colour-blind, belongs to the extreme red of the spectrum. At the other end Young took violet for the primitive colour, while Maxwell considers that it is more properly blue. The question is still an open one: according to J.J. Müller's experiments (*Archiv für Ophthalmologie*, XV.2. p. 208) violet is more probable. The fluorescence of the retina is here a source of difficulty.

the result is the sensation of white light.

The phenomena that occur in red-blindness must be referred to a condition in which the one kind of nerves, which are sensitive to red rays, are incapable of excitation. It is possible that this class of fibres are wanting, or at least very sparingly distributed, along the edge of the retina, even in the normal human eye.

It must be confessed that both in men and in quadrupeds we have at present no anatomical basis for this theory of colours; but Max Schultze has discovered a structure in birds and reptiles which manifestly corresponds with what we should expect to find. In the eyes of many of this group of animals there are found among the rods of the retina a number which contain a red drop of oil in their anterior end, that namely which is turned towards the light; while other rods contain a yellow drop, and others none at all. Now there can be no doubt that red light will reach the rods with a red drop much better than light of any other colour, while yellow and green light, on the contrary, will find easiest entrance to the rods with the yellow drop. Blue light would be shut off almost completely from both, but would affect the colourless rods all the more effectually. We may therefore with great probability regard these rods as the terminal organs of those nervous fibres which respectively convey impressions of red, of yellow, and of blue light.

I have myself subsequently found a similar hypothesis very convenient and well fitted to explain in a most simple manner certain peculiarities which have been observed in the perception of musical notes, peculiarities as enigmatical as those we have been considering in the eye. In the *cochlea* of the internal ear the ends of the nerve fibres lie regularly spread out side by side, and provided with minute elastic appendages (the rods of Corti) arranged like the keys and hammers of a piano. My hypothesis is, that here each separate nerve fibre is constructed so as to take cognizance of a definite note, to which its elastic fibre vibrates in perfect consonance. This is not the place to describe the special characters of our sensations of musical tones which led me to frame this hypothesis. Its analogy with Young's theory of colours is obvious, and it refers the origin of overtones, the perception of the quality of sounds, the difference between consonance and dissonance, the formation of the musical scale and other acoustic phenomena to as simple a principle as that of Young. But in the case of the ear, I could point to a much more distinct anatomical foundation for such a hypothesis, and since that time, I have been able actually to demonstrate the relation supposed; not, it is true, in man or any vertebrate animals, whose labyrinth lies too deep for experiment, but in some of the marine Crustacea. These animals have external appendages to their organs of hearing which may be observed in the living animal, jointed filaments to which the fibres of the auditory nerve are distributed; and Hensen, of Kiel, has satisfied himself that some of these filaments are set in motion by certain notes, and others by different ones.

It remains to reply to an objection against Young's theory of colour. I mentioned above that the outline of the colour-disc, which marks the position of the most saturated colours (those of the spectrum), approaches to a triangle in form; but our conclusions upon the theory of the three primary colours depend upon a perfect rectilinear triangle inclosing the complete colour-system, for only in that case is it possible to produce all possible tints by various combinations of the three primary colours at the angles. It must, however, be remembered that the colour-disc only includes the entire series of colours which actually occur in nature, while our theory has to do with the analysis of our subjective sensations of colour. We need then only assume that actual coloured light does not produce sensations of absolutely pure colour; that red, for instance, even when completely freed from all admixture of white light, still does not excite those nervous fibres alone which are sensitive to impressions of red, but also, to a very slight degree, those which are sensitive to green, and perhaps to a still smaller extent those which are sensitive to violet rays. If this be so, then the sensation which the purest red light produces in the eye is still not the purest sensation of red which we can conceive of as possible. This sensation could only be called forth by a fuller, purer, more saturated red than has ever been seen in this world.

It is possible to verify this conclusion. We are able to produce artificially a sensation of the kind I have described. This fact is not only important as a complete answer to a possible objection to Young's theory, but is also, as will readily be seen, of the greatest importance for understanding the real value of our sensations of colour. In order to describe the experiment I must first give an account of a new series of phenomena.

The result of nervous action is fatigue, and this will be proportioned to the activity of the function performed, and the time of its continuance. The blood, on the other hand, which flows in through the arteries, is constantly performing its function, replacing used material by fresh, and thus carrying away the chemical results of functional activity; that is to say, removing the source of fatigue.

The process of fatigue as the result of nervous action, takes place in the eye as well as other organs. When the entire retina becomes tired, as when we spend some time in the open air in brilliant sunshine, it becomes insensible to weaker light, so that if we pass immediately into a dimly lighted room we see nothing at first; we are blinded, as we call it, by the previous brightness. After a time the eye recovers itself, and at last we are able to see, and even to read, by the same dim light which at first appeared complete darkness.

It is thus that fatigue of the entire retina shows itself. But it is possible for separate parts of that membrane to become exhausted, if they alone have received a strong light. If we look steadily for some time at any bright object, surrounded by a dark background—it is necessary to look steadily in order that the image may remain quiet upon the retina, and thus fatigue a sharply defined portion of its surface—and afterwards turn our eyes upon a uniform dark-grey surface, we see projected upon it an *after-image* of the bright object we were looking at just before, with the same outline but with reversed illumination. What was dark appears bright, and what was bright dark, like the first negative of a photographer. By carefully fixing the attention, it is possible to produce very elaborate after-images, so much so that occasionally even printing can be distinguished in them. This phenomenon is the result of a local fatigue of the retina. Those parts of the membrane upon which the bright light fell before, are now less sensitive to the light of the dark grey background than the neighbouring regions, and there now appears a dark spot upon the really uniform surface, corresponding in extent to the surface of the retina which before received the bright light.

(I may here remark that illuminated sheets of white paper are sufficiently bright to produce this after-image. If we look at much brighter objects—at flames, or at the sun itself—the effect becomes complicated. The strong excitement of the retina does not pass away immediately, but produces a direct or positive after-image, which at first unites with the negative or indirect one produced by the fatigue of the retina. Besides this, the effects of the different colours of white light differ both in duration and intensity, so that the after-images become coloured, and the whole phenomenon much more complicated.)

By means of these after-images it is easy to convince oneself that the impression produced by a bright surface begins to diminish after the first second, and that by the end of a single minute it has lost from a quarter to half of its intensity. The simplest form of experiment for this object is as follows. Cover half of a white sheet of paper with a black one, fix the eye upon some point of the white sheet near the margin of the black, and after 30 to 60 seconds draw the black sheet quickly away, without losing sight of the point. The half of the white sheet which is then exposed appears suddenly of the most brilliant brightness; and thus it becomes apparent how very much the first impression produced by the upper half of the sheet had become blunted and weakened, even in the short time taken by the experiment. And yet, what is also important to remark, the observer does not at all notice this fact, until the contrast brings it before him.

Lastly, it is possible to produce a partial fatigue of the retina in another way. We may tire it for certain colours only, by exposing either the entire retina, or a portion of it, for a certain time (from half a minute to five minutes) to one and the same colour. According to Young's theory, only one or two kinds of the optic nerve fibres will then be fatigued, those namely which are sensitive to impressions of the colour in question. All the rest will remain unaffected. The result is, that when the after-image appears, red, we will suppose, upon a grey background, the uniformly mixed light of the latter can only produce sensations of green and violet in the part of the retina which has become fatigued by red light. This part is made red-blind for the time. The after-image accordingly appears of a bluish green, the complementary colour to red.

It is by this means that we are able to produce in the retina the pure and primitive sensations of saturated colours. If, for instance, we wish to see pure red, we fatigue a part of our retina by the bluish green of the spectrum, which is the complementary colour of red. We thus make this part at once green-blind and violet-blind. We then throw the after-image upon the red of as perfect a prismatic spectrum as possible; the image immediately appears in full and burning red, while the red light of the spectrum which surrounds it, although the purest that the world can offer, now seems to the unfatigued part of the retina, less saturated than the after-image, and looks as if it were covered by a whitish mist.

These facts are perhaps enough. I will not accumulate further details, to understand which it would be necessary to enter upon lengthy descriptions of many separate experiments.

We have already seen enough to answer the question whether it is possible to maintain the natural and innate conviction that the quality of our sensations, and especially our sensations of sight, give us a true impression of corresponding qualities in the outer world. It is clear that they do not. The question was really decided by Johannes Müller's deduction from well ascertained facts of the law of specific nervous energy. Whether the rays of the sun appear to us as colour, or as warmth, does not at all depend upon their own properties, but simply upon whether they excite the fibres of the optic nerve, or those of the skin. Pressure upon the eyeball, a feeble current of electricity passed through it, a narcotic drug carried to the retina by the blood, are capable of exciting the sensation of light just as well as the sunbeams. The most complete difference offered by our several sensations, that namely between those of sight, of hearing, of taste, of smell, and of touch—this deepest of all distinctions, so deep that it is impossible to draw any comparison of likeness, or unlikeness, between the sensations of colour and of musical tones—does not, as we now see, at all depend upon the nature of the external object, but solely upon the central connections of the nerves which are affected.

We now see that the question whether within the special range of each particular sense it is possible to discover a coincidence between its objects and the sensations they produce is of only subordinate interest. What colour the waves of ether shall appear to us when they are perceived by the optic nerve depends upon their length. The system of naturally visible colours offers us a series of varieties in the composition of light, but the number of those varieties is wonderfully reduced from an unlimited number to only three. Inasmuch as the most important property of the eye is its minute appreciation of locality, and as it is so much more perfectly organised for this purpose than the ear, we may be well content that it is capable of recognising comparatively few differences in quality of light; the ear, which in the latter respect is so enormously better provided, has scarcely any power of appreciating differences of locality. But it is certainly matter for astonishment to any one who trusts to the direct information of his natural senses, that neither the limits within which the spectrum affects our eyes, nor the differences of colour which alone remain as the simplified effect of all the actual differences of light in kind, should have any other demonstrable import than for the sense of sight. Light which is precisely the same to our eyes, may in all other physical and chemical effects be completely different. Lastly, we find that the unmixed primitive elements of all our sensations of colour (the perception of the simple primary tints) cannot be produced by any kind of external light in the natural unfatigued condition of the eye. These elementary sensations of colour can only be called forth by artificial preparation of the organ, so that, in fact, they only exist as subjective phenomena. We see, therefore, that as to any correspondence in kind of external light with the sensations it produces, there is only one bond of connection between them, a bond which at first sight may seem slender enough, but is in fact quite sufficient to lead to an infinite number of most useful applications. This law of correspondence between what is subjective and objective in vision is as follows:—

Similar light produces under like conditions a like sensation of colour. Light which under like conditions excites unlike sensations of colour is dissimilar.

When two relations correspond to one another in this manner, the one is a *sign* for the other. Hitherto the notions of a 'sign' and of an 'image' or representation have not been carefully enough distinguished in the theory of perception; and this seems to me to have been the source of numberless mistakes and false hypotheses. In an 'image' the representation must be of the same kind as that which is represented. Indeed, it is only so far an image as it is like in kind. A statue is an image of a man, so far as its form reproduces his: even if it is executed on a smaller scale, every dimension will be represented in proportion. A picture is an image or representation of the original, first because it represents the colours of the latter by similar colours, secondly because it represents a part of its relations in space—those, namely, which belong to perspective—by corresponding relations in space.

Functional cerebral activity and the mental conceptions which go with it may be 'images' of actual occurrences in the outer world, so far as the former represent the sequence in time of the latter, so far as they represent likeness of objects by likeness of signs—that is, a regular arrangement by a regular arrangement.

This is obviously sufficient to enable the understanding to deduce what is constant from the varied changes of the external world, and to formulate it as a notion or a law. That it is also sufficient for all practical purposes we shall see in the next chapter. But not only uneducated persons, who are accustomed to trust blindly to their senses, even the educated, who know that their senses may be deceived, are inclined to demur to so complete a want of any closer correspondence in kind between actual objects and the sensations they produce than the law I have just expounded. For instance, natural philosophers long hesitated to admit the identity of the rays of light and of heat, and exhausted all possible means of escaping a conclusion which seemed to contradict the evidence of their senses.

Another example is that of Goethe, as I have endeavoured to show elsewhere. He was led to contradict Newton's theory of colours, because he could not persuade himself that white, which appears to our sensation as the purest manifestation of the brightest light, could be composed of darker colours. It was Newton's discovery of the composition of light that was the first germ of the modern doctrine of the true functions of the senses; and in the writings of his contemporary, Locke, were correctly laid down the most important principles on which the right interpretation of sensible qualities depends. But, however clearly we may feel that here lies the difficulty for a large number of people, I have never found the opposite conviction of certainty derived from the senses so distinctly expressed that it is possible to lay hold of the point of error; and the reason seems to me to lie in the fact that beneath the popular notions on the subject lie other and more fundamentally erroneous conceptions.

We must not be led astray by confounding the notions of a *phenomenon* and an *appearance*. The colours of objects are phenomena caused by certain real differences in their constitution. They are, according to the scientific as well as to the uninstructed view, no mere appearance, even though the way in which they appear depends chiefly upon the constitution of our nervous system. A 'deceptive appearance' is the result of the normal phenomena of one object being confounded with those of another. But the sensation of colour is by no means deceptive appearance. There is no other way in which colour can appear; so that there is nothing which we could describe as the normal phenomenon, in distinction from the impressions of colour received through the eye.

Here the principal difficulty seems to me to lie in the notion of *quality*. All difficulty vanishes as soon as we clearly understand that each quality or property of a thing is, in reality, nothing else but its capability of exercising certain effects upon other things. These actions either go on between similar parts of the same body, and so produce the differences of its aggregate condition; or they proceed from one body upon another, as in the case of chemical reactions; or they produce their effect on our organs of special sense, and are there recognised as *sensations*, as those of sight, with which we have now to do. Any of these actions is called a 'property,' when its object is understood without being expressly mentioned. Thus, when we speak of the 'solubility' of a substance, we mean its behavior toward *water*; when we speak of its 'weight,' we mean its attraction to *the earth*; and in the same way we may correctly call a substance 'blue,' understanding, as a tacit assumption, that we are only speaking of its action upon *a normal eye*.

But if what we call a property always implies an action of one thing on another, then a property or quality can never depend upon the nature of one agent alone, but exists only in relation to, and dependent on, the nature of some second object, which is acted upon. Hence, there is really no meaning in talking of properties of light which belong to it absolutely, independent of all other objects, and which we may expect to find represented in the sensations of the human eye. The notion of such properties is a contradiction in itself. They cannot possibly exist, and therefore we cannot expect to find any coincidence of our sensations of colour with qualities of light.

These considerations have naturally long ago suggested themselves to thoughtful minds; they may be found clearly expressed in the writings of Locke and Herbart,[24] and they are completely in accordance with Kant's philosophy. But in former times, they demanded a more than usual power of abstraction, in order that their truth should be understood; whereas now the facts which we have laid before the reader illustrate them in the clearest manner.

After this excursion into the world of abstract ideas, we return once more to the subject of colour, and will now examine it as a sensible 'sign' of certain external qualities, either of light itself or of the objects which reflect it.

It is essential for a good sign to be constant—that is, the same sign must always denote the same object. Now, we have already seen that in this particular our sensations of colour are imperfect; they are not quite uniform over the entire field of the retina. But the constant movement of the eye

---

24. Johann Friedrich Herbart, born 1776, died 1841, professor of philosophy at Königsberg and Göttingen, author of *Psychologie als Wissenschaft, neugegrundet auf Erfahrung, Metaphysik und Mathematik.*—TR.

supplies this imperfection, in the same way as it makes up for the unequal sensitiveness of the different parts of the retina to form.

We have also seen that when the retina becomes tired, the intensity of the impression produced on it rapidly diminishes, but here again the usual effect of the constant movements of the eye is to equalise the fatigue of the various parts, and hence we rarely see after-images. If they appear at all, it is in the case of brilliant objects like very bright flames, or the sun itself. And, so long as the fatigue of the entire retina is uniform, the relative brightness and colour of the different objects in sight remains almost unchanged, so that the effect of fatigue is gradually to weaken the apparent illumination of the entire field of vision.

This brings us to consider the differences in the pictures presented by the eye, which depend on different degrees of illumination. Here again we meet with instructive facts. We look at external objects under light of very different intensity, varying from the most dazzling sunshine to the pale beams of the moon; and the light of the full moon is 150,000 times less than that of the sun.

Moreover, the colour of the illumination may vary greatly. Thus, we sometimes employ artificial light, and this is always more or less orange in colour; or the natural daylight is altered, as we see it in the green shade of an arbour, or in a room with coloured carpets and curtains. As the brightness and the colour of the illumination changes, so of course will the brightness and colour of the light which the illuminated objects reflect to our eyes, since all differences in local colour depend upon different bodies reflecting and absorbing various proportions of the several rays of the sun. Cinnabar reflects the rays of great wave length without any obvious loss, while it absorbs almost the whole of the other rays. Accordingly, this substance appears of the same red colour as the beams which it throws back into the eye. If it is illuminated with light of some other colour, without any mixture of red, it appears almost black.

These observations teach what we find confirmed by daily experience in a hundred ways, that the apparent colour and brightness of illuminated objects varies with the colour and brightness of the illumination. This is a fact of the first importance for the painter, for many of his finest effects depend on it.

But what is most important practically is for us to be able to recognise surrounding objects when we see them: it is only seldom that, for some artistic or scientific purpose, we turn our attention to the way in which they are illuminated. Now what is constant in the colour of an object is not the brightness and colour of the light which it reflects, but the relation between the intensity of the different coloured constituents of this light, on the one hand, and that of the corresponding constituents of the light which illuminates it on the other. This proportion alone is the expression of a constant property of the object in question.

Considered theoretically, the task of judging of the colour of a body under changing illumination would seem to be impossible; but in practice we soon find that we are able to judge of local colour without the least uncertainty or hesitation, and under the most different conditions. For instance, white paper in full moonlight is darker than black satin in daylight, but we never find any difficulty in recognising the paper as white and the satin as black. Indeed, it is much more difficult to satisfy ourselves that a dark object with the sun shining on it reflects light of exactly the same colour, and perhaps the same brightness, as a white object in shadow, than that the proper colour of a white paper in shadow is the same as that of a sheet of the same kind lying close to it in the sunlight. Grey seems to us something altogether different from white, and so it is, regarded as a *proper* colour;[25] for anything which only reflects half the light it receives must have a different surface from one which reflects it all. And yet the impression upon the retina of a grey surface under illumination may be absolutely identical with that of a white surface in the shade. Every painter represents a white object in shadow by means of grey pigment, and if he has correctly imitated nature, it appears pure white. In order to convince one's self of the identity in this respect—i.e. as *illumination* colours—of grey and white, the following experiment may be tried. Cut out a circle in grey paper, and concentrate a strong beam of light upon it with a lens, so that the limits of the illumination exactly correspond with those of the grey circle. It will then be impossible to tell that there is any artificial illumination at all. The grey looks white.[26]

We may assume, and the assumption is justified by certain phenomena of contrast, that illumination of the brightest white we can produce, gives a true criterion for judging of the darker objects in the neighbourhood, since, under ordinary circumstances, the brightness of any proper colour diminishes in proportion as the illumination is diminished, or the fatigue of the retina increased.

This relation holds even for extreme degrees of illumination, so far as the objective intensity of the light is concerned, but not for our sensation. Under illumination so brilliant as to approach what would be blinding, degrees of brightness of light-coloured objects become less and less distinguishable; and, in the same way, when the illumination is very feeble, we are unable

---

25. The local or proper colour of an object (*Körperfarbe*) is that which it shows in common white light, while the 'illumination-colour,' as I have translated *Lichtfarbe*, is that which is produced by coloured light. Thus the red of some sandstone rocks seen by common white light is their proper colour, that of a snow mountain in the rays of the setting sun is an illumination-colour.—TR.

26. The demonstration is more striking if the grey disk is placed on a sheet of white paper in diffused light.—TR.

to appreciate slight differences in the amount of light reflected by dark objects. The result is that in sunshine local colours of moderate brightness approach the brightest, whereas in moonlight they approach the darkest. The painter utilises this difference in order to represent noonday or midnight scenes, although pictures, which are usually seen in uniform daylight, do not really admit of any difference of brightness approaching that between sunshine and moonlight. To represent the former, he paints the objects of moderate brightness almost as bright as the brightest; for the latter, he makes them almost as dark as the darkest.

The effect is assisted by another difference in the sensation produced by the same actual conditions of light and colour. If the brightness of various colours is equally increased, that of red and yellow becomes apparently stronger than that of blue. Thus, if we select a red and a blue paper which appear of the same brightness in ordinary daylight, the red seems much brighter in full sunlight, the blue in moonlight or starlight. This peculiarity in our perception is also made use of by painters; they make yellow tints predominate when representing landscapes in full sunshine, while every object of a moonlight scene is given a shade of blue. But it is not only local colour which is thus affected; the same is true of the colours of the spectrum.

These examples show very plainly how independent our judgment of colours is of their actual amount of illumination. In the same way, it is scarcely affected by the colour of the illumination. We know, of course, in a general way that candle-light is yellowish compared with daylight, but we only learn to appreciate how much the two kinds of illumination differ in colour when we bring them together of the same intensity—as, for example, in the experiment of *coloured shadows*. If we admit light from a cloudy sky through a narrow opening into a dark room, so that it falls sideways on a horizontal sheet of white paper, while candle-light falls on it from the other side, and if we then hold a pencil vertically upon the paper, it will of course throw two shadows: the one made by the daylight will be orange, and looks so; the other made by the candle-light is really white, but appears blue by contrast. The blue and the orange of the two shadows are both colours which we call white, when we see them by daylight and candle-light respectively. Seen together, they appear as two very different and tolerably saturated colours, yet we do not hesitate a moment in recognising white paper by candle-light as white, and very different from orange.[27]

The most remarkable of this series of facts is that we can separate the colour of any transparent medium from that of objects seen through it. This is proved by a number of experiments contrived to illustrate the effects of contrast. If we look through a green veil at a field of snow, although the

27. This experiment with diffused white day-light may also be made with moonlight.

light reflected from it must really have a greenish tint when it reaches our eyes, yet it appears, on the contrary, of a reddish tint, from the effect of the indirect after-image of green. So completely are we able to separate the light which belongs to the transparent medium from that of the objects seen through it.[28]

The changes of colour in the two last experiments are known as phenomena of contrast. They consist in mistakes as to local colour, which for the most part depend upon imperfectly defined after-images.[29] This effect is known as *successive contrast*, and is experienced when the eye passes over a series of coloured objects. But a similar mistake may result from our custom of judging of local colour according to the brightness and colour of the various objects seen at the same time. If these relations happen to be different from what is usual, contrast phenomena ensue. When, for example, objects are seen under two different coloured illuminations, or through two different coloured media (whether real or apparent), these conditions produce what is called *simultaneous contrast*. Thus in the experiment described above of coloured shadows thrown by daylight and candle-light, the doubly illuminated surface of the paper being the brightest object seen, gives a false criterion for white. Compared with it, the really white but less bright light of the shadow thrown by the candle looks blue. Moreover, in these curious effects of contrast, we must take into account that differences in sensation which are easily apprehended appear to us greater than those less obvious. Differences of colour which are actually before our eyes are more easily apprehended than those which we only keep in memory, and contrasts between objects which are close to one another in the field of vision are more easily recognised than when they are at a distance. All this contributes to the effect. Indeed, there are a number of subordinate circumstances affecting the result which it would be very interesting to follow out in detail, for they throw great light upon the way in which we judge of local colour: but we must not pursue the inquiry further here. I will only remark that all these effects of contrast are not less interesting for the scientific painter than for the physiologist, since he must often exaggerate the natural phenomena of contrast, in order to produce the impression of greater varieties of light and greater fulness of colour than can be actually produced by artificial pigments.

Here we must leave the theory of the Sensations of Sight. This part of our inquiry has shown us that the qualities of these sensations can only be

---

28. A number of similar experiments will be found described in the author's *Handbuch der physiologischen Optik*, pp. 398–411.

29. These after-images have been described as 'accidental images,' positive when of the same colour as the original colour, negative when of the complementary colour.—TR.

regarded as *signs* of certain different qualities, which belong sometimes to light itself, sometimes to the bodies it illuminates, but that there is not a single actual quality of the objects seen which precisely corresponds to our sensations of sight. Nay, we have seen that, even regarded as signs of real phenomena in the outer world, they do not possess the one essential requisite of a complete system of signs—namely, constancy—with anything like completeness; so that all that we can say of our sensations of sight is, that 'under similar conditions, the qualities of this sensation appear in the same way for the same objects.'

And yet, in spite of all this imperfection, we have also found that by means of so inconstant a system of signs, we are able to accomplish the most important part of our task—to recognise the same proper colours wherever they occur; and, considering the difficulties in the way, it is surprising how well we succeed. Out of this inconstant system of brightness and of colours, varying according to the illumination, varying according to the fatigue of the retina, varying according to the part of it affected, we are able to determine the proper colour of any object, the one constant phenomenon which corresponds to a constant quality of its surface; and this we can do, not after long consideration, but by an instantaneous and involuntary decision.

The inaccuracies and imperfections of the eye as an optical instrument, and those which belong to the image on the retina, now appear insignificant in comparison with the incongruities which we have met with in the field of sensation. One might almost believe that Nature had here contradicted herself on purpose, in order to destroy any dream of a pre-existing harmony between the outer and the inner world.

And what progress have we made in our task of explaining Sight? It might seem that we are farther off than ever; the riddle only more complicated, and less hope than ever of finding out the answer. The reader may perhaps feel inclined to reproach Science with only knowing how to break up with fruitless criticism the fair world presented to us by our senses, in order to annihilate the fragments.

> Woe! woe!
> Thou hast destroyed
> The beautiful world
> With powerful fist;
> In ruin 'tis hurled,
> By the blow of a demigod shattered.
> The scattered
> Fragments into the void we carry,
> Deploring

The beauty perished beyond restoring.[30]

and may feel determined to stick fast to the 'sound common sense' of mankind, and believe his own senses more than physiology.

But there is still a part of our investigation which we have not touched—that into our conceptions of space. Let us see whether, after all, our natural reliance upon the accuracy of what our senses teach us, will not be justified even before the tribunal of Science.

III. THE PERCEPTION OF SIGHT

The colours which have been the subject of the last chapter are not only an ornament we should be sorry to lose, but are also a means of assisting us in the distinction and recognition of external objects. But the importance of *colour* for this purpose is far less than the means which the rapid and far-reaching power of the eye gives us of distinguishing the various relations of *locality*. No other sense can be compared with the eye in this respect. The sense of touch, it is true, can distinguish relations of space, and has the special power of judging of all matter within reach, at once as to resistance, volume, and weight; but the range of touch is limited, and the distinction it can make between small distances is not nearly so accurate as that of sight. Yet the sense of touch is sufficient, as experiments upon persons born blind have proved, to develop complete notions of space. This proves that the possession of sight is not necessary for the formation of these conceptions, and we shall soon see that we are continually controlling and correcting the notions of locality derived from the eye by the help of the sense of touch, and always accept the impressions on the latter sense as decisive. The two senses, which really have the same task, though with very different means of accomplishing it, happily supply each other's deficiencies. Touch is a trustworthy and experienced servant, but enjoys only a limited range, while sight rivals the boldest flights of fancy in penetrating to illimitable distances.

30. Bayard Taylor's translation of the passage in *Faust*:—

> Du hast sie zerstört
> Die schöne Welt
> Mit mächtiger Faust;
> Sie stürzt, sie zerfällt,
> Ein Halbgott hat sie zerschlagen.
> Wir tragen
> Die Trümmern ins Nichts hinüber,
> Und klagen
> Ueber die verlorne Schöne.

This combination of the two senses is of great importance for our present task; for, since we have here only to do with vision, and since touch is sufficient to produce complete conceptions of locality, we may assume these conceptions to be already complete, at least in their general outline, and may, accordingly, confine our investigation to ascertaining the common point of agreement between the visual and tactile perceptions of space. The question how it is possible for any conception of locality to arise from either or both of these sensations, we will leave till last.

It is obvious, from a consideration of well-known facts, that the distribution of our sensations among nervous structures separated from one another does not at all necessarily bring with it the conception that the causes of these sensations are locally separate. For example, we may have sensations of light, of warmth, of various notes of music, and also perhaps of an odour, in the same room, and may recognise that all these agents are diffused through the air of the room at the same time, and without any difference of locality. When a compound colour falls upon the retina, we are conscious of three separate elementary impressions, probably conveyed by separate nerves, without any power of distinguishing them. We hear in a note struck on a stringed instrument or in the human voice, different tones at the same time, one fundamental, and a series of harmonic overtones, which also are probably received by different nerves, and yet we are unable to separate them in space. Many articles of food produce a different impression of taste upon different parts of the tongue, and also produce sensations of odour by their volatile particles ascending into the nostrils from behind. But these different sensations, recognised by different parts of the nervous system, are usually completely and inseparably united in the compound sensation which we call taste.

No doubt, with a little attention it is possible to ascertain the parts of the body which receive these sensations, but, even when these are known to be locally separate, it does not follow that we must conceive of the sources of these sensations as separated in the same way.

We find a corresponding fact in the physiology of sight—namely, that we see only a single object with our two eyes, although the impression is conveyed by two distinct nerves. In fact, both phenomena are examples of a more universal law.

Hence, when we find that a plane optical image of the objects in the field of vision is produced on the retina, and that the different parts of this image excite different fibres of the optic nerve, this is not a sufficient ground for our referring the sensations thus produced to locally distinct regions of our field of vision. Something else must clearly be added to produce the notion of separation in space.

The sense of touch offers precisely the same problem. When two different parts of the skin are touched at the same time, two different sensitive nerves are excited, but the local separation between these two nerves is not a sufficient ground for our recognition of the two parts which have been touched as distinct, and for the conception of two different external objects which follows. Indeed, this conception will vary according to circumstances. If we touch the table with two fingers, and feel under each a grain of sand, we suppose that there are two separate grains of sand; but if we place the two fingers one against the other, and a grain of sand between them, we may have the same sensations of touch in the same two nerves as before, and yet, under these circumstances, we suppose that there is only a single grain. In this case, our consciousness of the position of the fingers has obviously an influence upon the result at which the mind arrives. This is further proved by the experiment of crossing two fingers one over the other, and putting a marble between them, when the single object will produce in the mind the conception of two.

What, then, is it which comes to help the anatomical distinction in locality between the different sensitive nerves, and, in cases like those I have mentioned, produces the notion of separation in space? In attempting to answer this question, we cannot avoid a controversy which has not yet been decided.

Some physiologists, following the lead of Johannes Müller, would answer that the retina or skin, being itself an organ which is extended in space, perceives impressions which carry with them this quality of extension in space; that this conception of locality is innate; and that impressions derived from external objects are transmitted of themselves to corresponding local positions in the image produced in the sensitive organ. We may describe this as the *Innate* or *Intuitive* Theory of conceptions of Space. It obviously cuts short all further enquiry into the origin of these conceptions, since it regards them as something original, inborn, and incapable of further explanation.

The opposing view was put forth in a more general form by the early English philosophers of the sensational school—by Molyneux,[31] Locke, and Jurin.[32] Its application to special physiological problems has only become possible in very modern times, particularly since we have gained more accurate knowledge of the movements of the eye. The invention of the stereoscope by Wheatstone (p. 183) made the difficulties and imperfections of the Innate Theory of sight much more obvious than before, and led to

31. William Molyneux, author of *Dioptrica Nova*, was born in Dublin, 1656, and died in the same city, 1698.
32. James Jurin, M.D., Sec. R.S., physician to Guy's Hospital, and President of the Royal College of Physicians, was born in 1684, and died in 1750. Beside works on the Contraction of the Heart, on *Vis viva*, &c., he published, in 1738, a treatise on *Distinct and Indistinct Vision.*—Tr.

another solution which approached much nearer to the older view, and which we will call the *Empirical* Theory of Vision. This assumes that none of our sensations give us anything more than 'signs' for external objects and movements, and that we can only learn how to interpret these signs by means of experience and practice. For example, the conception of differences in locality can only be attained by means of movement, and, in the field of vision, depends upon our experience of the movements of the eye. Of course, this Empirical Theory must assume a difference between the sensations of various parts of the retina, depending upon their local difference. If it were not so, it would be impossible to distinguish any local difference in the field of vision. The sensation of red, when it falls upon the right side of the retina, must in some way be different from the sensation of the same red when it affects the left side; and, moreover, *this* difference between the two sensations must be of another kind from that which we recognise when the same spot in the retina is successively affected by two different shades of red. Lotze[33] has named this difference between the sensations which the same colour excites when it affects different parts of the retina, the *local sign* of the sensation. We are for the present ignorant of the nature of this difference, but I adopt the name given by Lotze as a convenient expression. While it would be premature to form any further hypothesis as to the nature of these 'local signs,' there can be no doubt of their existence, for it follows from the fact that we are able to distinguish local differences in the field of vision.

The difference, therefore, between the two opposing views is as follows. The Empirical Theory regards the local signs (whatever they really may be) as signs the signification of which must be learnt, and is actually learnt, in order to arrive at a knowledge of the external world. It is not at all necessary to suppose any kind of correspondence between these local signs and the actual differences of locality which they signify. The Innate Theory, on the other hand, supposes that the local signs are nothing else than direct conceptions of differences in space as such, both in their nature and their magnitude.

The reader will see how the subject of our present enquiry involves the consideration of that far-reaching opposition between the system of philosophy which assumes a pre-existing harmony of the laws of mental operations with those of the outer world, and the system which attempts to derive all correspondence between mind and matter from the results of experience.

So long as we confine ourselves to the observation of a field of two dimensions, the individual parts of which offer no, or, at any rate, no

---

33. Rudolf Hermann Lotze, Professor in the University of Göttingen, originally a disciple of Herbart (v. *supra*), author of *Allgemeine Physiologie des menschlichen Körpers*, 1851.—TR.

recognisable, difference in their distances from the eye—so long, for instance, as we only look at the sky and distant parts of the landscape, both the above theories practically offer an equally good explanation of the way in which we form conceptions of local relations in the field of vision. The extension of the retinal image corresponds to the extension of the actual image presented by the objects before us; or, at all events, there are no incongruities which may not be reconciled with the Innate Theory of sight without any very difficult assumptions or explanations.

The first of these incongruities is that in the retinal picture the top and bottom and the right and left of the actual image are inverted. This is seen in Fig. 4 to result from the rays of light crossing as they enter the pupil; the point *a* is the retinal image of *A*, *b* of *B*. This has always been a difficulty in the theory of vision, and many hypotheses have been invented to explain it. Two of these have survived. We may, with Johannes Müller, regard the conception of upper and lower as only a relative distinction, so far as sight is concerned—that is, as only affecting the relation of the one to the other; and we must further suppose that the feeling of correspondence between what is upper in the sense of sight and in the sense of touch is only acquired by experience, when we see the hands, which feel, moving in the field of vision. Or, secondly, we may assume with Fick[34] that, since all impressions upon the retina must be conveyed to the brain in order to be there perceived, the nerves of sight and those of feeling are so arranged in the brain as to produce a correspondence between the notions they suggest of upper and under, right and left. This supposition has, however, no pretense of any anatomical facts to support it.

The second difficulty for the Intuitive Theory is that, while we have two retinal pictures, we do not see double. This difficulty was met by the assumption that both retinae when they are excited produce only a single sensation in the brain, and that the several points of each retina correspond with each other, so that each pair of corresponding or 'identical' points produces the sensation of a single one. Now there is an actual anatomical arrangement which might perhaps support this hypothesis. The two optic nerves cross before entering the brain, and thus become united. Pathological observations make it probable that the nerve-fibres from the right-hand halves of both retinae pass to the right cerebral hemisphere, those from the left halves to the left hemisphere.[35] But although corresponding nerve-fibres

34. Ludwig Fick, late Professor of Medicine in the University of Marburg, the brother of Prof. Adolf Fick, of Zürich.

35. We may compare the arrangement to that of the reins of a pair of horses: the inner fibres only of each optic nerve cross, so that those which run to the right half of the brain are the outer fibres of the right and the inner of the left retina, while those which run to the left cerebral hemisphere are the outer fibres of the left and the inner of the right retina: just as the inner reins of both horses cross, so that the outer rein of the off horse and the inner of the near one run

would thus be brought close together, it has not yet been shown that they actually unite in the brain.

These two difficulties do not apply to the Empirical Theory, since it only supposes that the actual sensible 'sign,' whether it be simple or complex, is recognised as the sign of that which it signifies. An uninstructed person is as sure as possible of the notions he derives from his eyesight, without ever knowing that he has two retinae, that there is an inverted picture on each, or that there is such a thing as an optic nerve to be excited, or a brain to receive the impression. He is not troubled by his retinal images being inverted and double. He knows what impression such and such an object in such and such a position makes on him through his eyesight, and governs himself accordingly. But the possibility of learning the signification of the local signs which belong to our sensations of sight, so as to be able to recognise the actual relations which they denote, depends, first, on our having movable parts of our own body within sight; so that, when we once know by means of touch what relation in space and what movement is, we can further learn what changes in the impressions on the eye correspond to the voluntary movements of a hand which we can see. In the second place, when we move our eyes while looking at a field of vision filled with objects at rest, the retina, as it moves, changes its relation to the almost unchanged position of the retinal picture. We thus learn what impression the same object makes upon different parts of the retina. An unchanged retinal picture, passing over the retina as the eye turns, is like a pair of compasses which we move over a drawing in order to measure its parts. Even if the 'local signs' of sensation were quite arbitrary, thrown together without any systematic arrangement (a supposition which I regard as improbable), it would still be possible by means of the movements of the hand and of the eye, as just described, to ascertain which signs go together, and which correspond in different regions of the retina to points at similar distances in the two dimensions of the field of vision. This is in accordance with experiments by Fechner,[36] Volkmann[37] and myself, which prove that even the fully developed eye of an adult can only accurately compare the size of those lines or angles in the field of vision, the images of which can be thrown one after another upon precisely the same spot of the retina by means of the ordinary movements of the eye.

Moreover, we may convince ourselves by a simple experiment that the harmonious results of the perceptions of feeling and of sight depend, even

together to the driver's right hand, while the inner rein of the off and the outer of the near horse pass to his left hand.—TR.

36. Gustav Theodor Fechner, author of *Elemente der Psychophysik*, 1860; also known as a satirist.—TR.

37. Alfred Wilhelm Volkmann, successively Professor of Physiology at Leipzig, Dorpat, and Halle; author of *Physiologische Untersuchungen im Gebiete der Optik*, 1864, &c.—TR.

in the adult, upon a constant comparison of the two, by means of the retinal pictures of our hands as they move. If we put on a pair of spectacles with prismatic glasses, the two flat surfaces of which converge towards the right, all objects appear to be moved over to the right. If we now try to touch anything we see, taking care to shut the eyes before the hand appears in sight, it passes to the right of the object; but if we follow the movement of the hand with the eye, we are able to touch what we intend, by bringing the retinal image of the hand up to that of the object. Again, if we handle the object for one or two minutes, watching it all the time, a fresh correspondence is formed between the eye and the hand, in spite of the deceptive glass, so that we are now able to touch the object with perfect certainty, even when the eyes are shut. And we can even do the same with the other hand without seeing it, which proves that it is not the perception of touch which has been rectified by comparison with the false retinal images, but, on the contrary, the perception of sight which has been corrected by that of touch. But, again, if, after trying this experiment several times, we take off the spectacles and then look at any object, taking care not to bring our hands into the field of vision, and now try to touch it with our eyes shut, the hand will pass beyond it on the opposite side—that is, to the left. The new harmony which was established between the perceptions of sight and of touch continues its effects, and thus leads to fresh mistakes when the normal conditions are restored.

In preparing objects with needles under a compound microscope, we must learn to harmonise the inverted microscopical image with our muscular sense; and we have to get over a similar difficulty in shaving before a looking-glass, which changes right to left.

These instances, in which the image presented in the two dimensions of the field of vision is essentially of the same kind as the retinal images, and resembles them, can be equally well explained (or nearly so) by the two opposite theories of vision to which I have referred. But it is quite another matter when we pass to the observation of near objects of three dimensions. In this case there is a thorough and complete incongruity between our retinal images on the one hand, and, on the other, the actual condition of the objects as well as the correct impression of them which we receive. Here we are compelled to choose between the two opposite theories, and accordingly this department of our subject—the explanation of our Perception of Solidity or *Depth* in the field of vision, and that of binocular vision on which the former chiefly depends—has for many years become the field of much investigation and no little controversy. And no wonder, for we have already learned enough to see that the questions which have here to be decided are of fundamental importance, not only for the physiology of sight, but for a correct understanding of the true nature and limits of human knowledge generally.

Each of our eyes projects a plane image upon its own retina. However we may suppose the conducting nerves to be arranged, the two retinal images when united in the brain can only reappear as a plane image. But instead of the two plane retinal images, we find that the actual impression on our mind is a solid image of three dimensions. Here, again, as in the system of colours, the outer world is richer than our sensation by one dimension; but in this case the conception formed by the mind completely represents the reality of the outer world. It is important to remember that this perception of depth is fully as vivid, direct, and exact as that of the plane dimensions of the field of vision. If a man takes a leap from one rock to another, his life depends just as much upon his rightly estimating the distance of the rock on which he is to alight, as upon his not misjudging its position, right or left; and, as a matter of experience, we find that we can do the one just as quickly and as surely as the other.

In what way can this appreciation of what we call depth, solidity, and direct distance come about? Let us first ascertain what are the facts.

At the outset of the enquiry we must bear in mind that the perception of the solid form of objects and of their relative distance from us is not quite absent, even when we look at them with only one eye and without changing our position. Now the means which we possess in this case are just the same as those which the painter can employ in order to give the figures on his canvas the appearance of being solid objects, and of standing at different distances from the spectator. It is part of a painter's merit for his figures to stand out boldly. Now how does he produce the illusion? We shall find, in the first place, that in painting a landscape he likes to have the sun near the horizon, which gives him strong shadows; for these throw objects in the foreground into bold relief. Next he prefers an atmosphere which is not quite clear, because slight obscurity makes the distance appear far off. Then he is fond of bringing in figures of men and cattle, because, by help of these objects of known size, we can easily measure the size and distance of other parts of the scene. Lastly, houses and other regular productions of art are also useful for giving a clue to the meaning of the picture, since they enable us easily to recognise the position of horizontal surfaces. The representation of solid forms by drawings in correct perspective is most successful in the case of objects of regular and symmetrical shape, such as buildings, machines, and implements of various kinds. For we know that all of these are chiefly bounded either by planes which meet at a right angle or by spherical and cylindrical surfaces; and this is sufficient to supply what the drawing does not directly show. Moreover, in the case of figures of men or animals, our knowledge that the two sides are symmetrical further assists the impression conveyed.

But objects of unknown and irregular shape, as rocks or masses of ice, baffle the skill of the most consummate artist; and even their representation in the most complete and perfect manner possible, by means of photography, often shows nothing but a confused mass of black and white. Yet, when we have these objects in reality before our eyes, a single glance is enough for us to recognise their form.

The first who clearly showed in what points it is impossible for any picture to represent actual objects was the great master of painting, Leonardo da Vinci,[38] who was almost as distinguished in natural philosophy as in art. He pointed out in his *Trattato della Pittura*, that the views of the outer world presented by each of our eyes are not precisely the same. Each eye sees in its retinal image a perspective view of the objects which lie before it; but, inasmuch as it occupies a somewhat different position in space from the other, its point of view and so its whole perspective image is different. If I hold up my finger and look at it first with the right and then with the left eye, it covers, in the picture seen by the latter, a part of the opposite wall of the room which is more to the right than in the picture seen by the right eye. If I hold up my right hand with the thumb towards me, I see with the right eye more of the back of the hand, with the left more of the palm; and the same effect is produced whenever we look at bodies of which the several parts are at different distances from our eyes. But when I look at a hand represented in the same position in a painting, the right eye will see exactly the same figure as the left, and just as much of either the palm or the back of it. Thus we see that actual solid objects present different pictures to the two eyes, while a painting shows only the same. Hence follows a difference in the impression made upon the sight which the utmost perfection in a representation on a flat surface cannot supply.

The clearest proof that seeing with two eyes, and the difference of the pictures presented by each, constitute the most important cause of our perception of a third dimension in the field of vision, has been furnished by Wheatstone's invention of the stereoscope.[39] I may assume that this instrument and the peculiar illusion which it produces are well known. By its help we see the solid shape of the objects represented on the stereoscopic slide, with the same complete evidence of the senses with which we should look at the real objects themselves. This illusion is produced by presenting somewhat different pictures to the two eyes—to the right, one which represents the object in perspective as it would appear to that eye, and to the left one as it would appear to the left. If the pictures are otherwise exact and drawn from two different points of view corresponding to the position of

38. Born at Vinci, near Florence, 1452; died at Cloux, near Amboise, 1519. Mr. Hallam says of his scientific writings, that they are 'more like revelations of physical truths vouchsafed to a single mind, than the superstructure of its reasoning upon any established basis. . . . He first laid down the grand principle of Bacon, that experiment and observation must be the guides to just theory in the investigation of nature.'—TR.

39. Described in the *Philosophical Transactions* for 1838.—TR.

the two eyes, as can be easily done by photography, we receive on looking into the stereoscope precisely the same impression in black and white as theobject itself would give.

Anyone who has sufficient control over the movements of his eyes does not need the help of an instrument in order to combine the two pictures on a stereoscopic slide into a single solid image. It is only necessary so to direct the eyes, that each of them shall at the same time see corresponding points in the two pictures; but it is easier to do so by help of an instrument which will apparently bring the two pictures to the same place.

In Wheatstone's original stereoscope, represented in Fig. 9, the observer looked with the right eye into the mirror *b*, and with the left into the mirror *a*. Both mirrors were placed at an angle to the observer's line of sight, and the two pictures were so placed at *k* and *g* that their reflected images appeared at the same place behind the two mirrors; but the right eye saw the picture *g* in the mirror *b*, while the left saw the picture *k* in the mirror *a*.

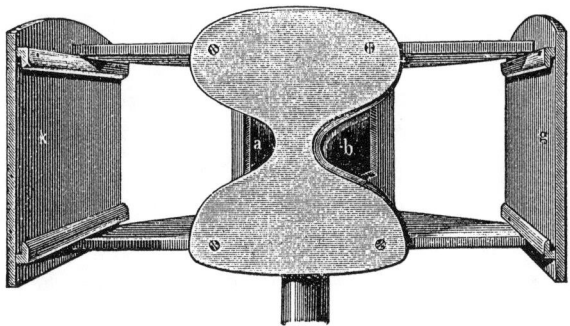

Fig. 9.

A more convenient instrument, though it does not give such sharply defined effects, is the ordinary stereoscope of Brewster,[40] shown in Fig. 10. Here the two pictures are placed on the same slide and laid in the lower part of the stereoscope, which is divided by a partition s. Two slightly prismatic glasses with convex surfaces are fixed at the top of the instrument which show the pictures somewhat further off, somewhat magnified, and at the same time overlapping each other, so that both appear to be in the middle of the instrument. The section of the double eye-piece shown in Fig. 11 exhibits the position and shape of the right and left prisms. Thus both pictures are apparently brought to the same spot, and each eye sees only the one which belongs to it.

40. Sir David Brewster, Professor of Mathematics at Edinburgh, born 1781, died 1868.—TR.

The illusion produced by the stereoscope is most obvious and striking when other means of recognising the form of an object fail. This is the case with geometrical outlines of solid figures, such as diagrams of crystals, and

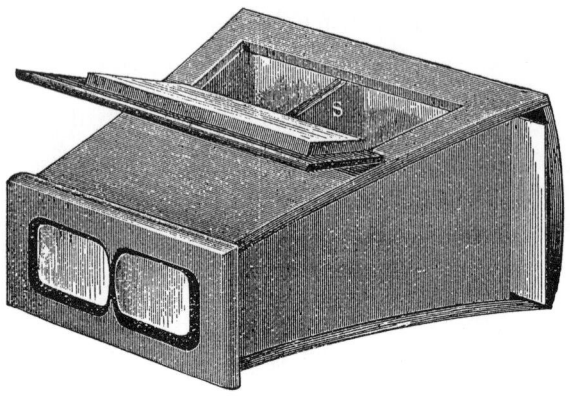

Fig. 10.

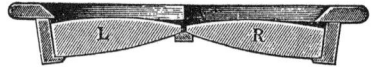

Fig. 11.

also with representations of irregular objects, especially when they are transparent, so that the shadows do not fall as we are accustomed to see them in opaque objects. Thus glaciers in stereoscopic photographs often appear to the unassisted eye an incomprehensible chaos of black and white, but when seen through a stereoscope the clear transparent ice, with its fissures and polished surfaces, comes out as if it were real. It has often happened that when I have seen for the first time buildings, cities or landscapes, with which I was familiar from stereoscopic pictures, they seemed familiar to me; but I never experienced this impression after seeing any number of ordinary pictures, because these but so imperfectly represent the real effect upon the senses.

The accuracy of the stereoscope is no less wonderful. Dove[41] has contrived an ingenious illustration of this. Take two pieces of paper printed with the same type, or from the same copper-plate, and hence exactly alike, and put them in the stereoscope in place of the two ordinary photographs.

41. Heinrich Wilhelm Dove, Professor in the University of Berlin, author of *Optische Studien* (1859); also eminent for his researches in meteorology and electricity.

They will then unite into a single completely flat image, because, as we have seen above, the two retinal images of a flat picture are identical. But no human skill is able to copy the letters of one copperplate on to another so perfectly that there shall not be some difference between them. If, therefore, we print off the same sentence from the original plate and a copy of it, or the same letters with different specimens of the same type, and put the two pieces of paper into the stereoscope, some lines will appear nearer and some further off than the rest. This is the easiest way of detecting spurious bank notes. A suspected one is put in a stereoscope along with a genuine specimen of the same kind, and it is then at once seen whether all the marks in the combined image appear on the same plane. This experiment is also important for the theory of vision, since it teaches us in a most striking manner how vivid, sure, and minute is our judgment as to depth derived from the combination of the two retinal images.

We now come to the question how is it possible for two different flat perspective images upon the retina, each of them representing only two dimensions, to combine so as to present a solid image of three dimensions. We must first make sure that we are really able to distinguish between the two flat images offered us by our eyes. If I hold my finger up and look towards the opposite wall, it covers a different part of the wall to each eye, as I mentioned above. Accordingly I see the finger twice, in front of two different places on the wall; and if I see a single image of the wall I must see a double image of the finger.

Now in ordinary vision we try to recognise the solid form of surrounding objects, and either do not notice this double image at all, or only when it is unusually striking. In order to see it we must look at the field of vision in another way—in the way that an artist does who intends to draw it. He tries to forget the actual shape, size, and distance of the objects that he represents. One would think that this is the more simple and original way of seeing things; and hitherto most physiologists have regarded it as the kind of vision which results most directly from sensation, while they have looked on ordinary solid vision as a secondary way of seeing things, which has to be learned as the result of experience. But every draughtsman knows how much harder it is to appreciate the apparent form in which objects appear in the field of vision, and to measure the angular distance between them, than to recognise what is their actual form and comparative size. In fact, the knowledge of the true relations of surrounding objects of which the artist cannot divest himself, is his greatest difficulty in drawing from nature.

Accordingly, if we look at the field of vision with both eyes, in the way an artist does, fixing our attention upon the outlines, as they would appear if projected on a pane of glass between us and them, we then become at once aware of the difference between the two retinal images. We see those objects double which lie further off or nearer than the point at which we are looking, and are not too far removed from it laterally to admit of their position being sufficiently seen. At first we can only recognise double images of objects at very different distances from the eye, but by practice they will be seen with objects at nearly the same distance.

All these phenomena, and others like them, of double images of objects seen with both eyes, may be reduced to a simple rule which was laid down by Johannes Müller:—'For each point of one retina there is on the other a *corresponding* point.' In the ordinary flat field of vision presented by the two eyes, the images received by corresponding points as a rule coincide, while images received by those which do not correspond do not coincide. The corresponding points in each retina (without noticing slight deviations) are those which are situated at the same lateral and vertical distance from the point of the retina at which rays of light come to a focus when we fix the eye for exact vision, namely, the yellow spot.

The reader will remember that the intuitive theory of vision of necessity assumes a complete combination of those sensations which are excited by impressions upon *corresponding*, or, as Müller calls them, 'identical' points. This supposition was most fully expressed in the anatomical hypothesis, that two nerve fibres which arise from corresponding points of the two retinae actually unite so as to form a single fibre, either at the commissure of the optic nerves or in the brain itself. I may, however, remark that Johannes Müller did not definitely commit himself to this mechanical explanation, although he suggested its possibility. He wished his law of identical points to be regarded simply as an expression of facts, and only insisted that the position in the field of vision of the images they receive is always the same.

But a difficulty arose. The distinction between the double images is comparatively imperfect, whenever it is possible to combine them into a single view; a striking contrast to the extraordinary precision with which, as Dove has shown, we can judge of stereoscopic relief. Yet the latter power depends upon the same differences between the two retinal pictures which cause the phenomenon of double images. The slight difference of distance between the objects represented in the right and left half of a stereoscopic photograph, which suffices to produce the most striking effect of solidity, must be increased twenty or thirty-fold before it can be recognised in the production of a double image, even if we suppose the most careful observation by one who is well practised in the experiment.

Again, there are a number of other circumstances which make the recognition of double images either easy or difficult. The most striking instance of the latter is the effect of relief. The more vivid the impression of solidity, the more difficult are double images to see, so that it is easier to see them in stereoscopic pictures than in the actual objects they represent.

On the other hand, the observation of double images is facilitated by varying the colour and brightness of the lines in the two stereoscopic pictures, or by putting lines in both which exactly correspond, and so will make more evident by contrast the imperfect coalescence of the other lines. All these circumstances ought to have no influence, if the combination of the two images in our sensation depended upon any anatomical arrangement of the conducting nerves.

Again, after the invention of the stereoscope, a fresh difficulty arose in explaining our perceptions of solidity by the differences between the two retinal images. First, Brücke[42] called attention to a series of facts which apparently made it possible to reconcile the new phenomena discovered with the theory of the innate identity of the sensations conveyed by the two retinae. If we carefully follow the way in which we look at stereoscopic pictures or at real objects, we notice that the eye follows the different outlines one after another, so that we see the 'fixed point' at each moment single, while the other points appear double. But, usually, our attention is concentrated upon the fixed point, and we observe the double images so little that to many people they are a new and surprising phenomenon when first pointed out. Now since in following the outlines of these pictures, or of an actual image, we move the eyes unequally this way and that, sometimes they converge, and sometimes diverge, according as we look at points of the outline which are apparently nearer or further off; and these differences in movement may give rise to the impression of different degrees of distance of the several lines.

Now it is quite true, that by this movement of the eye while looking at stereoscopic outlines, we gain a much more clear and exact image of the raised surface they represent, than if we fix our attention upon a single point. Perhaps the simple reason is that when we move the eyes we look at every point of the figure in succession *directly*, and therefore see it much more sharply defined than when we see only one point directly and the others indirectly. But Brücke's hypothesis, that the perception of solidity is only produced by this movement of the eyes, was disproved by experiments made by Dove, which showed that the peculiar illusion of stereoscopic pictures is also produced when they are illuminated with an electric spark. The light then lasts for less than the four thousandth part of a second. In this time heavy bodies move so little, even at great velocities, that they seem to be at rest. Hence there cannot be the slightest movement of the eye, while the spark lasts, which can possibly be recognised; and yet we receive the complete impression of stereoscopic relief.

Secondly, such a combination of the sensations of the two eyes as the anatomical hypothesis assumes, is proved not to exist by the phenomenon

42. Professor of Physiology in the University of Vienna.

of *stereoscopic lustre*, which was also discovered by Dove. If the same surface is made white in one stereoscopic picture and black in another, the combined image appears to shine, though the paper itself is quite dull. Stereoscopic drawings of crystals are made so that one shows white lines on a black ground, and the other black lines on a white ground. When looked at through a stereoscope they give the impression of a solid crystal of shining graphite. By the same means it is possible to produce in stereoscopic photographs the still more beautiful effect of the sheen of water or of leaves.

The explanation of this curious phenomenon is as follows:—A dull surface, like unglazed white paper, reflects the light which falls on it equally in all directions, and, therefore, always looks equally bright, from whatever point it is seen; hence, of course, it appears equally bright to both eyes. On the other hand, a polished surface, beside the reflected light which it scatters equally in all directions, throws back other beams by *regular* reflection, which only pass in definite directions. Now one eye may receive this regularly reflected light and the other not; the surface will then appear much brighter to the one than to the other, and, as this can only happen with shining bodies, the effect of the black and white stereoscopic pictures appears like that of a polished surface.

Now if there were a complete combination of the impressions produced upon both retinae, the union of white and black would give grey. The fact, therefore, that when they are actually combined in the stereoscope they produce the effect of lustre, that is to say, an effect which cannot be produced by any kind of uniform grey surface, proves that the impressions on the two retinae are not combined into one sensation.

That, again, this effect of stereoscopic lustre does not depend upon an alternation between the perceptions of the two eyes, on what is called the 'rivalry of the retinae,' is proved by illuminating stereoscopic pictures for an instant with the electric spark. The same effect is perfectly produced.

In the third place, it can be proved, not only that the images received by the two eyes do not coalesce in our sensation, but that the two sensations which we receive from the two eyes are not exactly similar, that they can, on the contrary, be readily distinguished. For if the sensation given by the right eye were indistinguishably the same as that given by the left, it would follow that, at least in the case of the electric spark (when no movements of the eye can help us in distinguishing the two images), it would make no difference whether we saw the right hand stereoscopic picture with the right eye, and the left with the left, or put the two pictures into the stereoscope reversed, so as to see that intended for the right eye with the left, and that intended for the left eye with the right. But practically we find that it makes all the difference, for if we make the two pictures change places, the relief appears to be inverted: what should be further off seems nearer, what should stand out seems to fall back. Now, since when we look at objects by the momentary light of the electric spark, they always appear in their true relief and never reversed, it follows that the impression produced on the right eye is not indistinguishable from that on the left.

Lastly, there are some very curious and interesting phenomena seen when two pictures are put before the two eyes at the same time which cannot be combined so as to present the appearance of a single object. If, for example, we look with one eye at a page of print, and with the other at an engraving,[43] there follows what is called the 'rivalry' of the two fields of vision. The two images are not then seen at the same time, one covering the other; but at some points one prevails, and at others the other. If they are equally distinct, the places where one or the other appears usually change after a few seconds. But if the engraving presents anywhere in the field of vision a uniform white or black surface, then the printed letters which occupy the same position in the image presented to the other eye, will usually prevail exclusively over the uniform surface of the engraving. In spite, however, of what former observers have said to the contrary, I maintain that it is possible for the observer at any moment to control this rivalry by voluntary direction of his attention. If he tries to read the printed sheet, the letters remain visible, at least at the spot where for the moment he is reading. If, on the contrary, he tries to follow the outline and shadows of the engraving, then these prevail. I find, moreover, that it is possible to fix the attention upon a very feebly illuminated object, and make it prevail over a much brighter one, which coincides with it in the retinal image of the other eye. Thus, I can follow the watermarks of a white piece of paper and cease to see strongly-marked black figures in the other field. Hence the retinal rivalry is not a trial of strength between two sensations, but depends upon our fixing on failing to fix the attention. Indeed there is scarcely any phenomenon so well fitted for the study of the causes which are capable of determining the attention. It is not enough to form the conscious intention of seeing first with one eye and then with the other; we must form as clear a notion as possible of what we expect to see. Then it will actually appear. If, on the other hand, we leave the mind at liberty without a fixed intention to observe a definite object, that alternation between the two pictures ensues which is called retinal rivalry. In this case, we find that, as a rule, bright and strongly marked objects in one field of vision prevail over those which are darker and less distinct in the other, either completely or at least for a time.

We may vary this experiment by using a pair of spectacles with different coloured glasses. We shall then find, on looking at the same objects with

---

43. The practiced observer is able to do this without any apparatus, but most persons will find it necessary to put the two objects in a stereoscope or, at least, to hold a book, or a sheet of paper, or the hand in front of the face, to serve for the partition in the stereoscope.—TR.

both eyes at once, that there ensues a similar rivalry between the two colours. Everything appears spotted over first with one and then with the other. After a time, however, the vividness of both colours becomes weakened, partly by the elements of the retina which are affected by each of them being tired, and partly by the complementary after-images which result. The alternation then ceases, and there ensues a kind of mixture of the two original colours.

It is much more difficult to fix the attention upon a colour than upon such an object as an engraving. For the attention upon which, as we have seen, the whole phenomenon of 'rivalry' depends, fixes itself with constancy only upon such a picture as continually offers something new for the eye to follow. But we may assist this by reflecting on the side of the glasses next the eye letters or other lines upon which the attention can fix. These reflected images themselves are not coloured, but as soon as the attention is fixed upon one of them we become conscious of the colour of the corresponding glass.

These experiments on the rivalry of colours have given rise to a singular controversy among the best observers; and the possibility of such difference of opinion is an instructive hint as to the nature of the phenomenon itself. One party, including the names of Dove, Regnault,[44] Brücke, Ludwig,[45] Panum,[46] and Hering,[47] maintains that the result of a binocular view of two colours is the true combination-colour. Other observers, as Heinrich Meyer of Zürich, Volkmann, Meissner,[48] and Funke,[49] declare quite as positively that, under these conditions, they have never seen the combination-colour. I myself entirely agree with the latter, and a careful examination of the cases in which I might have imagined that I saw the combination-colour, has always proved to me that it was the result of phenomena of contrast. Each time that I brought the true combination-colour side by side with the binocular mixture of colours, the difference between the two was very apparent. On the other hand, there can of course be no doubt that the observers I first named really saw what they profess, so that there must here be great individual difference. Indeed with certain experiments which Dove recommends as particularly well fitted to prove the correctness of his conclusion, such as the binocular combination of complementary polarisation-colours into white, I could not myself see the slightest trace of a combination-colour.

44. The distinguished French chemist, father of the well-known painter who was killed in the second siege of Paris.
45. Professor of Physiology in the University of Leipzig.
46. Professor of Physiology in the University of Kiel.
47. Ewald Hering, Professor of Physiology in the University of Prague, lately in the Josephs-akademie of Vienna.
48. Professor of Physiology in the University of Göttingen.
49. Professor of Physiology in the University of Freiburg.—TR.

This striking difference in a comparatively simple observation seems to me to be of great interest. It is a remarkable confirmation of the supposition above made, in accordance with the empirical theory of vision, that in general only those sensations are perceived as separated in space, which can be separated one from another by voluntary movements. Even when we look at a compound colour with one eye, only three separate sensations are, according to Young's theory, produced together; but it is impossible to separate these by any movement of the eye, so that they always remain locally united. Yet we have seen that even in this case we may become conscious of a separation under certain circumstances; namely, when it is seen that part of the colour belongs to a transparent covering. When two corresponding points of the retinae are illuminated with different colours, it will be rare for any separation between them to appear in ordinary vision; if it does, it will usually take place in the part of the field of sight outside the region of exact vision. But there is always a possibility of separating the compound impression thus produced into its two parts, which will appear to some extent independent of each other, and will move with the movements of the eye; and it will depend upon the degree of attention which the observer is accustomed to give to the region of indirect vision and to double images, whether he is able to separate the colours which fall on both retinae at the same time. Mixed hues, whether looked at with one eye or with both, excite many simple sensations of colour at the same time, each having exactly the same position in the field of vision. The difference in the way in which such a compound-colour is regarded by different people depends upon whether this compound sensation is at once accepted as a coherent whole without any attempt at analysis, or whether the observer is able by practice to recognise the parts of which it is composed, and to separate them from one another. The former is our usual (though not constant) habit when looking with one eye, while we are more inclined to the latter when using both. But inasmuch as this inclination must chiefly depend upon practice in observing distinctions, gained by preceding observation, it is easy to understand how great individual peculiarities may arise.

If we carefully observe the rivalry which ensues when we try to combine two stereoscopic drawings, one of which is in black lines on a white ground and the other in white lines on black, we shall see that the white and black lines which affect nearly corresponding points of each retina always remain visible side by side—an effect which of course implies that the white and black grounds are also visible. By this means the brilliant surface, which seems to shine like black lead, makes a much more stable impression than that produced under the operation of retinal rivalry by entirely different drawings. If we cover the lower half of the white figure on a black ground with a sheet of printed paper, the upper half of the combined stereoscopic image shows the phenomenon of lustre, while in the lower we see Retinal Rivalry between the black lines of the figure and the black marks of the type. As long as the observer attends to the solid form of the object represented, the black and white outlines of the two stereoscopic drawings carry on in common the point of exact vision as it moves along them, and the effect can only be kept up by continuing to follow both. He must steadily keep his attention upon both drawings, and then the impression of each will be equally combined. There is no better way of preserving the combined effect of two stereoscopic pictures than this. Indeed it is possible to combine (at least partially and for a short time) two entirely different drawings when put into the stereoscope, by fixing the attention upon the way in which they cover each other, watching, for instance, the angles at which their lines cross. But as soon as the attention turns from the angle to follow one of the lines which makes it, the picture to which the other line belongs vanishes.

Let us now put together the results to which our inquiry into binocular vision has led us.

I. The excitement of corresponding points of the two retinae is not indistinguishably combined into a single impression; for, if it were, it would be impossible to see Stereoscopic Lustre. And we have found reason to believe that this effect is not a consequence of Retinal Rivalry, even if we admit the latter phenomenon to belong to sensation at all, and not rather to the degree of attention. On the contrary the appearance of lustre is associated with the *restriction* of this rivalry.

II. The sensations which are produced by the excitation of corresponding points of each retina are not indistinguishably the same; for otherwise we should not be able to distinguish the true from the inverted or 'pseudo-scopic' relief, when two stereoscopic pictures are illuminated by the electric spark.

III. The combination of the two different sensations received from corresponding retinal points is not produced by one of them being suppressed for a time; for, in the first place, the perception of solidity given by the two eyes depends upon our being at the same time conscious of the two different images, and, in the second, this perception of solidity is independent of any movement of the retinal images, since it is possible under momentary illumination.

We therefore learn that two distinct sensations are transmitted from the two eyes, and reach the consciousness at the same time and without coalescing; that accordingly the combination of these two sensations into the single picture of the external world of which we are conscious in ordinary vision is not produced by any anatomical mechanism of sensation, but by a mental act.

IV. Further, we find that there is, on the whole, complete, or at least nearly complete, coincidence as to localisation in the field of vision of impressions of sight received from corresponding points of the retinae; but that when we refer both impressions to the same object, their coincidence of localisation is much disturbed.

If this coincidence were the result of a direct function of sensation, it could not be disturbed by the mental operation which refers the two impressions to the same object. But we avoid the difficulty, if we suppose that the coincidence in localisation of the corresponding pictures received from the two eyes depends upon the power of measuring distances at sight which we gain by experience, that is, on an acquired knowledge of the meaning of the 'signs of localisation.' In this case it is simply one kind of experience opposing another; and we can then understand how the conclusion that two images belong to the same object should influence our estimation of their relative position by the measuring power of the eye, and how in consequence the distance of the two images from the fixed point in the field of vision should be regarded as the same, although it is not exactly so in reality.

But if the practical coincidence of corresponding points as to localisation in the two fields of vision does not depend upon sensation, it follows that the original power of comparing different distances in each separate field of vision cannot depend upon direct sensation. For, if it were so, it would follow that the coincidence of the two fields would be completely established by direct sensation, as soon as the observer had got his two fixed points to coincide and a single meridian of one eye to coincide with the corresponding one of the other.

The reader sees how this series of facts has driven us by force to the empirical theory of vision. It is right to mention that lately fresh attempts have been made to explain the origin of our perception of solidity and the phenomena of single and double binocular vision by the assumption of some ready-made anatomical mechanism. We cannot criticise these attempts here: it would lead us too far into details. Although many of these hypotheses are very ingenious (and at the same time very indefinite and elastic), they have hitherto always proved insufficient; because the actual world offers us far more numerous relations than the authors of these attempts could provide for. Hence, as soon as they have arranged one of their systems to explain any particular phenomenon of vision, it is found not to answer for any other. Then, in order to help out the hypothesis, the very doubtful assumption has to be made that, in these other cases, sensation is overcome and extinguished by opposing experience. But what confidence could we put in any of our perceptions if we were able to extinguish our sensations as we please, whenever they concern an object of our attention, for the sake of previous conceptions to which they are opposed? At any rate, it is clear that in every case where experience must finally decide, we shall succeed much better in forming a correct notion of what we see, if we have no opposing sensations to overcome, than if a correct judgment must be formed in spite of them.

It follows that the hypotheses which have been successively framed by the various supporters of intuitive theories of vision, in order to suit one phenomenon after another, are really quite unnecessary. No fact has yet been discovered inconsistent with the Empirical Theory: which does not assume any peculiar modes of physiological action in the nervous system, nor any hypothetical anatomical structures; which supposes nothing more than the well known association between the impressions we receive and the conclusions we draw from them, according to the fundamental laws of daily experience. It is true that we cannot at present offer any complete scientific explanation of the mental operations involved, and there is no immediate prospect of our doing so. But since these operations actually exist, and since hitherto every form of the intuitive theory has been obliged to fall back on their reality when all other explanation failed, these mysteries of the laws of thought cannot be regarded from a scientific point of view as constituting any deficiency in the empirical theory of vision.

It is impossible to draw any line in the study of our perceptions of space which shall sharply separate those which belong to direct Sensation from those which are the result of Experience. If we attempt to draw such a boundary, we find that experience proves more minute, more direct and more exact than supposed sensation, and in fact proves its own superiority by overcoming the latter. The only supposition which does not lead to any contradiction is that of the Empirical Theory, which regards all our perceptions of space as depending upon experience, and not only the qualities, but even the local signs of the sense of sight as nothing more than signs, the meaning of which we have to learn by experience.

We become acquainted with their meaning by comparing them with the result of our own movements, with the changes which we thus produce in the outer world. The infant first begins to play with its hands. There is a time when it does not know how to turn its eyes or its hands to an object which attracts its attention by its brightness or colour. When a little older, a child seizes whatever is presented to it, turns it over and over again, looks at it, touches it, and puts it in his mouth. The simplest objects are what a child likes best, and he always prefers the most primitive toy to the elaborate inventions of modern ingenuity. After he has looked at such a toy every day for weeks together, he learns at last all the perspective images which it presents; then he throws it away and wants a fresh toy to handle like the first. By this means the child learns to recognise the different views which the same object can afford, in connection with the movements which he is constantly giving it. The conception of the shape of any object, gained in this manner, is the result of associating all these visual images. When we have obtained an accurate conception of the form of any object, we are then able to imagine what appearance it would present, if we looked at it from some other point of view. All these different views are combined in the judgment we form as to the dimensions and shape of an object. And, consequently, when we are once acquainted with this, we can deduce from it the various images it would present to the sight when seen from different points of view, and the various movements which we should have to impress upon it in order to obtain these successive images.

I have often noticed a striking instance of what I have been saying in looking at stereoscopic pictures. If, for example, we look at elaborate outlines of complicated crystalline forms, it is often at first difficult to see what they mean. When this is the case, I look out two points in the diagram which correspond, and make them overlap by a voluntary movement of the eyes. But as long as I have not made out what kind of form the drawings are intended to represent, I find that my eyes begin to diverge again, and the two points no longer coincide. Then I try to follow the different lines of the figure, and suddenly I see what the form represented is. From that moment my two eyes pass over the outlines of the apparently solid body with the utmost ease, and without ever separating. As soon as we have gained a correct notion of the shape of an object, we have the rule for the movements of the eyes which are necessary for seeing it. In carrying out these movements, and thus receiving the visual impressions we expect, we retranslate the notion we have formed into reality, and by finding this retranslation agrees with the original, we become convinced of the accuracy of our conception.

This last point is, I believe, of great importance. The meaning we assign to our sensations depends upon experiment, and not upon mere observation of what takes place around us. We learn by experiment that the correspondence between two processes takes place at any moment that we choose, and under conditions which we can alter as we choose. Mere observation would not give us the same certainty, even though often repeated under different conditions. For we should thus only learn that the processes in question appear together frequently (or even always, as far as our experience goes); but mere observation would not teach us that they appear together at any moment we select.

Even in considering examples of scientific observation, methodically carried out, as in astronomy, meteorology, or geology, we never feel fully convinced of the causes of the phenomena observed until we can demonstrate the working of these same forces by actual experiment in the laboratory. So long as science is not experimental it does not teach us the knowledge of any new force.[50]

It is plain that, by the experience which we collect in the way I have been describing, we are able to learn as much of the meaning of sensible 'signs' as can afterwards be verified by further experience; that is to say, all that is real and positive in our conceptions.

It has been hitherto supposed that the sense of *touch* confers the notion of space and movement. At first of course the only direct knowledge we acquire is that we can produce, by an act of volition, changes of which we are cognisant by means of touch and sight. Most of these voluntary changes are movements, or changes in the relations of space; but we can also produce changes in an object itself. Now, can we recognise the movements of our hands and eyes as changes in the relations of space, without knowing it beforehand? and can we distinguish them from other changes which affect the properties of external objects? I believe we can. It is an essentially distinct character of the Relations of Space that they are *changeable relations between objects which do not depend on their quality or quantity*, while all other material relations between objects depend upon their properties. The perceptions of sight prove this directly and easily. A movement of the eye which causes the retinal image to shift its place upon the retina always produces the same series of changes as often as it is repeated, whatever objects the field of vision may contain. The effect is that the impressions which had before the local signs a_1, a_2, a_3, a_4, receive the new local signs b_1, b_2, b_3, b_4; and this may always occur in the same way, whatever be the quality of the impressions. By this means we learn to recognise such changes as belonging to the special phenomena which we call changes in space. This is enough for the object of Empirical Philosophy, and we need not further enter upon a discussion of the question, how much of universal conceptions of space is derived *a priori*, and how much *a posteriori*?[51]

An objection to the empirical Theory of Vision might be found in the fact that illusions of the senses are possible; for if we have learnt the meaning of our sensations from experience, they ought always to agree with experience. The explanation of the possibility of illusions lies in the fact that we transfer the notions of external objects, which would be correct under normal conditions, to cases in which unusual circumstances have altered the retinal pictures. What I call 'observation under normal

---

50. An interesting paper, applying this view of the 'experimental' character of progressive science to Zoology, has been published by M. Lacaze-Duthiers, in the first number of his *Archives de Zoologie.*—TR.

51. The question of the origin of our conceptions of space is discussed by Mr. Bain on empirical principles in his treatise on *The Senses and the Intellect*, pp. 114–118, 189–194, 245, 363–392, &c.—TR.

conditions' implies not only that the rays of light must pass in straight lines from each visible point to the cornea, but also that we must use our eyes in the way they should be used in order to receive the clearest and most easily distinguishable images. This implies that we should successively bring the images of the separate points of the outline of the objects we are looking at upon the centres of both retinae (the yellow spot), and also move the eyes so as to obtain the surest comparison between their various positions. Whenever we deviate from these conditions of normal vision, illusions are the result. Such are the long recognised effects of the refraction or reflection of rays of light before they enter the eye. But there are many other causes of mistake as to the position of the objects we see—defective accommodation when looking through one or two small openings, improper convergence when looking with one eye only, irregular position of the eye-ball from external pressure or from paralysis of its muscles. Moreover, illusions may come in from certain elements of sensation not being accurately distinguished; as, for instance, the degree of convergence of the two eyes, of which it is difficult to form an accurate judgment when the muscles which produce it become fatigued.

The simple rule for all illusions of sight is this: *we always believe that we see such objects as would, under conditions of normal vision, produce the retinal image of which we are actually conscious.* If these images are such as could not be produced by any normal kind of observation, we judge of them according to their nearest resemblance; and in forming this judgment, we more easily neglect the parts of sensation which are imperfectly than those which are perfectly apprehended. When more than one interpretation is possible, we usually waver involuntarily between them; but it is possible to end this uncertainty by bringing the idea of any of the possible interpretations we choose as vividly as possible before the mind by a conscious effort of the will.

These illusions obviously depend upon mental processes which may be described as false inductions. But there are, no doubt, judgments which do not depend upon our consciously thinking over former observations of the same kind, and examining whether they justify the conclusion which we form. I have, therefore, named these 'unconscious judgments;' and this term, though accepted by other supporters of the empirical theory, has excited much opposition, because, according to generally-accepted psychological doctrines, a *judgment*, or *logical conclusion*, is the culminating point of the conscious operations of the mind. But the judgments which play so great a part in the perceptions we derive from our senses cannot be expressed in the ordinary form of logically analysed conclusions, and it is necessary to deviate somewhat from the beaten paths of psychological analysis in order to convince ourselves that we really have here the same kind of mental operation as that involved in conclusions usually recognised as such. There appears to me to be in reality only a superficial difference between the 'conclusions' of logicians and those inductive conclusions of which we recognise the result in the conceptions we gain of the outer world through our sensations. The difference chiefly depends upon the former conclusions being capable of expression in words, while the latter are not; because, instead of words, they only deal with sensations and the memory of sensations. Indeed, it is just the impossibility of describing sensations, whether actual or remembered, in words, which makes it so difficult to discuss this department of psychology at all.

Beside the knowledge which has to do with Notions, and is, therefore, capable of expression in words, there is another department of our mental operations, which may be described as knowledge of the relations of those impressions on the senses which are not capable of direct verbal expression. For instance, when we say that we 'know'[52] a man, a road, a fruit, a perfume, we mean that we have seen, or tasted, or smelt, these objects. We keep the sensible impression fast in our memory, and we shall recognise it again when it is repeated, but we cannot describe the impression in words, even to ourselves. And yet it is certain that this kind of knowledge (*Kennen*) may attain the highest possible degree of precision and certainty, and is so far not inferior to any knowledge (*Wissen*) which can be expressed in words; but it is not directly communicable, unless the object in question can be brought actually forward, or the impression it produces can be otherwise represented—as by drawing the portrait of a man instead of producing the man himself.

It is an important part of the former kind of knowledge to be acquainted with the particular innervation of muscles, which is necessary in order to produce any effect we intend by moving our limbs. As children, we must learn to walk; we must afterwards learn how to skate or go on stilts, how to ride, or swim, or sing, or pronounce a foreign language. Moreover, observation of infants shows that they have to learn a number of things which afterwards they will know so well as entirely to forget that there was ever a time when they were ignorant of them. For example, everyone of us had to learn, when an infant, how to turn his eyes toward the light in order to see. This kind of 'knowledge' (*Kennen*) we also call 'being able' to do a thing (*können*), or 'understanding' how to do it (*verstehen*), as, 'I know how to ride,' 'I am able to ride,' or 'I understand how to ride.'[53]

It is important to notice that this 'knowledge' of the effort of the will to

---

52. In German this kind of knowledge is expressed by the verb *kennen* (*cognoscere, connaître*), to be acquainted with, while *wissen* (*scire, savoir*) means to be aware of. The former kind of knowledge is only applicable to objects directly cognisable by the senses, whereas the latter applies to notions or conceptions which can be formally stated as propositions.—TR.

53. The German word *können* is said to be of the same etymology as *kennen*, and so their likeness in form would be explained by their likeness in meaning.

be exerted must attain the highest possible degree of certainty, accuracy, and precision, for us to be able to maintain so artificial a balance as is necessary for walking on stilts or for skating, for the singer to know how to strike a note with his voice, or the violin-player with his finger, so exactly that its vibration shall not be out by a hundredth part.

Moreover, it is clearly possible, by using these sensible images of memory instead of words, to produce the same kind of combination which, when expressed in words, would be called a proposition or a conclusion. For example, I may know that a certain person with whose face I am familiar, has a peculiar voice, of which I have an equally lively recollection. I should be able with the utmost certainty to recognise his face and his voice among a thousand, and each would recall the other. But I cannot express this fact in words, unless I am able to add some other characters of the person in question which can be better defined. Then I should be able to resort to a syllogism and say, 'This voice which I now hear belongs to the man whom I saw then and there.' But universal, as well as particular conclusions, may be expressed in terms of sensible impressions, instead of words. To prove this I need only refer to the effect of works of art. The statue of a god would not be capable of conveying a notion of a definite character and disposition, if I did not know that the form of face and the expression it wears have usually or constantly a certain definite signification. And, to keep in the domain of the perceptions of the senses, if I know that a particular way of looking, for which I have learnt how to employ exactly the right kind of innervation, is necessary in order to bring into direct vision a point two feet off and so many feet to the right, this also is a universal proposition which applies to every case in which I have fixed a given point at that distance before, or may do so hereafter. It is a piece of knowledge which cannot be expressed in words, but is the result which sums up my previous successful experience. It may at any moment become the *major* premiss of a syllogism, whenever, in fact, I fix a point in the supposed position and feel that I do so by looking as that major proposition states. This perception of what I am doing is my *minor* proposition, and the '*conclusion*' is that the object I am looking for will be found at the spot in question.

Suppose that I employ the same way of looking, but look into a stereoscope. I am now aware that there is no real object before me at the spot I am looking at; but I have the same sensible impression as if one were there; and yet I am unable to describe this impression to myself or others, or to characterise it otherwise than as 'the same impression which would arise in the normal method of observation, if an object were really there.' It is important to notice this. No doubt the physiologist can describe the impression in other ways, by the direction of the eyes, the position of the retinal images, and so on; but there is no other way of directly defining and characterising the sensation which we experience. Thus we may recognise it as an illusion, but yet we cannot get rid of the sensation of this illusion; for we cannot extinguish our remembrance of its normal signification, even when we know that in the case before us this does not apply—just as little as we are able to drive out of the mind the meaning of a word in our mother tongue, when it is employed as a sign for an entirely different purpose.

These conclusions in the domain of our sensible perceptions appear as inevitable as one of the forces of nature, and hence their results seem to be directly apprehended, without any effort on our part; but this does not distinguish them from logical and conscious conclusions, or at least from those which really deserve the name. All that we can do by voluntary and conscious effort, in order to come to a conclusion, is, after all, only to supply complete materials for constructing the necessary premises. As soon as this is done, the conclusion forces itself upon us. Those conclusions which (it is supposed) may be accepted or avoided as we please, are not worth much.

The reader will see that these investigations have led us to a field of mental operations which has been seldom entered by scientific explorers. The reason is that it is difficult to express these operations in words. They have been hitherto most discussed in writings on aesthetics, where they play an important part as Intuition, Unconscious Ratiocination, Sensible Intelligibility, and such obscure designations. There lies under all these phrases the false assumption that the mental operations we are discussing take place in an undefined, obscure, half-conscious fashion; that they are, so to speak, mechanical operations, and thus subordinate to conscious thought, which can be expressed in language. I do not believe that any difference in kind between the two functions can be proved. The enormous superiority of knowledge which has become ripe for expression in language, is sufficiently explained by the fact that, in the first place, speech makes it possible to collect together the experience of millions of individuals and thousands of generations, to preserve them safely, and by continual verification to make them gradually more and more certain and universal; while, in the second place, all deliberately combined actions of mankind, and so the greatest part of human power, depend on language. In neither of these respects can mere familiarity with phenomena (*das Kennen*) compete with the knowledge of them which can be communicated by speech (*das Wissen*); and yet it does not follow of necessity that the one kind of knowledge should be of a different nature from the other, or less clear in its operation.

The supporters of Intuitive Theories of Sensation often appeal to the capabilities of new-born animals, many of which show themselves much more skilful than a human infant. It is quite clear that an infant, in spite of the greater size of its brain, and its power of mental development, learns with extreme slowness to perform the simplest tasks; as, for example, to direct its eyes to an object or to touch what it sees with its hands. Must we not conclude that a child has much more to learn than an animal which is safely guided, but also restricted, by its instincts? It is said that the calf sees the udder and goes after it, but it admits of question whether it does not simply smell it, and make those movements which bring it nearer to the scent.[54] At any rate, the child knows nothing of the meaning of the visual image presented by its mother's breast. It often turns obstinately away from it to the wrong size and tries to find it there. The young chicken very soon pecks at grains of corn, but it pecked while it was still in the shell, and when it hears the hen peck, it pecks again, at first seemingly at random. Then, when it has by chance hit upon a grain, it may, no doubt, learn to notice the field of vision which is at the moment presented to it. The process is all the quicker because the whole of the mental furniture which it requires for its life is but small.

We need, however, further investigations on the subject in order to throw light upon this question. As far as the observations with which I am acquainted go, they do not seem to me to prove that anything more than certain tendencies is born with animals. At all events one distinction between them and man lies precisely in this, that these innate or congenital tendencies, impulses or instincts are in him reduced to the smallest possible number and strength.[55]

There is a most striking analogy between the entire range of processes which we have been discussing, and another System of Signs, which is not given by nature but arbitrarily chosen, and which must undoubtedly be learned before it is understood. I mean the words of our mother tongue.

Learning how to speak is obviously a much more difficult task than acquiring a foreign language in after life. First, the child has to guess that the sounds it hears are intended to be signs at all; next, the meaning of each separate sound must be found out, by the same kind of induction as the meaning of the sensations of sight or touch; and yet we see children by the end of their first year already understanding certain words and phrases, even if they are not yet able to repeat them. We may sometimes observe the same in dogs.

Now this connection between Names and Objects, which demonstrably must be *learnt*, becomes just as firm and indestructible as that between Sensations and the Objects which produce them. We cannot help thinking of the usual signification of a word, even when it is used exceptionally in

---

54. See Darwin on the *Expression of the Emotions*, p. 47.—TR.
55. See on this subject Bain on the *Senses and the Intellect*, p. 293; also a paper on 'Instinct' in *Nature*, Oct. 10, 1872.

some other sense; we cannot help feeling the mental emotions which a fictitious narrative calls forth, even when we know that it is not true; just in the same way as we cannot get rid of the normal signification of the sensations produced by any illusion of the senses, even when we know that they are not real. There is one other point of comparison which is worth notice. The elementary signs of language are only twenty-six letters, and yet what wonderfully varied meanings can we express and communicate by their combination! Consider, in comparison with this, the enormous number of elementary signs with which the machinery of sight is provided. We may take the number of fibres in the optic nerves as two hundred and fifty thousand. Each of these is capable of innumerable different degrees of sensation of one, two, or three primary colours. It follows that it is possible to construct an immeasurably greater number of combinations here than with the few letters which build up our words. Nor must we forget the extremely rapid changes of which the images of sight are capable. No wonder, then, if our senses speak to us in language which can express far more delicate distinctions and richer varieties than can be conveyed by words.

This is the solution of the riddle of how it is possible to see; and, as far as I can judge, it is the only one of which the facts at present known admit. Those striking and broad incongruities between Sensations and Objects, both as to quality and to localisation, on which we dwelt, are just the phenomena which are most instructive; because they compel us to take the right road. And even those physiologists who try to save fragments of a pre-established harmony between sensations and their objects, cannot but confess that the completion and refinement of sensory perceptions depend so largely upon experience, that it must be the latter which finally decides whenever they contradict the supposed congenital arrangements of the organ. Hence the utmost significance which may still be conceded to any such anatomical arrangements is that they are possibly capable of helping the first practice of our senses.

The correspondence, therefore, between the external world and the Perceptions of Sight rests, either in whole or in part, upon the same foundation as all our knowledge of the actual world—on *experience*, and on constant *verification* of its accuracy by experiments which we perform with every movement of our body. It follows, of course, that we are only warranted in accepting the reality of this correspondence so far as these means of verification extend, which is really as far as for practical purposes we need.

Beyond these limits, as, for example, in the region of Qualities, we are in some instances able to prove conclusively that there is no correspondence at all between sensations and their objects.

Only the relations of time, of space, of equality, and those which are derived from them, of number, size, regularity of coexistence and of sequence—'mathematical relations' in short—are common to the outer and the inner world, and here we may indeed look for a complete correspondence between our conceptions and the objects which excite them.

But it seems to me that we should not quarrel with the bounty of nature because the greatness, and also the emptiness, of these abstract relations have been concealed from us by the manifold brilliance of a system of signs; since thus they can be the more easily surveyed and used for practical ends, while yet traces enough remain visible to guide the philosophical spirit aright, in its search after the meaning of sensible Images and Signs.

A Course of Lectures Delivered in Frankfort and Heidelberg, and Republished in the Preussische Jahrbücher, 1868.

7

On the Aim and Progress of Physical Science

IN accepting the honour you have done me in requesting me to deliver the first lecture at the opening meeting of this year's Association, it appears to me to be more in keeping with the import of the moment and the dignity of this assembly that, in place of dealing with any particular line of research of my own, I should invite you to cast a glance at the development of all the branches of physical science represented on these occasions. These branches include a vast area of special investigation, material of almost too varied a character for comprehension, the range and intrinsic value of which become greater with each year, while no bounds can be assigned to its increase. During the first half of the present century we had an Alexander von Humboldt, who was able to scan the scientific knowledge of his time in its details, and to bring it within one vast generalisation. At the present juncture, it is obviously very doubtful whether this task could be accomplished in a similar way, even by a mind with gifts so peculiarly suited for the purpose as Humboldt's was, and if all his time and work were devoted to the purpose.

We, however, working as we do to advance a single department of science, can devote but little of our time to the simultaneous study of the other branches. As soon as we enter upon any investigation, all our powers have to be concentrated on a field of narrowed limit. We have not only, like the philologian or historian, to seek out and search through books and gather from them what others have already determined about the subject under inquiry; that is but a secondary portion of our work. We have to attack the things themselves, and in doing so each offers new and peculiar difficulties of a kind quite different from those the scholar encounters; while in the majority of instances, most of our time and labour is consumed by secondary matters that are but remotely connected with the purpose of the investigation.

At one time, we have to study the errors of our instruments, with a view to their diminution, or, where they cannot be removed, to compass their detrimental influence; while at other times we have to watch for the moment when an organism presents itself under circumstances most favourable for research. Again, in the course of our investigation we learn for the first time of possible errors which vitiate the result, or perhaps merely raise a suspicion that it may be vitiated, and we find ourselves compelled to begin the work anew, till every shadow of doubt is removed. And it is only when the observer takes such a grip of the subject, so fixes all his thoughts and all his interest upon it that he cannot separate himself from it for weeks, for months, even for years, cannot force himself away from it, in short, till he has mastered every detail, and feels assured of all those results which must come in time, that a perfect and valuable piece of work is done. You are all aware that in every good research, the preparation, the secondary operations, the control of possible errors, and especially in the separation of the results attainable in the time from those that cannot be attained, consume far more time than is really required to make actual observations or experiments. How much more ingenuity and thought are expended in bringing a refractory piece of brass or glass into subjection, than in sketching out the plan of the whole investigation! Each of you will have experienced such impatience and over-excitement during work where all the thoughts are directed on a narrow range of questions, the import of which to an outsider appears trifling and contemptible because he does not see the end to which the preparatory work tends. I believe I am correct in thus describing the work and mental condition that precedes all those great results which hastened so much the development of science after its long inaction, and gave it so powerful an influence over every phase of human life.

The period of work, then, is no time for broad comprehensive survey. When, however, the victory over difficulties has happily been gained, and results are secured, a period of repose follows, and our interest is next directed to examining the bearing of the newly established facts, and once more venturing on a wider survey of the adjoining territory. This is essential, and those only who are capable of viewing it in this light can hope to find useful starting-points for further investigation.

The preliminary work is followed by other work, treating of other subjects. In the course of its different stages, the observer will not deviate far from a direction of more or less narrowed range. For it is not alone of importance to him that he may have collected information from books regarding the region to be explored. The human memory is, on the whole, proportionately patient, and can store up an almost incredibly large amount of learning. In addition, however, to the knowledge which the student of science acquires from lectures and books, he requires intelligence which only an ample and diligent perception can give him; he needs skill which comes only by repeated experiment and long practice. His senses must be sharpened for certain kinds of observation, to detect minute differences of form, colour, solidity, smell, &c., in the object under examination; his hand must be equally trained to the work of the blacksmith, the locksmith, and the carpenter, or the draughtsman and the violin-player, and, when operating with the microscope, must surpass the lace-maker in delicacy of handling the needle. Moreover, when he encounters superior destructive forces, or performs bloody operations upon man or beast, he must possess the courage and coolness of the soldier. Such qualities and capabilities, partly the result of natural aptitude, partly cultivated by long practice, are not so readily and so easily acquired as the mere massing of facts in the memory; and hence it happens that an investigator is compelled, during the entire labours of his life, to strictly limit his field, and to confine himself to those branches which suit him best.

We must not, however, forget that the more the individual worker is compelled to narrow the sphere of his activity, so much the more will his intellectual desires induce him not to sever his connection with the subject in its entirety. How shall he go stout and cheerful to his toilsome work, how feel confident that what has given him so much labour will not molder uselessly away, but remain a thing of lasting value, unless he keeps alive within himself the conviction that he also has added a fragment to the stupendous whole of Science which is to make the reasonless forces of nature subservient to the moral purposes of humanity?

An immediate practical use cannot generally be counted on *a priori*, for each particular investigation. Physical science, it is true, has by the practical realisation of its results transformed the entire life of modern humanity. But, as a rule, these applications appear under circumstances when they are least expected; to search in that direction generally leads to nothing unless certain points have already been definitely fixed, so that all that has to be done is to remove certain obstacles in the way of practical application. If we search the records of the most important discoveries, they are either, especially in earlier times, made by workmen who their whole lives through did but one kind of work, and, either by a happy accident, or by a searching, repeated, tentative experiment, hit upon some new method advantageous to their particular handicraft; others there are, and this is especially the case in most of the recent discoveries, which are the fruit of a matured scientific acquaintance with the subject in question, an acquaintance that in each instance had originally been acquired without any direct view to possible use.

Our Association represents the whole of natural science. To-day are assembled mathematicians, physicists, chemists and zoologists, botanists and geologists, the teacher of science and the physician, the technologist and the amateur who finds in scientific pursuits relaxation from other occupation. Here each of us hopes to meet with fresh impulse and encouragement for his peculiar work; the man who lives in a small country place hopes to meet with the recognition, otherwise unattainable, of having aided in the advance of science; he hopes by intercourse with men pursuing more or less the same object to mark the aim of new researches. We rejoice to find among us a goodly proportion of members representing the cultivated classes of the nation; we see influential statesmen among us. They all have an interest in our labours; they look to us for further progress in civilisation, further victories over the powers of nature. They it is who place at our disposal the actual means for carrying on our labours, and are therefore entitled to enquire into the results of those labours. It appears to me, therefore, appropriate to this occasion to take account of the progress of science as a whole, of the objects it aspires to, and the magnitude of the efforts made to attain them.

Such a survey is desirable; that it lies beyond the powers of any one man to accomplish with even an approximate completeness such a task as this is clear from what I have already said. If I stand here to-day with such a problem entrusted to me, my excuse must be that no other would attempt it, and I hold that an attempt to accomplish it, even if with small success, is better than none whatever. Besides, a physiologist has perhaps more than all others immediate occasion to maintain a clear and constant view of the entire field, for in the present state of things it is peculiarly the lot of the physiologist to receive help from all other branches of science and to stand in alliance with them. In physiology, in fact, the importance of the vast strides to which I shall allude, has been chiefly felt, while to physiology, and the leading controversies arising in it, some of the most valuable discoveries are directly due.

If I leave considerable gaps in my survey, my excuse must be the magnitude of the task, and the fact that the pressing summons of my friend the secretary of this Association reached me but recently, and that too in the course of my summer holiday in the mountains. The gaps which I may leave will at all events be abundantly filled up by the proceedings of the Sections.

Let us then proceed to our task. In discussing the progress of physical science as a whole, the first question which presents itself is, By what standard are we to estimate this progress?

To the uninitiated, this science of ours is an accumulation of a vast number of facts, some of which are conspicuous for their practical utility, while others are merely curiosities, or objects of wonder. And, if it were possible to classify this unconnected mass of facts, as was done in the Linnean system, or in encyclopaedias, so that each may be readily found when required, such knowledge as this would not deserve the name of science, nor satisfy either the scientific wants of the human mind, or the desire for progressive mastery over the powers of nature. For the former requires an intellectual grasp of the connection of ideas, the latter demands our anticipation of a result in cases yet untried, and under conditions that we propose to introduce in the course of our experiment. Both are obviously arrived at by a knowledge of the *law* of the phenomena.

Isolated facts and experiments have in themselves no value, however great their number may be. They only become valuable in a theoretical or practical point of view when they make us acquainted with the *law* of a series of uniformly recurring phenomena, or, it may be, only give a negative result showing an incompleteness in our knowledge of such a law, till then held to be perfect. From the exact and universal conformity to law of natural phenomena, a single observation of a condition that we may presume to be rigorously conformable to law, suffices, it is true, at times to establish a rule with the highest degree of probability; just as, for example, we assume our knowledge of the skeleton of a prehistoric animal to be complete if we find only one complete skeleton of a single individual. But we must not lose sight of the fact that the isolated observation is not of value in that it is isolated, but because it is an aid to the knowledge of the conformable regularity in bodily structure of an entire species of organisms. In like manner, the knowledge of the specific heat of one small fragment of a new metal is important because we have no grounds for doubting that any other pieces of the same metal subjected to the same treatment will yield the same result.

To find the *law* by which they are regulated is to *understand* phenomena. For law is nothing more than the general conception in which a series of similarly recurring natural processes may be embraced. Just as we include in the conception 'mammal' all that is common to the man, the ape, the dog, the lion, the hare, the horse, the whale, &c., so we comprehend in the law of refraction that which we observe to regularly recur when a ray of light of any colour passes in any direction through the common boundary of any two transparent media.

A law of nature, however, is not a mere logical conception that we have adopted as a kind of *memoria technica* to enable us to more readily remember facts. We of the present day have already sufficient insight to know that the laws of nature are not things which we can evolve by any speculative method. On the contrary, we have to *discover* them in the facts; we have to test them by repeated observation or experiment, in constantly new cases, under ever-varying circumstances; and in proportion only as they hold good under a constantly increasing change of conditions, in a constantly increasing number of cases and with greater delicacy in the means of observation, does our confidence in their trustworthiness rise.

Thus the laws of nature occupy the position of a power with which we are not familiar, not to be arbitrarily selected and determined in our minds, as one might devise various systems of animals and plants one after another, so long as the object is only one of classification. Before we can say that our knowledge of any one law of nature is complete, we must see that it *holds good without exception*, and make this the test of its correctness. If we can be assured that the conditions under which the law operates have presented themselves, the result must ensue without arbitrariness, without choice, without our co-operation, and from the very necessity which regulates the things of the external world as well as our perception. The law then takes the form of an objective power, and for that reason we call it *force*.

For instance, we regard the law of refraction objectively as a refractive force in transparent substances; the law of chemical affinity as the elective force exhibited by different bodies towards one another. In the same way, we speak of electrical force of contact of metals, of a force of adhesion, capillary force, and so on. Under these names are stated objectively laws which for the most part comprise small series of natural processes, the conditions of which are somewhat involved. In science our conceptions begin in this way, proceeding to generalizations from a number of well-established special laws. We must endeavour to eliminate the incidents of form and distribution in space which masses under investigation may present by trying to find from the phenomena attending large visible masses laws for the operation of infinitely small particles; or, expressed objectively, by resolving the forces of composite masses into the forces of their smallest elementary particles. But precisely in this, the simplest form of expression of force—namely, of mechanical force acting on a point of the mass—is it especially clear that force is only the law of action objectively expressed. The force arising from the presence of such and such bodies is equivalent to the acceleration of the mass on which it operates multiplied by this mass. The actual meaning of such an equation is that it expresses the following law: if such and such masses are present and no other, such and such acceleration of their individual points occurs. Its actual signification may be compared with the facts and tested by them. The abstract conception of force we thus introduce implies moreover, that we did not discover this law at random, that it is an essential law of phenomena.

Our desire to *comprehend* natural phenomena, in other words, to ascertain their *laws*, thus takes another form of expression—that is, we have to seek out the *forces* which are the *causes* of the phenomena. The conformity to law in nature must be conceived as a causal connection the moment we recognise that it is independent of our thought and will.

If then we direct our inquiry to the progress of physical science as a whole, we shall have to judge of it by the measure in which the recognition and knowledge of a causative connection embracing all natural phenomena has advanced.

On looking back over the history of our sciences, the first great example we find of the subjugation of a wide mass of facts to a comprehensive law, occurred in the case of theoretical mechanics, the fundamental conception of which was first clearly propounded by Galileo. The question then was to find the general propositions that to us now appear so self-evident, that all substance is inert, and that the magnitude of force is to be measured not by its velocity, but by changes in it. At first the operation of a continually acting force could only be represented as a series of small impacts. It was not till Leibnitz and Newton, by the discovery of the differential calculus, had dispelled the ancient darkness which enveloped the conception of the infinite, and had clearly established the conception of the Continuous and of continuous change, that a full and productive application of the newly-found mechanical conceptions made any progress. The most singular and most splendid instance of such an application was in regard to the motion of the planets, and I need scarcely remind you here how brilliant an example astronomy has been for the development of the other branches of science. In its case, by the theory of gravitation, a vast and complex mass of facts were first embraced in a single principle of great simplicity, and such a reconciliation of theory and fact established as has never been accomplished in any other department of science, either before or since. In supplying the wants of astronomy, have originated almost all the exact methods of measurement as well as the principal advances made in modern mathematics; the science itself was peculiarly fitted to attract the attention of the general public, partly by the grandeur of the objects under investigation, partly by its practical utility in navigation and geodesy, and the many industrial and social interests arising from them.

Galileo began with the study of terrestrial gravity. Newton extended the application, at first cautiously and hesitatingly, to the moon, then boldly to all the planets. And, in more recent times, we learn that these laws of the common inertia and gravitation of all ponderable masses hold good of the movements of the most distant double stars of which the light has yet reached us.

During the latter half of the last and the first half of the present century came the great progress of chemistry which conclusively solved the ancient problem of discovering the elementary substances, a task to which so much metaphysical speculation had been devoted. Reality has always far exceeded even the boldest and wildest speculation, and, in the place of the four primitive metaphysical elements—fire, water, air, and earth—we have now the sixty-five simple bodies of modern chemistry. Science has shown that these elements are really indestructible, unalterable in their mass, unalterable also in their properties; in short, that from every condition into which they may have been converted, they can invariably be isolated, and recover those qualities which they previously possessed in the free state. Through all the varied phases of the phenomena of animated and inanimate nature, so far as we are acquainted with them, in all the astonishing results of chemical decomposition and combination, the number and diversity of which the chemist with unwearied diligence augments from year to year, the one law of the *immutability of matter* prevails as a necessity that knows no exception. And chemistry has already pressed on into the depths of immeasurable space, and detected in the most distant suns or nebulae indications of well-known terrestrial elements, so that doubts respecting the prevailing homogeneity of the matter of the universe no longer exist, though certain elements may perhaps be restricted to certain groups of the heavenly bodies.

From this invariability of the elements follows another and wider consequence. Chemistry shows by actual experiment that all matter is made up of the elements which have been already isolated. These elements may exhibit great differences as regards combination or mixture, the mode of aggregation or molecular structure—that is to say, they may vary the mode of their *distribution in space*. In their *properties*, on the other hand, they are altogether unchangeable; in other words, when referred to the same compound, as regards isolation, and to the same state of aggregation, they invariably exhibit the same properties as before. If, then, all elementary substances are unchangeable in respect to their properties, and only changeable as regards their combination and their states of aggrega-tion—that is, in respect to their distribution in space—it follows that all changes in the world are changes in the local distribution of elementary matter, and are eventually brought about through *Motion*.

If, however, motion be the primordial change which lies at the root of all the other changes occurring in the world, every elementary force is a force of motion, and the ultimate aim of physical science must be to determine the movements which are the real causes of all other phenomena and discover the motive powers on which they depend; in other words, to merge itself into mechanics.

Though this is clearly the final consequence of the qualitative and quantitative immutability of matter, it is after all an ideal proposition, the realization of which is still very remote. The field is a prescribed one, in which we have succeeded in tracing back actually observed changes to motions and forces of motion of a definite kind. Besides astronomy, may be mentioned the purely mechanical part of physics, then acoustics, optics, and electricity; in the science of heat and in chemistry, strenuous endeavours are being made towards perfecting definite views respecting the nature of the motion and position of molecules, while physiology has scarcely made a definite step in this direction.

This renders all the more important, therefore, a noteworthy advance-ment of the most general importance made during the last quarter of a century in the direction we are considering. If all elementary forces are forces of motion, and all, consequently, of similar nature, they should all be measurable by the same standard, that is, the standard of the mechanical forces. And that this is actually the fact is now regarded as proved. The law expressing this is known under the name of *the law of the Conservation of Force.* For a small group of natural phenomena it had already been pronounced by Newton, then more definitely and in more general terms by D. Bernoulli, and so continued of recognised application in the greater part of the then known purely mechanical processes. Certain amplifications at times attracted attention, like those of Rumford, Davy, and Montgolfier. The first, however, to compass the clear and distinct idea of this law, and to venture to pronounce its absolute universality, was one whom we shall have soon the pleasure of hearing from this platform, Dr. Robert Mayer, of Heilbronn. While Dr. Mayer was led by physiological questions to the discovery of the most general form of this law, technical questions in mechanical engineering led Mr. Joule, of Manchester, simultaneously, and independently of him, to the same considerations; and it is to Mr. Joule that we are indebted for those important and laborious experimental researches in that department where the applicability of the law of the conservation of force appeared most doubtful, and where the greatest gaps in actual knowledge occurred, namely, in the production of work from heat, and of heat from work.

To state the law clearly it was necessary, in contradistinction to Galileo's conception of the *intensity of force*, that a new mechanical idea was elaborated, which we may term the conception of the *quantity of force*, and which has also been called *quality of work* or of *energy*.

A way to this conception of the quantity of force had been prepared partly, in theoretical mechanics, through the conception of the *amount of vis viva* of a moving body, and partly by practical mechanics through the conception of the *motive power* necessary to keep a machine at work. Practical machinists had already found a standard by which any motive power could be measured, in the determination of the number of pounds that it could lift one foot in a second; and, as is known, a horse-power was defined to be equivalent to the motive power required to lift seventy kilogrammes one metre in each second.

Machines, and the motive powers required for their movement, furnish, in fact, the most familiar illustrations of the uniformity of all natural forces expressed by the law of the conservation of force. Any machine which is to be set in motion requires a mechanical motive power. Whence this power is derived or what its form, is of no consequence, provided only it be sufficiently great and act continuously. At one time we employ a steam-engine, at another a water-wheel or turbine, here horses or oxen at a whim, there a windmill, or if but little power is required, the human arm, a raised weight, or an electromagnetic engine. The choice of the machine is merely dependent on the amount of power we would use, or the force of circumstance. In the watermill the weight of the water flowing down the hills is the agent; it is lifted to the hills by a meteorological process, and becomes the source of motive power for the mill. In the windmill it is the *vis viva* of the moving air which drives round the sails; this motion also is due to a meteorological operation of the atmosphere. In the steam-engine we have the tension of the heated vapour which drives the piston to and fro; this is engendered by the heat arising from the combustion of the coal in the fire-box, in other words, by a chemical process; and in this case the latter action is the source of the motive power. If it be a horse or the human arm which is at work, we have the muscles stimulated through the nerves, directly producing the mechanical force. In order, however, that the living body may generate muscular power it must be nourished and breathe. The food it takes separates again from it, after having combined with the oxygen inhaled from the air, to form carbonic acid and water. Here again, then, a chemical process is an essential element to maintain muscular power. A similar state of things is observed in the electro-magnetic machines of our telegraphs.

Thus, then, we obtain mechanical motive force from the most varied processes of nature in the most different ways; but it will also be remarked in only a limited quantity. In doing so we always *consume* something that nature supplies to us. In the watermill we use a quantity of water collected at an elevation, coal in the steam-engine, zinc and sulphuric acid in the electro-magnetic machine, food for the horse; in the windmill we use up the motion of the wind, which is arrested by the sails.

Conversely, if we have a motive force at our disposal we can develop with it forms of action of the most varied kind. It will not be necessary in this place to enumerate the countless diversity of industrial machines, and the varieties of work which they perform.

Let us rather consider the physical differences of the possible performance of a motive power. With its help we can raise loads, pump water to an elevation, compress gases, set a railway train in motion, and through friction generate heat. By its aid we can turn magneto-electric machines, and produce electric currents, and with them decompose water and other chemical compounds having the most powerful affinities, render wires incandescent, magnetise iron, &c.

Moreover, had we at our disposal a sufficient mechanical motive force we could restore all those states and conditions from which, as was seen above, we are enabled at the outset to derive mechanical motive power.

As, however, the motive power derived from any given natural process is limited, so likewise is there a limitation to the total amount of modifications which we may produce by the use of any given motive power.

These deductions, arrived at first in isolated instances from machines and physical apparatus, have now been welded into a law of nature of the widest validity. Every change in nature is equivalent to a certain development, or a certain consumption of motive force. If motive power be developed it may either appear as such, or be directly used up again to form other changes equivalent in magnitude. The leading determinations of this equivalency are founded on Joule's measurements of the mechanical equivalent of heat. When, by the application of heat, we set a steam-engine in motion, heat proportional to the work done disappears within it; in short, the heat which can warm a given weight of water one degree of the Centigrade scale is able, if converted into work, to lift the same weight of water to a height of 425 metres. If we convert work into heat by friction we again use, in heating a given weight of water one degree Centigrade, the motive force which the same quantity of water would have generated in flowing down from a height of 425 metres. Chemical processes generate heat in definite proportion, and in like manner we estimate the motive power equivalent to such chemical forces; and thus the energy of the chemical force of affinity is also measurable by the mechanical standard. The same holds true for all the other forms of natural forces, but it will not be necessary to pursue the subject further here.

It has actually been established, then, as a result of these investigations, that all the forces of nature are measurable by the same mechanical standard, and that all pure motive forces are, as regards performance of work, equivalent. And thus one great step towards the solution of the comprehensive theoretical task of referring all natural phenomena to motion has been accomplished.

Whilst the foregoing considerations chiefly seek to elucidate the logical value of the law of the conservation of force, its actual signification in the general conception of the processes of nature is expressed in the grand connection which it establishes between the entire processes of the universe, through all distances of place or time. The universe appears, according to this law, to be endowed with a store of energy which, through all the varied changes in natural processes, can neither be increased nor diminished, which is maintained therein in ever-varying phases, but, like matter itself, is from eternity to eternity of unchanging magnitude; *acting in space*, but not *divisible*, as matter is, with it. Every change in the world simply consists in a variation in the mode of appearance of this store of energy. Here we find one portion of it as the *vis viva* of moving bodies, there as regular oscillation in light and sound; or, again, as heat, that is to say, the irregular motion of invisible particles; at another point the energy appears in the form of the weight of two masses gravitating towards each other, then as internal tension and pressure of elastic bodies, or as chemical attraction, electrical tension, or magnetic distribution. If it disappear in one form, it reappears as surely in another; and whenever it presents itself in a new phase we are certain that it does so at the expense of one of its other forms.

Carnot's law of the mechanical theory of heat, as modified by Clausius, has, in fact, made it clear that this change moves in the main continuously onward in a definite direction, so that a constantly increasing amount of the great store of energy in the universe is being transformed into heat. We can, therefore, see with the mind's eye the original condition of things in which the matter composing the celestial bodies was still cold, and probably distributed as chaotic vapour or dust through space; we see that it must have developed heat when it collected together under the influence of gravity. Even at the present time spectrum analysis (a method the theoretical principles of which owe their origin to the mechanical theory of heat) enables us to detect remains of this loosely distributed matter in the nebulae; we recognise it in the meteor-showers and comets; the act of agglomeration and the development of heat still continue, though in our portion of the stellar system they have ceased to a great extent. The chief part of the primordial energy of the matter belonging to our system is now in the form of solar heat. This energy, however, will not remain locked up in our system for ever: portions of it are continually radiating from it, in the form of light and heat, into infinite space. Of this radiation our earth receives a share. It is these solar heat-rays which produce on the earth's surface the winds and the currents of the ocean, and lift the watery vapour from the tropical seas, which, distilling over hill and plain, returns as springs and rivers to the sea. The solar rays impart to the plant the power to separate from carbonic acid and water those combustible substances which serve as food for animals, and thus, in even the varied changes of organic life, the moving power is derived from the infinitely vast store of the universe.

This exalted picture of the connection existing between all the processes of nature has been often presented to us in recent times; it will suffice here that I direct attention to its leading features. If the task of physical science be to determine laws, a step of the most comprehensive significance towards that object has here been taken.

The application of the law of the conservation of force to the vital processes of animals and plants, which has just been discussed, leads us in another direction in which our knowledge of nature's conformity to law has made an advance. The law to which we referred is of the most essential importance in leading questions of physiology, and it was for this reason that Dr. Mayer and I were led on physiological grounds to investigations having especial reference to the conservation of force.

As regards the phenomena of inorganic nature all doubts have long since been laid to rest respecting the principles of the method. It was apparent that these phenomena had fixed laws, and examples enough were already known to make the finding of such laws probable.

In consequence, however, of the greater complexity of the vital processes, their connection with mental action, and the unmistakable evidence of adaptability to a purpose which organic structures exhibit, the existence of a settled conformity to law might well appear doubtful, and, in fact, physiology has always had to encounter this fundamental question: are all vital processes absolutely conformable to law? Or is there, perhaps, a range of greater or less magnitude within which an exception prevails? More or less obscured by words, the view of Paracelsus, Helmont, and Stahl, has been, and is at present, held, particularly outside Germany, that there exists a soul of life (*"Lebensseele"*) directing the organic processes which is endowed more or less with consciousness like the soul of man. The influence of the inorganic forces of nature on the organism was still recognised on the assumption that the soul of life only exercises power over matter by means of the physical and chemical forces of matter itself; so that without this aid it could accomplish nothing, but that it possessed the faculty of suspending or permitting the operation of the forces at pleasure.

After death, when no longer subject to the control of the soul of life or vital force, it was these very chemical forces of organic matter which brought about decomposition. In short, through all the different modes of expressing it, whether it was termed the Archäus, the *anima inscia*, or the *vital force* and the *restorative power of nature*, the faculty to build up the body according to system, and to suitably accommodate it to external circumstances, remained the most essential attribute of this hypothetically controlling principle of the vitalistic theory with which, therefore, by reason of its attributes, only the name of soul fully harmonised.

It is apparent, however, that this notion runs directly counter to the law of the conservation of force. If vital force were for a time to annul the gravity of a weight, it could be raised without labour to any desired height, and subsequently, if the action of gravity were again restored, could perform work of any desired magnitude. And thus work could be obtained out of nothing without expense. If vital force could for a time suspend the chemical affinity of carbon for oxygen, carbonic acid could be decomposed without work being employed for that purpose, and the liberated carbon and oxygen could perform new work.

In reality, however, no trace of such an action is to be met with as that of the living organism being able to generate an amount of work without an equivalent expenditure. When we consider the work done by animals, we find the operation comparable in every respect with that of the steam-engine. Animals, like machines, can only move and accomplish work by being continuously supplied with fuel (that is to say, food) and air containing oxygen; both give off again this material in a burnt state, and at the same time produce heat and work. All investigation, thus far, respecting the amount of heat which an animal produces when at rest is in no way at variance with the assumption that this heat exactly corresponds to the equivalent, expressed as work, of the forces of chemical affinity then in action.

As regards the work done by plants, a source of power in every way sufficient, exists in the solar rays which they require for the increase of the organic matter of their structures. Meanwhile it is true that exact quantitative determinations of the equivalents of force, consumed and produced in the vegetable as well as the animal kingdom, have still to be made in order to fully establish the exact accordance of these two values.

If, then, the law of the conservation of force hold good also for the living body, it follows that the physical and chemical forces of the material employed in building up the body are in continuous action without intermission and without choice, and that *their exact conformity to law never suffers a moment's interruption.*

Physiologists, then, must expect to meet with an unconditional conformity to law of the forces of nature in their inquiries respecting the vital processes; they will have to apply themselves to the investigation of the physical and chemical processes going on within the organism. It is a task of vast complexity and extent; but the workers, in Germany especially, are both numerous and enthusiastic, and we may already affirm that their labours have not been unrewarded, inasmuch as our knowledge of the vital phenomena has made greater progress during the last forty years than in the two preceding centuries.

Assistance, that cannot be too highly valued, towards the elucidation of the fundamental principles of the doctrine of life, has been rendered on the part of descriptive natural history, through Darwin's theory of the evolution of organic forms, by which the possibility of an entirely new interpretation of organic adaptability is furnished.

The adaptability in the construction of the functions of the living body, most wonderful at any time, and with the progress of science becoming still more so, was doubtless the chief reason that provoked a comparison of the vital processes with the actions of a principle acting like a soul. In the whole external world we know of but one series of phenomena possessing similar characteristics, we mean the actions and deeds of an intelligent human being; and we must allow that in innumerable instances the organic adaptability appears to be so extraordinarily superior to the capacities of the human intelligence that we might feel disposed to ascribe to it a higher rather than a lower character.

Before the time of Darwin only two theories respecting organic adaptability were in vogue, both of which pointed to the interference of free intelligence in the course of natural processes. On the one hand it was held, in accordance with the vitalistic theory, that the vital processes were continuously directed by a living soul; or, on the other, recourse was had to an act of supernatural intelligence to account for the origin of every living species. The latter view indeed supposes that the causal connection of natural phenomena had been broken less often, and allows of a strict scientific examination of the processes observable in the species of human beings now existing; but even it is not able to entirely explain away those exceptions to the law of causality, and consequently it enjoyed no considerable favour as opposed to the vitalistic view, which was powerfully supported, by apparent evidence, that is, by the natural desire to find similar causes behind similar phenomena.

Darwin's theory contains an essentially new creative thought. It shows how adaptability of structure in organisms can result from a blind rule of a law of nature without any intervention of intelligence. I allude to the law of transmission of individual peculiarities from parent to offspring, a law long known and recognised, and only needing a more precise definition. If both parents have individual peculiarities in common, the majority of their offspring also possess them; and if among the offspring there are some which present these peculiarities in a less marked degree, there will, on the other hand, always be found among a great number, others in which the same peculiarities have become intensified. If, now, these be selected to propagate offspring, a greater and greater intensification of these peculiarities may be attained and transmitted. This is, in fact, the method employed in cattlebreeding and gardening, in order with greater certainty to obtain new breeds and varieties, with well-marked different characters. The experience of artificial breeding is to be regarded, from a scientific point of view, as an experimental confirmation of the law under discussion; and, in fact, this experiment has proved successful, and is still doing so, with species of every class of the animal kingdom, and, with respect to the most different organs of the body, in a vast number of instances.

After the general application of the law of transmission had been established in this way, it only remained for Darwin to discuss the bearings of the question as regards animals and plants in the wild state. The result which has been arrived at is that those individuals which are distinguished in the struggle for existence by some advantageous quality, are the most likely to produce offspring, and thus transmit to them their advantageous qualities. And in this way from generation to generation a gradual adjustment is arrived at in the adaptation of each species of living creation to the conditions under which it has to live until the type has reached such a degree of perfection that any substantial variation from it is a disadvantage. It will then remain unchanged so long as the external conditions of its existence remain materially unaltered. Such an almost absolutely fixed condition appears to be attained by the plants and animals now living, and thus the continuity of the species, at least during historic times, is found to prevail.

An animated controversy, however, still continues, concerning the truth or probability of the Darwinian theory, for the most part respecting the limits that should be assigned to the variation of species. The opponents of this view would hardly deny that, as assumed by Darwin, hereditary differences of race could have arisen in one and the same species; or, in other words, that many of the forms hitherto regarded as distinct species of the same genus have been derived from the same primitive form. Whether we must restrict our view to this, or whether, perhaps, we venture to derive all mammals from one original marsupial, or, again, all vertebrates from a primitive lancelet, or all plants and animals together from the slimy protoplasm of a protiston, depends at the present moment rather on the leanings of individual observers than on facts. Fresh links, connecting classes of apparently irreconcilable type, are always presenting themselves; the actual transition of forms, into others widely different, has already been traced in regularly deposited geological strata, and has come to be beyond question; and since this line of research has been taken up, how numerous are the facts which fully accord with Darwin's theory, and give special effect to it in detail!

At the same time, we should not forget the clear interpretation Darwin's grand conception has supplied of the till then mysterious notions respecting natural affinity, natural systems, and homology of organs in various animals; how by its aid the remarkable recurrence of the structural peculiarities of lower animals in the embryos of others higher in the scale, the special kind of development appearing in the series of palaeontological forms, and the peculiar conditions of affinity of the faunas and floras of limited areas have, one and all, received elucidation. Formerly natural affinity appeared to be a mere enigmatical, and altogether groundless similarity of forms; now it has become a matter for actual consanguinity. The natural system certainly forced itself as such upon the mind, although theory strictly disavowed any real significance to it; at present it denotes an actual genealogy of organisms. The facts of palaeontological and embryological evolution and of geographical distribution were enigmatical wonders so long as each species was regarded as the result of an independent act of creation, and cast a scarcely favourable light on the strange tentative method which was ascribed to the Creator. Darwin has raised all these isolated questions from the condition of a heap of enigmatical wonders to a great consistent system of development, and established definite ideas in the place of such a fanciful hypothesis as, among the first, had occurred to Goethe, respecting the facts of the comparative anatomy and the morphology of plants.

This renders possible a definite statement of problems for further inquiry, a great gain in any case, even should it happen that Darwin's theory does not embrace the whole truth, and that, in addition to the influences which he has indicated, there should be found to be others which operate in the modification of organic forms.

While the Darwinian theory treats exclusively of the gradual modification of species after a succession of generations, we know that a single individual may adapt itself, or become accustomed, in a certain degree, to the circumstances under which it has to live; and that even during the single life of an individual a distinct progress towards a higher development of organic adaptability may be attained. And it is more especially in those forms of organic life where the adaptability in structure has reached the highest grade and excited the greatest admiration, namely, in the region of mental perception, that, as the latest results of physiology teach us, this individual adaptation plays a most prominent part.

Who has not marvelled at the fidelity and accuracy of the information which our senses convey to us from the surrounding world, more especially those of the far-reaching eye? The information so gained furnishes the premises for the conclusions which we come to, the acts that we perform; and unless our senses convey to us correct impressions, we cannot expect to act accurately, so that results shall correspond with our expectations. By the success or failure of our acts we again and again test the truth of the information with which our senses supply us, and experience, after millions of repetitions, shows us that this fidelity is exceedingly great, in fact, almost free from exceptions. At all events, these exceptions, the so-called illusions of the senses, are rare, and are only brought about by very special and unusual circumstances.

Whenever we stretch forth the hand to lay hold of something, or advance the foot to step upon some object, we must first form an accurate optical image of the position of the object to be touched, its form, distance, &c., or we shall fail. The certainty and accuracy of our perception by the senses must at least equal the certainty and accuracy which our actions have attained after long practice; and the belief, therefore, in the trustworthiness of our senses is no blind belief, but one, the accuracy of which has been tested and verified again and again by numberless experiments.

Were this harmony between the perceptions through the senses and the objects causing them, in other words, this basis of all our knowledge, a direct product of the vital principle, its formative power would, in fact, then have attained the highest degree of perfection. But an examination of the actual facts at once destroys in the most merciless manner all belief in a preordained harmony of the inner and external world.

I need not call to mind the startling and unexpected results of ophthalmometrical and optical research which have proved the eye to be a by no means more perfect optical instrument than those constructed by human hands; but, on the contrary, to exhibit, in addition to the faults inseparable from any dioptric instrument, others that in an artificial instrument we should severely condemn; nor need I remind you that the ear conveys to us sounds from without in no wise in the ratio of their actual intensity, but strangely resolves them and modifies them, intensifying or weakening them in very different degrees, according to their varieties of pitch.

These anomalies, however, are as nothing compared with those to be met with in examining the nature of the sensations by which we become acquainted with the various properties of the objects surrounding us. Here it can at once be proved that no kind and no degree of similarity exists between the quality of a sensation and the quality of the agent inducing it, and portrayed by it.

In its leading features this was demonstrated by Johannes Müller in his law of the Specific Action of the Senses. According to him, each nerve of sense possesses a peculiar kind of sensation. A nerve, we know, can be rendered active by a vast number of exciting agents, and the same agent may likewise affect different organs of sense; but however it be brought about, we never have in nerves of sight any other sensation than that of light; in the nerves of the ear any other than a sensation of sound; in short, in each individual nerve of sense only that sensation which corresponds to its peculiar specific action. The most marked differences in the qualities of sensation, in other words, those between the sensations of different senses, are, then, in no way dependent on the nature of the exciting agent, but only on that of the nerve apparatus under operation.

The bearing of Müller's law has been extended by later research. It appears highly probable that even the sensations of different colours and different pitch, as well as qualitative peculiarities of luminous sensations *inter se*, and of sonorous sensations *inter se*, also depend on the excitation of systems of fibres, with distinct character and endowed with different specific energy, of nerves of sight and hearing respectively. The infinitely more varied diversity of composite light is in this way referable to sensations of only threefold heterogeneous character, in other words, to mixtures of the three primary colours. From this reduction in the number of possible differences it follows that very different composite light may appear the same. In this case it has been shown that no kind of physical similarity whatever corresponds to the subjective similarity of different composite light of the same colour. By these and similar facts we are led to the very important conclusion that our sensations are, as regards their quality, only *signs* of external objects, and in no sense *images* of any degree of resemblance. An image must, in certain respects, be *analogous* to the original object; a statue, for instance, has the same corporeal form as the human being after which it is made; a picture the same colour and perspective projection. For a *sign* it is sufficient that it become apparent as often as the occurrence to be depicted makes its appearance, the conformity between them being restricted to their presenting themselves simultaneously; and the correspondence existing between our sensations and the objects producing them is precisely of this kind. They are signs which we have *learned to decipher*, and a language given us with our organisation by which external objects discourse to us—a language, however, like our mother tongue, that we can only learn by practice and experience.

Moreover, what has been said holds good not only for the qualitative differences of sensations, but also, in any case, for the greatest and most important part, if not the whole, of our various perceptions of extension in space. In their bearings on this question the new doctrine of binocular vision and the invention of the stereoscope have been of importance. All that the sensation of the two eyes could convey to us directly, and without psychical aid was, at the most, two somewhat different flat pictures of two dimensions as they lay on the two retinae; instead of this we perceive a representation with three dimensions of the things around us. We are sensible as well of the *distance* of objects not too far removed from us as of their perspective *juxtaposition*, and compare the actual magnitude of two objects of apparently unequal size at different distances from us with greater certainty than the apparent equal magnitudes of a finger, say, and the moon.

One explanation only of our perception of extension in space, which stands the test of each separate fact, can in my judgment be brought forward by our assuming with Lotze that to the sensations of nerve-fibres, differently situated in space, certain differences, *local signs*, attach themselves, the significations of which, as regards space, we have to learn. That a knowledge of their signification may be attained by these hypotheses, and with the help of the movements of our body, and that we can at the same time learn which are the right movements to bring about a desired result, and become conscious of having arrived at it, has in many ways been established.

That experience exercised an enormous influence over the signification of visual pictures, and, in cases of doubt, is generally the final arbiter, is allowed even by those physiologists who wish to save as much as possible of the innate harmony of the senses with the external world. The controversy is at present almost entirely confined to the question of the proportion at birth of the innate impulses that can facilitate training in the understanding of sensations. The assumption of the existence of impulses of this kind is unnecessary, and renders difficult instead of elucidating an interpretation of well-observed phenomena in adults.[1]

It follows, then, that this subtile and most admirable harmony existing

---

1. A further exposition of these conditions will be found in the lectures on the Recent Progress of the Theory of Vision, pp. 127 *et seq.*

between our sensations and the objects causing them is substantially, and with but few doubtful exceptions, a conformity individually acquired, a result of experience, of training, the recollection of former acts of a similar kind.

This completes the circle of our observations, and lands us at the spot whence we set out. We found at the beginning, that what physical science strives after is the knowledge of laws, in other words, the knowledge how at different times under the same conditions the same results are brought about; and we found in the last instance how all laws can be reduced to laws of motion. We now find, in conclusion, that our sensations are merely signs of changes taking place in the external world, and can only be regarded as pictures in that they represent succession in time. For this very reason they are in a position to show directly the *conformity to law*, in regard to succession in time, of natural phenomena. If, under the same natural circumstances, the same action take place, a person observing it under the same conditions will find the same series of impressions regularly recur. That which our organs of sense perform is clearly sufficient to meet the demands of science as well as the practical ends of the man of business who must rely for support on the knowledge of natural laws, acquired, partly involuntarily by daily experience, and partly purposely by the study of science.

Having now completed our survey, we may, perhaps, strike a not unsatisfactory balance. Physical science has made active progress, not only in this or that direction, but as a vast whole, and what has been accomplished may warrant the attainment of further progress. Doubts respecting the entire conformity to law of nature are more and more dispelled; laws more general and more comprehensive have revealed themselves. That the direction which scientific study has taken is a healthy one its great practical issues have clearly demonstrated; and I may here be permitted to direct particular attention to the branch of science more especially my own. In physiology particularly scientific work had been crippled by doubts respecting the necessary conformity to law, which means, as we have shown, the intelligibility of vital phenomena, and this naturally extended itself to the practical science directly dependent on physiology, namely, medicine. Both have received an impetus, such as had not been felt for thousands of years, from the time that they seriously adopted the method of physical science, the exact observation of phenomena and experiment. As a practicing physician, in my earlier days, I can personally bear testimony to this. I was educated at a period when medicine was in a transitional stage, when the minds of the most thoughtful and exact were filled with despair. It was not difficult to recognise that the old predominant theorising methods of practicing medicine were altogether untenable; with these theories, however, the facts on which they had actually been founded had become so inextricably entangled that they also were mostly thrown overboard. How a science should be built up anew had already been seen in the case of the other sciences; but the new task assumed colossal proportions; few steps had been taken towards accomplishing it, and these first efforts were in some measure but crude and clumsy. We need feel no astonishment that many sincere and earnest men should at that time have abandoned medicine as unsatisfactory, or on principle given themselves over to an exaggerated empiricism.

But well directed efforts produced the right result more quickly even than many had hoped for. The application of the mechanical ideas to the doctrine of circulation and respiration, the better interpretation of thermal phenomena, the more refined physiological study of the nerves, soon led to practical results of the greatest importance; microscopic examination of parasitic structures, the stupendous development of pathological anatomy, irresistibly led from nebulous theories to reality. We found that we now possessed a much clearer means of distinguishing, and a clearer insight into the mechanism of the process of disease than the beats of the pulse, the urinary deposit, or the fever type of older medical science had ever given us. If I might name one department of medicine in which the influence of the scientific method has been, perhaps, most brilliantly displayed, it would be in ophthalmic medicine. The peculiar constitution of the eye enables us to apply physical modes of investigation as well in functional as in anatomical derangements of the living organ. Simple physical expedients, spectacles, sometimes spherical, sometimes cylindrical or prismatic, suffice, in many cases, to cure disorders which in earlier times left the organ in a condition of chronic incapacity; a great number of changes on the other hand, which formerly did not attract notice till they induced incurable blindness, can now be detected and remedied at the outset. From the very reason of its presenting the most favourable ground for the application of the scientific method, ophthalmology has proved attractive to a peculiarly large number of excellent investigators, and rapidly attained its present position, in which it sets an example to the other departments of medicine, of the actual capabilities of the true method, as brilliant as that which astronomy for long had offered to the other branches of physical science.

Though in the investigation of inorganic nature the several European nations showed a nearly uniform advancement, the recent progress of physiology and medicine is pre-eminently due to Germany. I have already spoken of the obstacles which formerly delayed progress in this direction. Questions respecting the nature of life are closely bound up with psychological and ethical inquiries. It demands, moreover, that we bestow on it unwearied diligence for purely ideal purposes, without any approaching prospect of the pure science becoming of practical value. And we may make it our boast that this exalted and self-denying assiduity, this labour for inward satisfaction, not for external success, has at all times peculiarly distinguished the scientific men of Germany.

What has, after all, determined the state of things in the present instance is in my opinion another circumstance, namely, that we are more fearless than others of the consequences of the entire and perfect truth. Both in England and France we find excellent investigators who are capable of working with thorough energy in the proper sense of the scientific methods; hitherto, however, they have almost always had to bend to social or ecclesiastical prejudices, and could only openly express their convictions at the expense of their social influence and their usefulness.

Germany has advanced with bolder step: she has had the full confidence, which has never been shaken, that truth fully known brings with it its own remedy for the danger and disadvantage that may here and there attend a limited recognition of what is true. A labour-loving, frugal, and moral people may exercise such boldness, may stand face to face with truth; it has nothing to fear though hasty or partial theories be advocated, even if they should appear to trench upon the foundations of morality and society.

We have met here on the southern frontier of our country. In science, however, we recognise no political boundaries, for our country reaches as far as the German tongue is heard, wherever German industry and German intrepidity in striving after truth find favour. And that it finds favour here is shown by our hospitable reception, and the inspiriting words with which we have been greeted. A new medical faculty has been established here. We will wish it in its career rapid progress in the cardinal virtues of German science, for then it will not only find remedies for bodily suffering, but become an active centre to strengthen intellectual independence, steadfastness to conviction and love of truth, and at the same time be the means of deepening the sense of unity throughout our country.

An Opening Address Delivered at the Naturforscher Versammlung, in Innsbrück, 1869.

8

On the Origin and Significance of Geometrical Axioms

THE fact that a science can exist and can be developed as has been the case with geometry, has always attracted the closest attention among those who are interested in questions relating to the bases of the theory of cognition. Of all branches of human knowledge, there is none which, like it, has sprung as a completely armed Minerva from the head of Jupiter; none before whose death-dealing Aegis doubt and inconsistency have so little dared to raise their eyes. It escapes the tedious and troublesome task of collecting experimental facts, which is the province of the natural sciences in the strict sense of the word; the sole form of its scientific method is deduction. Conclusion is deduced from conclusion, and yet no one of common sense doubts but that these geometrical principles must find their practical application in the real world about us. Land surveying, as well as architecture, the construction of machinery no less than mathematical physics, are continually calculating relations of space of the most varied kind by geometrical principles; they expect that the success of their constructions and experiments shall agree with these calculations; and no case is known in which this expectation has been falsified, provided the calculations were made correctly and with sufficient data.

Indeed, the fact that geometry exists, and is capable of all this, has always been used as a prominent example in the discussion on that question, which forms, as it were, the centre of all antitheses of philosophical systems, that there can be a cognition of principles destitute of any bases drawn from experience. In the answer to Kant's celebrated question, 'How are synthetical principles *a priori*, possible?' geometrical axioms are certainly those examples which appear to show most decisively that synthetical principles are *a priori* possible at all. The circumstance that such principles exist, and force themselves on our conviction, is regarded as a proof that space is an *a priori* mode of all external perception. It appears thereby to postulate, for this *a priori* form, not only the character of a purely formal scheme of itself quite unsubstantial, in which any given result experience would fit; but also to include certain peculiarities of the scheme, which bring it about that only a certain content, and one which, as it were, is strictly defined, could occupy it and be apprehended by us.[1]

It is precisely this relation of geometry to the theory of cognition which emboldens me to speak to you on geometrical subjects in an assembly of those who for the most part have limited their mathematical studies to the ordinary instruction in schools. Fortunately, the amount of geometry taught in our gymnasia will enable you to follow, at any rate the tendency, of the principles I am about to discuss.

I intend to give you an account of a series of recent and closely connected mathematical researches which are concerned with the geometrical axioms, their relations to experience, with the question whether it is logically possible to replace them by others.

Seeing that the researches in question are more immediately designed to furnish proofs for experts in a region which, more than almost any other, requires a higher power of abstraction, and that they are virtually inaccessible to the non-mathematician, I will endeavour to explain to such a one the question at issue. I need scarcely remark that my explanation will give no proof of the correctness of the new views. He who seeks this proof must take the trouble to study the original researches.

Anyone who has entered the gates of the first elementary axioms of geometry, that is, the mathematical doctrine of space, finds on his path that unbroken chain of conclusions of which I just spoke, by which the ever more varied and more complicated figures are brought within the domain of law. But even in their first elements certain principles are laid down, with respect to which geometry confesses that she cannot prove them, and can only assume that anyone who understands the essence of these principles will at once admit their correctness. These are the so-called axioms.

For example, the proposition that if the shortest line drawn between two points is called a *straight* line, there can be only one such straight line. Again, it is an axiom that through any three points in space, not lying in a straight line, a plane may be drawn, i.e. a surface which will wholly include every straight line joining any two of its points. Another axiom, about which there has been much discussion, affirms that through a point lying without a straight line only one straight line can be drawn parallel to the first; two straight lines that lie in the same plane and never meet, however

---

1. In his book, *On the Limits of Philosophy*, Mr. W. Tobias maintains that axioms of a kind which I formerly enunciated are a misunderstanding of Kant's opinion. But Kant specially adduces the axioms, that the straight line is the shortest *(Kritik der reinen Vernunft,* Introduction, v. 2nd ed. p. 16); that space has three dimensions *(Ibid.* part i. sect. i. § 3, p. 41); that only one straight line is possible between two points *(Ibid.* part ii. sect. i. 'On the Axioms of Intuition'), as axioms which express *a priori* the conditions of intuition by the senses. It is not here the question, whether these axioms were originally given as intuition of space, or whether they are only the starting-points from which the understanding can develop such axioms *a priori* on which my critic insists.

far they may be produced, being called parallel. There are also axioms that determine the number of dimensions of space and its surfaces, lines and points, showing how they are continuous; as in the propositions, that a solid is bounded by a surface, a surface by a line and a line by a point, that the point is indivisible, that by the movement of a point a line is described, by that of a line a line or a surface, by that of a surface a surface or a solid, but by the movement of a solid a solid and nothing else is described.

Now what is the origin of such propositions, unquestionably true yet incapable of proof in a science where everything else is reasoned conclusion? Are they inherited from the divine source of our reason as the idealistic philosophers think, or is it only that the ingenuity of mathematicians has hitherto not been penetrating enough to find the proof? Every new votary, coming with fresh zeal to geometry, naturally strives to succeed where all before him have failed. And it is quite right that each should make the trial afresh; for, as the question has hitherto stood, it is only by the fruitlessness of one's own efforts that one can be convinced of the impossibility of finding a proof. Meanwhile solitary inquirers are always from time to time appearing who become so deeply entangled in complicated trains of reasoning that they can no longer discover their mistakes and believe they have solved the problem. The axiom of parallels especially has called forth a great number of seeming demonstrations.

The main difficulty in these inquiries is, and always has been, the readiness with which results of everyday experience become mixed up as apparent necessities of thought with the logical processes, so long as Euclid's method of constructive intuition is exclusively followed in geometry. It is in particular extremely difficult, on this method, to be quite sure that in the steps prescribed for the demonstration we have not involuntarily and unconsciously drawn in some most general results of experience, which the power of executing certain parts of the operation has already taught us practically. In drawing any subsidiary line for the sake of his demonstration, the well-trained geometer always asks if it is possible to draw such a line. It is well known that problems of construction play an essential part in the system of geometry. At first sight, these appear to be practical operations, introduced for the training of learners; but in reality they establish the existence of definite figures. They show that points, straight lines, or circles such as the problem requires to be constructed are possible under all conditions, or they determine any exceptions that there may be. The point on which the investigations turn, that we are about to consider, is essentially of this nature. The foundation of all proof by Euclid's method consists in establishing the congruence of lines, angles, plane figures, solids, &c. To make the congruence evident, the geometrical figures are supposed to be applied to one another, of course without changing their form and dimensions. That this is in fact possible we have all experienced from our earliest youth. But, if we proceed to build necessities of thought upon this assumption of the free translation of fixed figures, with unchanged form, to every part of space, we must see whether the assumption does not involve some presupposition of which no logical proof is given. We shall see later on that it does indeed contain one of the most serious import. But if so, every proof by congruence rests upon a fact which is obtained from experience only.

I offer these remarks, at first only to show what difficulties attend the complete analysis of the presuppositions we make, in employing the common constructive method. We evade them when we apply, to the investigation of principles, the analytical method of modern algebraical geometry. The whole process of algebraical calculation is a purely logical operation; it can yield no relation between the quantities submitted to it that is not already contained in the equations which give occasion for its being applied. The recent investigations in question have accordingly been conducted almost exclusively by means of the purely abstract methods of analytical geometry.

However, after discovering by the abstract method what are the points in question, we shall best get a distinct view of them by taking a region of narrower limits than our own world of space. Let us, as we logically may, suppose reasoning beings of only two dimensions to live and move on the surface of some solid body. We will assume that they have not the power of perceiving anything outside this surface, but that upon it they have perceptions similar to ours. If such beings worked out a geometry, they would of course assign only two dimensions to their space. They would ascertain that a point in moving describes a line, and that a line in moving describes a surface. But they could as little represent to themselves what further spatial construction would be generated by a surface moving out of itself, as we can represent what would be generated by a solid moving out of the space we know. By the much-abused expression 'to represent' or 'to be able to think how some thing happens' I understand—and I do not see how anything else can be understood by it without loss of all meaning—the power of imagining the whole series of sensible impressions that would be had in such a case. Now as no sensible impression is known relating to such an unheard-of event, as the movement to a fourth dimension would be to us, or as a movement to our third dimension would be to the inhabitants of a surface, such a 'representation' is as impossible as the 'representation' of colours would be to one born blind, if a description of them in general terms could be given to him.

Our surface-beings would also be able to draw shortest lines in their superficial space. These would not necessarily be straight lines in our sense, but what are technically called *geodetic lines* of the surface on which they live; lines such as are described by a *tense* thread laid along the surface, and which can slide upon it freely. I will henceforth speak of such lines as the *straightest* lines of any particular surface or given space, so as to bring out their analogy with the straight line in a plane. I hope by this expression to make the conception more easy for the apprehension of my non-mathematical hearers without giving rise to misconception.

Now if beings of this kind lived on an infinite plane, their geometry would be exactly the same as our planimetry. They would affirm that only one straight line is possible between two points; that through a third point lying without this line only one line can be drawn parallel to it; that the ends of a straight line never meet though it is produced to infinity, and so on. Their space might be infinitely extended, but even if there were limits to their movement and perception, they would be able to represent to themselves a continuation beyond these limits; and thus their space would appear to them infinitely extended, just as ours does to us, although our bodies cannot leave the earth, and our sight only reaches as far as the visible fixed stars.

But intelligent beings of the kind supposed might also live on the surface of a sphere. Their shortest or straightest line between two points would then be an arc of the great circle passing through them. Every great circle, passing through two points, is by these divided into two parts; and if they are unequal, the shorter is certainly the shortest line on the sphere between the two points, but also the other or larger arc of the same great circle is a geodetic or straightest line, *i.e.* every smaller part of it is the shortest line between its ends. Thus the notion of the geodetic or straightest line is not quite identical with that of the shortest line. If the two given points are the ends of a diameter of the sphere, every plane passing through this diameter cuts semicircles, on the surface of the sphere, all of which are shortest lines between the ends; in which case there is an equal number of equal shortest lines between the given points. Accordingly, the axiom of there being only one shortest line between two points would not hold without a certain exception for the dwellers on a sphere.

Of parallel lines the sphere-dwellers would know nothing. They would maintain that any two straightest lines, sufficiently produced, must finally cut not in one only but in two points. The sum of the angles of a triangle would be always greater than two right angles, increasing as the surface of the triangle grew greater. They could thus have no conception of geometrical similarity between greater and smaller figures of the same kind, for with them a greater triangle must have different angles from a smaller one. Their space would be unlimited, but would be found to be finite or at least represented as such.

It is clear, then, that such beings must set up a very different system of geometrical axioms from that of the inhabitants of a plane, or from ours with our space of three dimensions, though the logical powers of all were the same; nor are more examples necessary to show that geometrical axioms must vary according to the kind of space inhabited by beings whose powers of reason are quite in conformity with ours. But let us proceed still farther. Let us think of reasoning beings existing on the surface of an egg-shaped body. Shortest lines could be drawn between three points of such a surface and a triangle constructed. But if the attempt were made to construct congruent triangles at different parts of the surface, it would be found that two triangles, with three pairs of equal sides, would not have their angles equal. The sum of the angles of a triangle drawn at the sharper pole of the body would depart farther from two right angles than if the triangle were drawn at the blunter pole or at the equator. Hence it appears that not even such a simple figure as a triangle can be moved on such a surface without change of form. It would also be found that if circles of equal radii were constructed at different parts of such a surface (the length of the radii being always measured by shortest lines along the surface) the periphery would be greater at the blunter than at the sharper end.

We see accordingly that, if a surface admits of the figures lying on it being freely moved without change of any of their lines and angles as measured along it, the property is a special one and does not belong to every kind of surface. The condition under which a surface possesses this important property was pointed out by Gauss in his celebrated treatise on the curvature of surfaces.[2] The 'measure of curvature,' as he called it, *i.e.* the reciprocal of the product of the greatest and least radii of curvature, must be everywhere equal over the whole extent of the surface.

Gauss showed at the same time that this measure of curvature is not changed if the surface is bent without distension or contraction of any part of it. Thus we can roll up a flat sheet of paper into the form of a cylinder, or of a cone, without any change in the dimensions of the figures taken along the surface of the sheet. Or the hemispherical fundus of a bladder may be rolled into a spindle-shape without altering the dimensions on the surface. Geometry on a plane will therefore be the same as on a cylindrical surface; only in the latter case we must imagine that any number of layers of this surface, like the layers of a rolled sheet of paper, lie one upon another, and that after each entire revolution round the cylinder a new layer is reached different from the previous ones.

These observations are necessary to give the reader a notion of a kind of surface the geometry of which is on the whole similar to that of the plane, but in which the axiom of parallels does not hold good. This is a kind of curved surface which is, as it were, geometrically the counterpart of a sphere, and which has therefore been called the *pseudospherical surface* by

2. Gauss, *Werke*, Bd. IV. p. 215, first published in *Commentationes Soc. Reg. Scientt. Gottengensis recentiores*, vol. vi., 1828.

the distinguished Italian mathematician E. Beltrami, who has investigated its properties.[3] It is a saddle-shaped surface of which only limited pieces or strips can be connectedly represented in our space, but which may yet be thought of as infinitely continued in all directions, since each piece lying at the limit of the part constructed can be conceived as drawn back to the middle of it and then continued. The piece displaced must in the process change its flexure but not its dimensions, just as happens with a sheet of paper moved about a cone formed out of a plane rolled up. Such a sheet fits the conical surface in every part, but must be more bent near the vertex and cannot be so moved over the vertex as to be at the same time adapted to the existing cone and to its imaginary continuation beyond.

Like the plane and the sphere, pseudospherical surfaces have their measure of curvature constant, so that every piece of them can be exactly applied to every other piece, and therefore all figures constructed at one place on the surface can be transferred to any other place with perfect congruity of form, and perfect equality of all dimensions lying in the surface itself. The measure of curvature as laid down by Gauss, which is positive for the sphere and zero for the plane, would have a constant negative value for pseudospherical surfaces, because the two principal curvatures of a saddle-shaped surface have their concavity turned opposite ways.

A strip of a pseudospherical surface may, for example, be represented by the inner surface (turned towards the axis) of a solid anchor-ring. If the plane figure *aabb* (Fig. 1) is made to revolve on its axis of symmetry AB, the two arcs *ab* will describe a pseudospherical concave-convex surface like that of the ring. Above and below, towards *aa* and *bb*, the surface will turn outwards with ever-increasing flexure, till it becomes perpendicular to the axis, and ends at the edge with one curvature infinite. Or, again, half of a pseudospherical surface may be rolled up into the shape of a champagne-glass (Fig. 2), with tapering stem infinitely prolonged. But the surface is always necessarily bounded by a sharp edge beyond which it cannot be directly continued. Only by supposing each single piece of the edge cut loose and drawn along the surface of the ring or glass, can it be brought to places of different flexure, at which farther continuation of the piece is possible.

In this way too the straightest lines of the pseudospherical surface may be infinitely produced. They do not, like those on a sphere, return upon themselves, but, as on a plane, only one shortest line is possible between the

---

3 *Saggio di Interpretazione della Geometria Non-Euclidea*, Napoli, *1868.—Teoria fondamentale degli Spazii di Curvatura costante, Annali di Matematica*, Ser. II. Tom. II. pp. 232–55. Both have been translated into French by J. Hoüel, *Annales Scientifiques de l'Ecole Normale*, Tom V., 1869.

two given points. The axiom of parallels does not, however, hold good. If a straightest line is given on the surface and a point without it, a whole pencil of straightest lines may pass through the point, no one of which, though infinitely produced, cuts the first line; the pencil itself being limited by two straightest lines, one of which intersects one of the ends of the given line at an infinite distance, the other the other end.

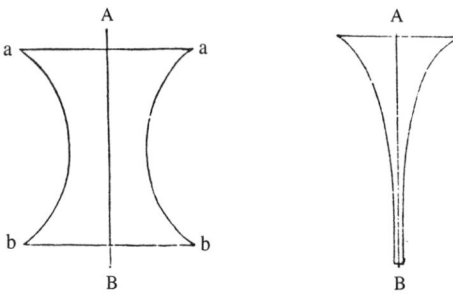

Figs. 1 and 2.

Such a system of geometry, which excluded the axiom of parallels, was devised on Euclid's synthetic method, as far back as the year 1829, by N. J. Lobachevsky, professor of mathematics at Kasan,[4] and it was proved that this system could be carried out as consistently as Euclid's. It agrees exactly with the geometry of the pseudospherical surfaces worked out recently by Beltrami.

Thus we see that in the geometry of two dimensions a surface is marked out as a plane, or a sphere, or a pseudospherical surface, by the assumption that any figure may be moved about in all directions without change of dimensions. The axiom, that there is only one shortest line between any two points, distinguishes the plane and the pseudospherical surface from the sphere, and the axiom of parallels marks off the plane from the pseudo-sphere. These three axioms are in fact necessary and sufficient, to define as a plane the surface to which Euclid's planimetry has reference, as distinguished from all other modes of space in two dimensions.

The difference between plane and spherical geometry has been long evident, but the meaning of the axiom of parallels could not be understood till Gauss had developed the notion of surfaces flexible without dilatation, and consequently that of the possibly infinite continuation of pseudo-spherical surfaces. Inhabiting, as we do, a space of three dimensions and endowed with organs of sense for their perception, we can represent to ourselves the various cases in which beings on a surface might have to

---

4. *Principien der Geometrie*, Kasan, 1829–30.

develop their perception of space; for we have only to limit our own perceptions to a narrower field. It is easy to think away perceptions that we have; but it is very difficult to imagine perceptions to which there is nothing analogous in our experience. When, therefore, we pass to space of three dimensions, we are stopped in our power of representation, by the structure of our organs and the experiences got through them which correspond only to the space in which we live.

There is however another way of treating geometry scientifically. All known space-relations are measurable, that is, they may be brought to determination of magnitudes (lines, angles, surfaces, volumes). Problems in geometry can therefore be solved, by finding methods of calculation for arriving at unknown magnitudes from known ones. This is done in *analytical geometry*, where all forms of space are treated only as quantities and determined by means of other quantities. Even the axioms themselves make reference to magnitudes. The straight line is defined as the *shortest* between two points, which is a determination of quantity. The axiom of parallels declares that if two straight lines in a plane do not intersect (are parallel), the alternate angles, or the corresponding angles, made by a third line intersecting them, are equal; or it may be laid down instead that the sum of the angles of any triangle is equal to two right angles. These, also, are determinations of quantity.

Now we may start with this view of space, according to which the position of a point may be determined by measurements in relation to any given figure (system of co-ordinates), taken as fixed, and then inquire what are the special characteristics of our space as manifested in the measurements that have to be made, and how it differs from other extended quantities of like variety. This path was first entered by one too early lost to science, B. Riemann of Göttingen.[5] It has the peculiar advantage that all its operations consist in pure calculation of quantities, which quite obviates the danger of habitual perceptions being taken for necessities of thought.

The number of measurements necessary to give the position of a point, is equal to the number of dimensions of the space in question. In a line the distance from one fixed point is sufficient, that is to say, one quantity; in a surface the distances from two fixed points must be given; in space, the distances from three; or we require, as on the earth, longitude, latitude, and height above the sea, or, as is usual in analytical geometry, the distances from three co-ordinate planes. Riemann calls a system of differences in which one thing can be determined by *n* measurements an '*n*fold extended aggregate' or an 'aggregate of *n* dimensions.' Thus the space in which we live is a threefold, a surface is a twofold, and a line is a simple extended

---

5. Ueber die Hypothesen welche der Geometrie zu Grunde liegen, Habilitationsschrift vom 10 Juni 1854. (*Abhandl. der königl. Gesellsch. zu Göttingen*, Bd. XIII.)

aggregate of points. Time also is an aggregate of one dimension. The system of colours is an aggregate of three dimensions, inasmuch as each colour, according to the investigations of Thomas Young and of Clerk Maxwell,[6] may be represented as a mixture of three primary colours, taken in definite quantities. The particular mixtures can be actually made with the colour-top.

In the same way we may consider the system of simple tones[7] as an aggregate of two dimensions, if we distinguish only pitch and intensity, and leave out of account differences of timbre. This generalisation of the idea is well suited to bring out the distinction between space of three dimensions and other aggregates. We can, as we know from daily experience, compare the vertical distance of two points with the horizontal distance of two others, because we can apply a measure first to the one pair and then to the other. But we cannot compare the difference between two tones of equal pitch and different intensity, with that between two tones of equal intensity and different pitch. Riemann showed, by considerations of this kind, that the essential foundation of any system of geometry, is the expression that it gives for the distance between two points lying in any direction towards one another, beginning with the infinitesimal interval. He took from analytical geometry the most general form for this expression, that, namely, which leaves altogether open the kind of measurements by which the position of any point is given.[8] Then he showed that the kind of free mobility without change of form which belongs to bodies in our space can only exist when certain quantities yielded by the calculation[9]—quantities that coincide with Gauss's measure of surface-curvature when they are expressed for surfaces—have everywhere an equal value. For this reason Riemann calls these quantities, when they have the same value in all directions for a particular spot, the measure of curvature of the space at this spot. To prevent misunderstanding,[10] I will once more observe that this so-called measure of space-curvature is a quantity obtained by purely analytical calculation, and that its introduction involves no suggestion of relations that would have a meaning only for sense-perception. The name is merely taken, as a short expression for a complete relation, from the one case in which the quantity designated admits of sensible representation.

Now whenever the value of this measure of curvature in any space is

6. Helmholtz's "The Recent Progress of the Theory of Vision," pp. 156–57.

7. Helmholtz's, "On the Physiological Causes of Harmony in Music," p. 62.

8. For the square of the distance of two infinitely near points the expression is a homogeneous quadric function of the differentials of their co-ordinates.

9. They are algebraical expressions compounded from the coefficients of the various terms in the expression for the square of the distance of two contiguous points and from their differential quotients.

10. As occurs, for instance, in the above-mentioned work of Tobias, pp. 70, etc.

everywhere zero, that space everywhere conforms to the axioms of Euclid; and it may be called a *flat* (*homaloid*) space in contradistinction to other spaces, analytically constructible, that may be called *curved*, because their measure of curvature has a value other than zero. Analytical geometry may be as completely and consistently worked out for such spaces as ordinary geometry can for our actually existing homaloid space.

If the measure of curvature is positive we have *spherical* space, in which straightest lines return upon themselves and there are no parallels. Such a space would, like the surface of a sphere, be unlimited but not infinitely great. A constant negative measure of curvature on the other hand gives *pseudo-spherical* space, in which straightest lines run out to infinity, and a pencil of straightest lines may be drawn, in any flattest surface, through any point which does not intersect another given straightest line in that surface.

Beltrami[11] has rendered these last relations imaginable by showing that the points, lines, and surfaces of a pseudospherical space of three dimensions, can be so portrayed in the interior of a sphere in Euclid's homaloid space, that every straightest line or flattest surface of the pseudospherical space is represented by a straight line or a plane, respectively, in the sphere. The surface itself of the sphere corresponds to the infinitely distant points of the pseudospherical space; and the different parts of this space, as represented in the sphere, become smaller, the nearer they lie to the spherical surface, diminishing more rapidly in the direction of the radii than in that perpendicular to them. Straight lines in the sphere, which only intersect beyond its surface, correspond to straightest lines of the pseudospherical space which never intersect.

Thus it appeared that space, considered as a region of measurable quantities, does not at all correspond with the most general conception of an aggregate of three dimensions, but involves also special conditions, depending on the perfectly free mobility of solid bodies without change of form to all parts of it and with all possible changes of direction; and, further, on the special value of the measure of curvature which for our actual space equals, or at least is not distinguishable from, zero. This latter definition is given in the axioms of straight lines and parallels.

Whilst Riemann entered upon this new field from the side of the most general and fundamental questions of analytical geometry, I myself arrived at similar conclusions,[12] partly from seeking to represent in space the system of colours, involving the comparison of one threefold extended aggregate with another, and partly from inquiries on the origin of our ocular measure for distances in the field of vision. Riemann starts by assuming the

11. *Teoria fondamentale, &c., ut sup.*

12. Ueber die Thatsachen die der Geometrie zum Grunde liegen (*Nachrichten von der königl. Ges. d. Wiss. zu Göttingen*, Juni 3, 1868).

above-mentioned algebraical expression which represents in the most general form the distance between two infinitely near points, and deduces therefrom, the conditions of mobility of rigid figures. I, on the other hand, starting from the observed fact that the movement of rigid figures is possible in our space, with the degree of freedom that we know, deduce the necessity of the algebraic expression taken by Riemann as an axiom. The assumptions that I had to make as the basis of the calculation were the following.

First, to make algebraical treatment at all possible, it must be assumed that the position of any point A can be determined, in relation to certain given figures taken as fixed bases, by measurement of some kind of magnitudes, as lines, angles between lines, angles between surfaces, and so forth. The measurements necessary for determining the position of A are known as its co-ordinates. In general, the number of co-ordinates necessary for the complete determination of the position of a point, marks the number of the dimensions of the space in question. It is further assumed that with the movement of the point A, the magnitudes used as co-ordinates vary continuously.

Secondly, the definition of a solid body, or rigid system of points, must be made in such a way as to admit of magnitudes being compared by congruence. As we must not, at this stage, assume any special methods for the measurement of magnitudes, our definition can, in the first instance, run only as follows. Between the co-ordinates of any two points belonging to a solid body, there must be an equation which, however the body is moved, expresses a constant spatial relation (proving at last to be the distance) between the two points, and which is the same for congruent pairs of points, that is to say, such pairs as can be made successively to coincide in space with the same fixed pair of points.

However indeterminate in appearance, this definition involves most important consequences, because with increase in the number of points, the number of equations increases much more quickly than the number of co-ordinates which they determine. Five points, A, B, C, D, E, give ten different pairs of points

$$AB, AC, AD, AE,$$
$$BC, BD, BE,$$
$$CD, CE,$$
$$DE,$$

and therefore ten equations, involving in space of three dimensions fifteen variable co-ordinates. But of these fifteen, six must remain arbitrary, if the system of five points is to admit of free movement and rotation, and thus the ten equations can determine only nine co-ordinates as functions of the six variables. With six points we obtain fifteen equations for twelve quantities, with seven points twenty-one equations for fifteen, and so on. Now from n independent equations we can determine n contained quantities, and if we have more than n equations, the superfluous ones must be deducible from the first n. Hence it follows that the equations which subsist between the co-ordinates of each pair of points of a solid body must have a special character, seeing that, when in space of three dimensions they are satisfied for nine pairs of points as formed out of any five points, the equation for the tenth pair follows by logical consequence. Thus our assumption for the definition of solidity, becomes quite sufficient to determine the kind of equations holding between the co-ordinates of two points rigidly connected.

Thirdly, the calculation must further be based on the fact of a peculiar circumstance in the movement of solid bodies, a fact so familiar to us that but for this inquiry it might never have been thought of as something that need not be. When in our space of three dimensions two points of a solid body are kept fixed, its movements are limited to rotations round the straight line connecting them. If we turn it completely round once, it again occupies exactly the position it had at first. This fact, that rotation in one direction always brings a solid body back into its original position, needs special mention. A system of geometry is possible without it. This is most easily seen in the geometry of a plane. Suppose that with every rotation of a plane figure its linear dimensions increased in proportion to the angle of rotation, the figure after one whole rotation through 360 degrees would no longer coincide with itself as it was originally. But any second figure that was congruent with the first in its original position might be made to coincide with it in its second position by being also turned though 360 degrees. A consistent system of geometry would be possible upon this supposition, which does not come under Riemann's formula.

On the other hand I have shown that the three assumptions taken together form a sufficient basis for the starting-point of Riemann's investigation, and thence for all his further results relating to the distinction of different spaces according to their measure of curvature.

It still remained to be seen whether the laws of motion, as dependent on moving forces, could also be consistently transferred to spherical or pseudospherical space. This investigation has been carried out by Professor Lipschitz of Bonn.[13] It is found that the comprehensive expression for all the laws of dynamics, Hamilton's principle, may be directly transferred to spaces of which the measure of curvature is other than zero. Accordingly, in this respect also, the disparate systems of geometry lead to no contradiction.

---

13. 'Untersuchungen über die ganzen homogenen Functionen von n Differentialen' (Borchardt's *Journal für Mathematik*, Bd. lxx. 3, 71; lxxiii. 3, 1); 'Untersuchung eines Problems der Variationsrechnung' (*Ibid.* Bd. lxxiv.).

We have now to seek an explanation of the special characteristics of our own flat space, since it appears that they are not implied in the general notion of an extended quantity of three dimensions and of the free mobility of bounded figures therein. *Necessities of thought*, such as are involved in the conception of such a variety, and its measurability, or from the most general of all ideas of a solid figure contained in it, and of its free mobility, they undoubtedly are not. Let us then examine the opposite assumption as to their origin being empirical, and see if they can be inferred from facts of experience and so established, or if, when tested by experience, they are perhaps to be rejected. If they are of empirical origin, we must be able to represent to ourselves connected series of facts, indicating a different value for the measure of curvature from that of Euclid's flat space. But if we can imagine such spaces of other sorts, it cannot be maintained that the axioms of geometry are necessary consequences of an *a priori* transcendental form of intuition, as Kant thought.

The distinction between spherical, pseudospherical, and Euclid's geometry depends, as was above observed, on the value of a certain constant called, by Riemann, the measure of curvature of the space in question. The value must be zero for Euclid's axioms to hold good. If it were not zero, the sum of the angles of a large triangle would differ from that of the angles of a small one, being larger in spherical, smaller in pseudospherical, space. Again, the geometrical similarity of large and small solids or figures is possible only in Euclid's space. All systems of practical mensuration that have been used for the angles of large rectilinear triangles, and especially all systems of astronomical measurement which make the parallax of the immeasurably distant fixed stars equal to zero (in pseudo-spherical space the parallax even of infinitely distant points would be positive), confirm empirically the axiom of parallels, and show the measure of curvature of our space thus far to be indistinguishable from zero. It remains, however, a question, as Riemann observed, whether the result might not be different if we could use other than our limited base-lines, the greatest of which is the major axis of the earth's orbit.

Meanwhile, we must not forget that all geometrical measurements rest ultimately upon the principle of congruence. We measure the distance between points by applying to them the compass, rule, or chain. We measure angles by bringing the divided circle or theodolite to the vertex of the angle. We also determine straight lines by the path of rays of light which in our experience is rectilinear; but that light travels in shortest lines as long as it continues in a medium of constant refraction would be equally true in space of a different measure of curvature. Thus all our geometrical measurements depend on our instruments being really, as we consider them, invariable in form, or at least on their undergoing no other than the small changes we know of, as arising from variation of temperature, or from gravity acting differently at different places.

In measuring, we only employ the best and surest means we know of to determine, what we otherwise are in the habit of making out by sight and touch or by pacing. Here our own body with its organs is the instrument we carry about in space. Now it is the hand, now the leg, that serves for a compass, or the eye turning in all directions is our theodolite for measuring arcs and angles in the visual field.

Every comparative estimate of magnitudes or measurement of their spatial relations proceeds therefore upon a supposition as to the behaviour of certain physical things, either the human body or other instruments employed. The supposition may be in the highest degree probable and in closest harmony with all other physical relations known to us, but yet it passes beyond the scope of pure space-intuition.

It is in fact possible to imagine conditions for bodies apparently solid such that the measurements in Euclid's space become what they would be in spherical or pseudospherical space. Let me first remind the reader that if all the linear dimensions of other bodies, and our own, at the same time were diminished or increased in like proportion, as for instance to half or double their size, we should with our means of space-perception be utterly unaware of the change. This would also be the case if the distension or contraction were different in different directions, provided that our own body changed in the same manner, and further that a body in rotating assumed at every moment, without suffering or exerting mechanical resistance, the amount of dilatation in its different dimensions corresponding to its position at the time. Think of the image of the world in a convex mirror. The common silvered globes set up in gardens give the essential features, only distorted by some optical irregularities. A well-made convex mirror of moderate aperture represents the objects in front of it as apparently solid and in fixed positions behind its surface. But the images of the distant horizon and of the sun in the sky lie behind the mirror at a limited distance, equal to its focal length. Between these and the surface of the mirror are found the images of all the other objects before it, but the images are diminished and flattened in proportion to the distance of their objects from the mirror. The flattening, or decrease in the third dimension, is relatively greater than the decrease of the surface-dimensions. Yet every straight line or every plane in the outer world is represented by a straight line or a plane in the image. The image of a man measuring with a rule a straight line from the mirror would contract more and more the farther he went, but with his shrunken rule the man in the image would count out exactly the same number of centimetres as the real man. And, in general, all geometrical measurements of lines or angles made with regularly varying images of real instruments would yield exactly the same results as in the outer world, all congruent bodies would coincide on being applied to one another in the mirror as in the outer world, all lines of sight in the outer world would be represented by straight lines of sight in the mirror. In short I do not see how men in the mirror are to discover that their bodies are not rigid solids and their experiences good examples of the correctness of Euclid's axioms. But if they could look out upon our world as we can look into theirs, without overstepping the boundary, they must declare it to be a picture in a spherical mirror, and would speak of us just as we speak of them; and if two inhabitants of the different worlds could communicate with one another, neither, so far as I can see, would be able to convince the other that he had the true, the other the distorted, relations. Indeed I cannot see that such a question would have any meaning at all, so long as mechanical considerations are not mixed up with it.

Now Beltrami's representation of pseudospherical space in a sphere of Euclid's space, is quite similar, except that the background is not a plane as in the convex mirror, but the surface of a sphere, and that the proportion in which the images as they approach the spherical surface contract, has a different mathematical expression.[14] If we imagine then, conversely, that in the sphere, for the interior of which Euclid's axioms hold good, moving bodies contract as they depart from the centre like the images in a convex mirror, and in such a way that their representatives in pseudospherical space retain their dimensions unchanged,—observers whose bodies were regularly subjected to the same change would obtain the same results from the geometrical measurements they could make as if they lived in pseudospherical space.

We can even go a step further, and infer how the objects in a pseudospherical world, were it possible to enter one, would appear to an observer, whose eye-measure and experiences of space had been gained like ours in Euclid's space. Such an observer would continue to look upon rays of light or the lines of vision as straight lines, such as are met with in flat space, and as they really are in the spherical representation of pseudospherical space. The visual image of the objects in pseudospherical space would thus make the same impression upon him as if he were at the centre of Beltrami's sphere. He would think he saw the most remote objects round about him at a finite distance,[15] let us suppose a hundred feet off. But as he approached these distant objects, they would dilate before him, though more in the third dimension than superficially, while behind him they would contract. He would know that his eye judged wrongly. If he saw two straight lines which in his estimate ran parallel for the hundred feet to his world's end, he would find on following them that the farther he advanced

14. Compare the Appendix at the end of this Lecture.
15. The reciprocal of the square of this distance, expressed in negative quantity, would be the measure of curvature of the pseudospherical space.

the more they diverged, because of the dilatation of all the objects to which he approached. On the other hand, behind him, their distance would seem to diminish, so that as he advanced they would appear always to diverge more and more. But two straight lines which from his first position seemed to converge to one and the same point of the background a hundred feet distant, would continue to do this however far he went, and he would never reach their point of intersection.

Now we can obtain exactly similar images of our real world, if we look through a large convex lens of corresponding negative focal length, or even through a pair of convex spectacles if ground somewhat prismatically to resemble pieces of one continuous larger lens. With these, like the convex mirror, we see remote objects as if near to us, the most remote appearing no farther distant than the focus of the lens. In going about with this lens before the eyes, we find that the objects we approach dilate exactly in the manner I have described for pseudospherical space. Now any one using a lens, were it even so strong as to have a focal length of only sixty inches, to say nothing of a hundred feet, would perhaps observe for the first moment that he saw objects brought nearer. But after going about a little the illusion would vanish, and in spite of the false images he would judge of the distances rightly. We have every reason to suppose that what happens in a few hours to any one beginning to wear spectacles would soon enough be experienced in pseudospherical space. In short, pseudospherical space would not seem to us very strange, comparatively speaking; we should only at first be subject to illusions in measuring by eye the size and distance of the more remote objects.

There would be illusions of an opposite description, if, with eyes practised to measure in Euclid's space, we entered a spherical space of three dimensions. We should suppose the more distant objects to be more remote and larger than they are, and should find on approaching them that we reached them more quickly than we expected from their appearance. But we should also see before us objects that we can fixate only with diverging lines of sight, namely, all those at a greater distance from us than the quadrant of a great circle. Such an aspect of things would hardly strike us as very extraordinary, for we can have it even as things are if we place before the eye a slightly prismatic glass with the thicker side towards the nose: the eyes must then become divergent to take in distant objects. This excites a certain feeling of unwonted strain in the eyes, but does not perceptibly change the appearance of the objects thus seen. The strangest sight, however, in the spherical world would be the back of our own head, in which all visual lines not stopped by other objects would meet again, and which must fill the extreme background of the whole perspective picture.

At the same time it must be noted that as a small elastic flat disk, say of india-rubber, can only be fitted to a slightly curved spherical surface with relative contraction of its border and distension of its centre, so our bodies, developed in Euclid's flat space, could not pass into curved space without undergoing similar distensions and contractions of their parts, their coherence being of course maintained only in as far as their elasticity permitted their bending without breaking. The kind of distension must be the same as in passing from a small body imagined at the centre of Beltrami's sphere to its pseudospherical or spherical representation. For such passage to appear possible, it will always have to be assumed that the body is sufficiently elastic and small in comparison with the real or imaginary radius of curvature of the curved space into which it is to pass.

These remarks will suffice to show the way in which we can infer from the known laws of our sensible perceptions the series of sensible impressions which a spherical or pseudospherical world would give us, if it existed. In doing so, we nowhere meet with inconsistency or impossibility any more than in the calculation of its metrical proportions. We can represent to ourselves the look of a pseudospherical world in all directions just as we can develop the conception of it. Therefore it cannot be allowed that the axioms of our geometry depend on the native form of our perceptive faculty, or are in any way connected with it.

It is different with the three dimensions of space. As all our means of sense-perception extend only to space of three dimensions, and a fourth is not merely a modification of what we have, but something perfectly new, we find ourselves by reason of our bodily organisation quite unable to represent a fourth dimension.

In conclusion, I would again urge that the axioms of geometry are not propositions pertaining only to the pure doctrine of space. As I said before, they are concerned with quantity. We can speak of quantities only when we know of some way by which we can compare, divide, and measure them. All space-measurements, and therefore in general all ideas of quantities applied to space, assume the possibility of figures moving without change of form or size. It is true we are accustomed in geometry to call such figures purely geometrical solids, surfaces, angles, and lines, because we abstract from all the other distinctions, physical and chemical, of natural bodies; but yet one physical quality, rigidity, is retained. Now we have no other mark of rigidity of bodies or figures but congruence, whenever they are applied to one another at any time or place, and after any revolution. We cannot, however, decide by pure geometry, and without mechanical considerations, whether the coinciding bodies may not both have varied in the same sense.

If it were useful for any purpose, we might with perfect consistency look upon the space in which we live as the apparent space behind a convex mirror with its shortened and contracted background; or we might consider a bounded sphere of our space, beyond the limits of which we perceive nothing further, as infinite pseudospherical space. Only then we should have to ascribe to the bodies which appear to us to be solid, and to our own body at the same time, corresponding distensions and contractions, and we should have to change our system of mechanical principles entirely; for even the proposition that every point in motion, if acted upon by no force, continues to move with unchanged velocity in a straight line, is not adapted to the image of the world in the convex-mirror. The path would indeed be straight, but the velocity would depend upon the place.

Thus the axioms of geometry are not concerned with space-relations only but also at the same time with the mechanical deportment of solidest bodies in motion. The notion of rigid geometrical figure might indeed be conceived as transcendental in Kant's sense, namely, as formed independently of actual experience, which need not exactly correspond therewith, any more than natural bodies do ever in fact correspond exactly to the abstract notion we have obtained of them by induction. Taking the notion of rigidity thus as a mere ideal, a strict Kantian might certainly look upon the geometrical axioms as propositions given, *a priori*, by transcendental intuition, which no experience could either confirm or refute, because it must first be decided by them whether any natural bodies can be considered as rigid. But then we should have to maintain that the axioms of geometry are not synthetic propositions, as Kant held them; they would merely define what qualities and deportment a body must have to be recognised as rigid.

But if to the geometrical axioms we add propositions relating to the mechanical properties of natural bodies, were it only the axiom of inertia, or the single proposition, that the mechanical and physical properties of bodies and their mutual reactions are, other circumstances remaining the same, independent of place, such a system of propositions has a real import which can be confirmed or refuted by experience, but just for the same reason can also be gained by experience. The mechanical axiom, just cited, is in fact of the utmost importance for the whole system of our mechanical and physical conceptions. That rigid solids, as we call them, which are really nothing else than elastic solids of great resistance, retain the same form in every part of space if no external force affects them, is a single case falling under the general principle.

In conclusion, I do not, of course, maintain that mankind first arrived at space-intuitions, in agreement with the axioms of Euclid, by any carefully executed systems of exact measurement. It was rather a succession of everyday experiences, especially the perception of the geometrical similarity of great and small bodies, only possible in flat space, that led to the rejection, as impossible, of every geometrical representation at variance with this fact. For this no knowledge of the necessary logical connection between the observed fact of geometrical similarity and the axioms was needed; but only an intuitive apprehension of the typical relations between lines, planes, angles, &c., obtained by numerous and attentive observations—an intuition of the kind the artist possesses of the objects he is to represent, and by means of which he decides with certainty and accuracy whether a new combination, which he tries, will correspond or not with their nature. It is true that we have no word but *intuition* to mark this; but it is knowledge empirically gained by the aggregation and reinforcement of similar recurrent impressions in memory, and not a transcendental form given before experience. That other such empirical intuitions of fixed typical relations, when not clearly comprehended, have frequently enough been taken by metaphysicians for *a priori* principles, is a point on which I need not insist.

To sum up, the final outcome of the whole inquiry may be thus expressed:—

(1.) The axioms of geometry, taken by themselves out of all connection with mechanical propositions, represent no relations of real things. When thus isolated, if we regard them with Kant as forms of intuition transcendentally given, they constitute a form into which any empirical content whatever will fit, and which therefore does not in any way limit or determine beforehand the nature of the content. This is true, however, not only of Euclid's axioms, but also of the axioms of spherical and pseudospherical geometry.

(2.) As soon as certain principles of mechanics are conjoined with the axioms of geometry, we obtain a system of propositions which has real import, and which can be verified or overturned by empirical observations, just as it can be inferred from experience. If such a system were to be taken as a transcendental form of intuition and thought, there must be assumed a pre-established harmony between form and reality.

APPENDIX

THE elements of the geometry of spherical space are most easily obtained by putting for space of four dimensions the equation for the sphere

$$x^2 + y^2 + z^2 + t^2 = R^2 \quad\text{............ (1.)}$$

and for the distance ds between the points (x, y, z, t) and $[(x + dx)\,(y + dy)\,(z + dz)\,(t + dt)]$ the value

$$ds^2 = dx^2 + dy^2 + dz^2 + dt^2 \quad\text{...... (2.)}$$

It is easily found by means of the methods used for three dimensions that the shortest lines are given by equations of the form

$$\left.\begin{array}{c} ax + by + cz + ft = 0 \\ \alpha x + \beta y + \gamma z + \phi t = 0 \end{array}\right\} \quad\text{......... (3.)}$$

in which $a, b, c, f,$ as well as $\alpha, \beta, \gamma, \phi$, are constants.

The length of the shortest arc, s, between the points (x, y, z, t), and (ξ, η, ζ, τ) follows, as in the sphere, from the equation

$$\cos\frac{s}{R} = \frac{x\xi + y\eta + z\zeta + t\tau}{R^2} \quad\text{....... (4.)}$$

One of the co-ordinates may be eliminated from the values given in 2 to 4, by means of equation 1, and the expressions then apply to space of three dimensions.

If we take the distances from the points

$$\xi = \eta = \zeta = 0$$

from which equation 1 gives $r = R$, then,

$$\sin\left(\frac{s_0}{R}\right) = \frac{\sigma}{R}$$

in which

$$\sigma = \sqrt{x^2 + y^2 + z^2}$$

or,

$$s_0 = R \cdot \arc\sin\left(\frac{\sigma}{R}\right) = R \cdot \arc\tan\left(\frac{\sigma}{t}\right) \dots (5.)$$

In this, s_0 is the distance of the point x, y, z, measured from the centre of the co-ordinates.

If now we suppose the point x, y, z, of spherical space, to be projected in a point of plane space whose co-ordinates are respectively

$$X = \frac{Rx}{t} \quad Y = \frac{Ry}{t} \quad Z = \frac{Rz}{t}$$

$$X^2 + Y^2 + Z^2 = r^2 = \frac{R^2\sigma^2}{t^2}$$

then in the plane space the equations 3, which belong to the straightest lines of spherical space, are equations of the straight line. Hence the shortest lines of spherical space are represented in the system of X, Y, Z by straight lines. For very small values of x, y, z, $t = R$, and

$$X = x, \ Y = y, \ Z = z.$$

Immediately about the centre of the co-ordinates, the measurements of both spaces coincide. On the other hand, we have for the distances from the centre

$$s_0 = R \cdot \arc\tan\left(\pm\frac{r}{R}\right) \dots\dots (6.)$$

In this, r may be infinite; but every point of plane space must be the projection of two points of the sphere, one for which $s_0 < \frac{1}{2} R\pi$, and one for which $s_0 > \frac{1}{2} R\pi$. The extension in the direction of r is then

$$\frac{ds_0}{dr} = \frac{R^2}{R^2 + r^2}.$$

In order to obtain corresponding expressions for pseudospherical space, let R and t be imaginary; that is, $R = \Re i$, and $t = \mathfrak{t}i$. Equation 6 gives then

$$\tan\frac{s_0}{i\Re} = \pm\frac{r}{i\Re},$$

from which, eliminating the imaginary form, we get

$$s_0 = \tfrac{1}{2}\Re\log.\,nat.\,\frac{\Re + r}{\Re - r}.$$

Here s_0 has real values only as long as $r = R$; for $r = \Re$ the distance s_0 in pseudospherical space is infinite. The image in plane space is, on the contrary, contained in the sphere of radius R, and every point of this sphere forms only one point of the infinite pseudospherical space. The extension in the direction of r is

$$\frac{ds_0}{dr} = \frac{\Re^2}{\Re^2 - r^2}.$$

For linear elements, on the contrary, whose direction is at right angles to r, and for which t is unchanged, we have in both cases

$$\frac{\sqrt{dx^2 + dy^2 + dz^2}}{\sqrt{dX^2 + dY^2 + dZ^2}} = \frac{t}{R} = \frac{\mathfrak{t}}{\Re} = \frac{\sigma}{r} = \frac{\sqrt{x^2 + y^2 + z^2}}{\sqrt{X^2 + Y^2 + Z^2}}.$$

Lecture delivered in the Docenten Verein in Heidelberg, in the year 1870.

9

On the Origin of the Planetary System

IT is my intention to bring a subject before you to-day which has been much discussed—that is, the hypothesis of Kant and Laplace as to the formation of the celestial bodies, and more especially of our planetary system. The choice of the subject needs no apology. In popular lectures, like the present, the hearers may reasonably expect from the lecturer, that he shall bring before them well-ascertained facts, and the complete results of investigation, and not unripe suppositions, hypotheses, or dreams.

Of all the subjects to which the thought and imagination of man could turn, the question as to the origin of the world has, since remote antiquity, been the favourite arena of the wildest speculation. Beneficent and malignant deities, giants, Kronos who devours his children, Niflheim, with the ice-giant Ymir, who is killed by the celestial Asas,[1] that out of him the world may be constructed—these are all figures which fill the cosmogonic systems of the more cultivated of the peoples. But the universality of the fact, that each people develops its own cosmogonies, and sometimes in great detail, is an expression of the interest, felt by all, in knowing what is our own origin, what is the ultimate beginning of the things about us. And with the question of the beginning is closely connected that of the end of all things; for that which may be formed, may also pass away. The question about the end of things is perhaps of greater practical interest than that of the beginning.

Now, I must premise that the theory which I intend to discuss to-day was first put forth by a man who is known as the most abstract of philosophical thinkers; the originator of transcendental idealism and of the Categorical Imperative, Immanuel Kant. The work in which he developed this, the *General Natural Philosophy and Theory of the Heavens*, is one of his first publications, having appeared in his thirty-first year. Looking at the writings of this first period of his scientific activity, which lasted to about

---

1. Cox's *Aryan Mythology*, vol. i. 372. Longmans.

his fortieth year, we find that they belong mostly to Natural Philosophy, and are far in advance of their times with a number of the happiest ideas. His philosophical writings at this period are but few, and partly like his introductory lecture, directly originating in some adventitious circumstance; at the same time the matter they contain is comparatively without originality, and they are only important from a destructive and partially sarcastic criticism. It cannot be denied that the Kant of early life was a natural philosopher by instinct and by inclination; and that probably only the power of external circumstances, the want of the means necessary for independent scientific research, and the tone of thought prevalent at the time, kept him to philosophy, in which it was only much later that he produced anything original and important; for the *Kritik der reinen Vernunft* appeared in his fifty-seventh year. Even in the later periods of his life, between his great philosophical works, he wrote occasional memoirs on natural philosophy, and regularly delivered a course of lectures on physical geography. He was restricted in this to the scanty measure of knowledge and of appliances of his time, and of the out-of-the-way place where he lived; but with a large and intelligent mind he strove after such more general points of view as Alexander von Humboldt afterwards worked out. It is exactly an inversion of the historical connection, when Kant's name is occasionally misused, to recommend that natural philosophy shall leave the inductive method, by which it has become great, to revert to the windy speculations of a so-called 'deductive method.' No one would have attacked such a misuse, more energetically and more incisively, than Kant himself if he were still among us.

The same hypothesis as to the origin of our planetary system was advanced a second time, but apparently quite independently of Kant, by the most celebrated of French astronomers, Simon, Marquis de Laplace. It formed, as it were, the final conclusion of his work on the mechanism of our system, executed with such gigantic industry and great mathematical acuteness. You see from the names of these two men, whom we meet as experienced and tried leaders in our course, that in a view in which they both agree, we have not to deal with a mere random guess, but with a careful and well-considered attempt to deduce conclusions as to the unknown past from known conditions of the present time.

It is in the nature of the case, that a hypothesis as to the origin of the world which we inhabit, and which deals with things in the most distant past, cannot be verified by direct observation. It may, however, receive direct confirmation, if, in the progress of scientific knowledge, new facts accrue to those already known, and like them are explained on the hypothesis; and particularly if survivals of the processes, assumed to have taken place in the formation of the heavenly bodies, can be proved to exist in the present.

Such direct confirmations of various kinds have, in fact, been formed for the view we are about to discuss, and have materially increased its probability.

Partly this fact, and partly the fact that the hypothesis in question has recently been mentioned in popular and scientific books, in connection with philosophical, ethical, and theological questions, have emboldened me to speak of it here. I intend not so much to tell you anything substantially new in reference to it, as to endeavor to give, as connectedly as possible, the reasons which have led to, and have confirmed it.

These apologies which I must premise, only apply to the fact that I treat a theme of this kind as a popular lecture. Science is not only entitled, but is indeed beholden, to make such an investigation. For her it is a definite and important question—the question, namely, as to the existence of limits to the validity of the laws of nature, which rule all that now surrounds us; the question whether they have always held in the past, and whether they will always hold in the future; or whether, on the supposition of an everlasting uniformity of natural laws, our conclusions from present circumstances as to the past, and as to the future, imperatively lead to an impossible state of things; that is, to the necessity of an infraction of natural laws, of a beginning which could not have been due to processes known to us. Hence, to begin such an investigation as to the possible or probable primeval history of our present world, is, considered as a question of science, no idle speculation, but a question as to the limits of its methods, and as to the extent to which existing laws are valid.

It may perhaps appear rash that we, restricted as we are, in the circle of our observations in space, by our position on this little earth, which is but as a grain of dust in our milky way; and limited in time by the short duration of the human race; that we should attempt to apply the laws which we have deduced from the confined circle of facts open to us, to the whole range of infinite space, and of time from everlasting to everlasting. But all our thought and our action, in the greatest as well as in the least, is based on our confidence in the unchangeable order of nature, and this confidence has hitherto been the more justified, the deeper we have penetrated into the interconnections of natural phenomena. And that the general laws, which we have found, also hold for the most distant vistas of space, has acquired strong actual confirmation during the past half-century.

In the front rank of all, then, is the law of gravitation. The celestial bodies, as you all know, float and move in infinite space. Compared with the enormous distances between them, each of us is but as a grain of dust. The nearest fixed stars, viewed even under the most powerful magnification, have no visible diameter; and we may be sure that even our sun, looked at from the nearest fixed stars, would only appear as a single luminous point; seeing that the masses of those stars, in so far as they have been determined, have not been found to be materially different from that of the sun. But, notwithstanding these enormous distances, there is an invisible tie between them which connects them together, and brings them in mutual interdependence. This is the force of gravitation, with which all heavy masses attract each other. We know this force as gravity, when it is operative between an earthly body and the mass of our earth. The force which causes a body to fall to the ground is none other than that which continually compels the moon to accompany the earth in its path round the sun, and which keeps the earth itself from fleeing off into space, away from the sun.

You may realise, by means of a simple mechanical model, the course of planetary motion. Fasten to the branch of a tree, at a sufficient height, or to a rigid bar, fixed horizontally in the wall, a silk cord, and at its end a small heavy body—for instance, a lead ball. If you allow this to hang at rest, it stretches the thread. This is the position of equilibrium of the ball. To indicate this, and keep it visible, put in the place of the ball any other solid body—for instance, a large terrestrial globe on a stand. For this purpose the ball must be pushed aside, but it presses against the globe, and, if taken away, it still tends to come back to it, because gravity impels it towards its position of equilibrium, which is in the centre of the sphere. And upon whatever side it is drawn, the same thing always happens. This force, which drives the ball towards the globe, represents in our model the attraction which the earth exerts on the moon, or the sun on the planets. After you have convinced yourselves of the accuracy of these facts, try to give the ball, when it is a little away from the globe, a slight throw in a lateral direction. If you have accurately hit the strength of the throw, the small ball will move round the large one in a circular path, and may retain this motion for some time; just as the moon persists in its course round the earth, or the planets about the sun. Now, in our model, the circles described by the lead ball will be continually narrower, because the opposing forces, the resistance of the air, the rigidity of the thread, friction, cannot be eliminated, in this case, as they are excluded in the planetary system.

If the path about the attracting centre is exactly circular, the attracting force always acts on the planets, or on the lead sphere, with equal strength. In this case, it is immaterial according to what law the force would increase or diminish at other distances from the centre in which the moving body does not come. If the original impulse has not been of the right strength in both cases, the paths will not be circular but elliptical, of the form of the curved line in Fig. 1. But these ellipses lie in both cases differently as regards the attracting centre. In our model, the attracting force is stronger, the further the lead sphere is removed from its position of equilibrium. Under these circumstances, the ellipse of the path has such a position in reference to the attracting centre, that this is in the centre, c, of the ellipse.

For planets, on the contrary, the attracting force is feebler the further it is removed from the attracting body, and this is the reason that an ellipse is described, one of whose foci lies in the centre of attraction. The two foci, *a* and *b*, are two points which lie symmetrically towards the ends of the ellipse, and are characterised by the property that the sum of their distances, *am* + *bm*, is the same from any given points.

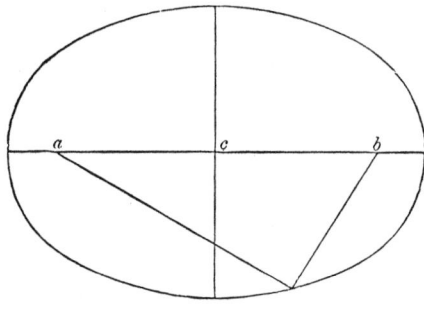

Fig. 1.

Kepler had found that the paths of the planets are ellipses of this kind; and since, as the above example shows, the form and position of the orbit depend on the law according to which the magnitude of the attracting force alters, Newton could deduce from the form of the planetary orbits the well-known law of the force of gravitation, which attracts the planets to the sun, according to which this force decreases with increase of distance as the square of that distance. Terrestrial gravity must obey this law, and Newton had the wonderful self-denial to refrain from publishing his important discovery until it had acquired a direct confirmation; this followed from the observations, that the force which attracts the moon towards the earth, bears towards the gravity of a terrestrial body the ratio required by the above law.

In the course of the eighteenth century the power of mathematical analysis, and the methods of astronomical observation, increased so far that all the complicated actions, which take place between all the planets, and all their satellites, in consequence of the mutual action of each upon each, and which astronomers call disturbances—disturbance, that is to say, of the simpler elliptical motions about the sun, which each one would produce if the others were absent—that all these could be theoretically predicted from Newton's law, and be accurately compared with what actually takes place in the heavens. The development of this theory of planetary motion in detail was, as has been said, the merit of Laplace. The agreement between this theory, which was developed from the simple law of gravitation, and the extremely complicated and manifold phenomena which follow therefrom, was so complete and so accurate, as had never previously been attained in any other branch of human knowledge. Emboldened by this agreement, the next step was to conclude that where slight defects were still constantly found, unknown causes must be at work. Thus, from Bessel's calculation of the discrepancy between the actual and the calculated motion of Uranus, it was inferred that there must be another planet. The position of this planet was calculated by Leverrier and Adams, and thus Neptune, the most distant of all known at that time, was discovered.

But it was not merely in the region of the attraction of our sun that the law of gravitation was found to hold. With regard to the fixed stars, it was found that double stars moved about each other in elliptical paths, and that therefore the same law of gravitation must hold for them as for our planetary system. The distance of some of them could be calculated. The nearest of them, α, in the constellation of the Centaur, is 270,000 times further from the sun than the earth. Light, which has a velocity of 186,000 miles a second, which traverses the distance from the sun to the earth in eight minutes, would take four years to travel from α Centauri to us. The more delicate methods of modern astronomy have made it possible to determine distances which light would take thirty-five years to traverse; as, for instance, the Pole Star; but the law of gravitation is seen to hold, ruling the motion of the double stars, at distances in the heavens, which all the means we possess have hitherto utterly failed to measure.

The knowledge of the law of gravitation has here also led to the discovery of new bodies, as in the case of Neptune. Peters of Altona found, confirming therein a conjecture of Bessel, that Sirius, the most brilliant of the fixed stars, moves in an elliptical path about an invisible centre. This must have been due to an unseen companion, and when the excellent and powerful telescope of the University of Cambridge, in the United States, had been set up, this was discovered. It is not quite dark, but its light is so feeble that it can only be seen by the most perfect instruments. The mass of Sirius is found to be 13·76, and that of its satellite 6·71, times the mass of the sun; their mutual distance is equal to thirty-seven times the radius of the earth's orbit, and is therefore somewhat larger than the distance of Neptune from the sun.

Another fixed star, Procyon, is in the same case as Sirius, but its satellite has not yet been discovered.

You thus see that in gravitation we have discovered a property common to all matter, which is not confined to bodies in our system, but extends, as far in the celestial space, as our means of observation have hitherto been able to penetrate.

But not merely is this universal property of all mass shared by the most distant celestial bodies, as well as by terrestrial ones; but spectrum analysis has taught us that a number of well-known terrestrial elements are met with in the atmospheres of the fixed stars, and even of the nebulae.

You all know that a fine bright line of light, seen through a glass prism, appears as a coloured band, red and yellow at one edge, blue and violet at the other, and green in the middle. Such a coloured image is called a spectrum—the rainbow is such a one, produced by the refraction of light, though not exactly by a prism; and it exhibits therefore the series of colours into which white sunlight can thus be decomposed. The formation of the prismatic spectrum depends on the fact that the sun's light, and that of most ignited bodies, is made up of various kinds of light which appear of different colours to our eyes, and the rays of which are separated from each other when refracted by a prism.

Now if a solid or a liquid is heated to such an extent that it becomes incandescent, the spectrum which its light gives is, like the rainbow, a broad coloured band without any breaks, with the well-known series of colours, red, yellow, green, blue, and violet, and in no wise characteristic of the nature of the body which emits the light.

The case is different if the light is emitted by an ignited gas, or by an ignited vapour—that is, a substance vaporised by heat. The spectrum of such a body consists, then, of one or more, and sometimes even a great number, of entirely distinct bright lines, whose position and arrangement in the spectrum is characteristic for the substances of which the gas or vapour consists, so that it can be ascertained, by means of spectrum analysis, what is the chemical constitution of the ignited gaseous body. Gaseous spectra of this kind are shown in the heavenly space by many nebulae; for the most part they are spectra which show the bright line of ignited hydrogen and oxygen, and along with it a line which, as yet, has never been again found in the spectrum of any terrestrial element. Apart from the proof of two well-known terrestrial elements, this discovery was of the utmost importance, since it furnished the first unmistakable proof that the cosmical nebulae are not, for the most part, small heaps of fine stars, but that the greater part of the light which they emit is really due to gaseous bodies.

The gaseous spectra present a different appearance when the gas is in front of an ignited solid whose temperature is far higher than that of the gas. The observer sees then a continuous spectrum of a solid, but traversed by fine dark lines, which are just visible in the places in which the gas alone, seen in front of a dark background, would show bright lines. The solar spectrum is of this kind, and also that of a great number of fixed stars. The dark lines of the solar spectrum, originally discovered by Wollaston, were first investigated and measured by Fraunhofer, and are hence known as Fraunhofer's lines.

Far more powerful apparatus was afterwards used by Kirchhoff, and then by Angström, to push the decomposition of light as far as possible. Fig. 2 represents an apparatus with four prisms, constructed by Steinheil for Kirchhoff. At the further end of the telescope B is a screen with a fine slit, representing a fine slice of light, which can be narrowed or widened by the small screw, and by which the light under investigation can be allowed to enter. It then passes through the telescope B, afterwards through the four

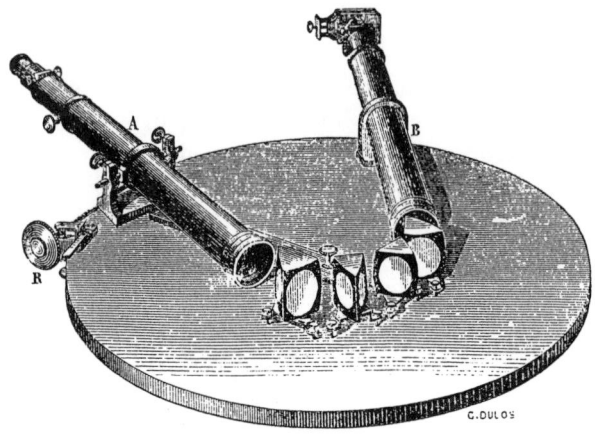

Fig. 2.

prisms, and finally through the telescope A, from which it reaches the eye of the observer. Figs. 3, 4, and 5 represent small portions of the solar spectrum as mapped by Kirchhoff, taken from the green, yellow, and golden-yellow, in which the chemical symbols below—Fe (iron), Ca calcium, Na (sodium), Pb (lead)—and the affixed lines, indicate the positions in which the vapours of these metals, when made incandescent, either in the flames or in the electrical spark, would show bright lines. The numbers above them show how far these fractions of Kirchhoff's map of the whole system are apart from each other. Here, also, we see a predominance of iron lines. In the whole spectrum Kirchhoff found not less than 450.

It follows from this, that the solar atmosphere contains an abundance of the vapours of iron, which, by the way, justifies us in concluding what an enormously high temperature must prevail there. It shows, moreover, how our figs. 3, 4, and 5 indicate iron, calcium, and sodium, and also the presence of hydrogen, of zinc, of copper, and of the metals of magnesia, alumina, baryta, and other terrestrial elements. Lead, on the other hand, is wanting, as well as gold, silver, mercury, antimony, arsenic, and some others.

The spectra of several fixed stars are similarly constituted; they show systems of fine lines which can be identified with those of terrestrial elements. In the atmosphere of Aldebaran in Taurus there is, again, hydrogen, iron, magnesium, calcium, sodium, and also mercury, antimony, and bismuth; and, according to H. C. Vogel, there is in α Orionis the rare metal thallium; and so on.

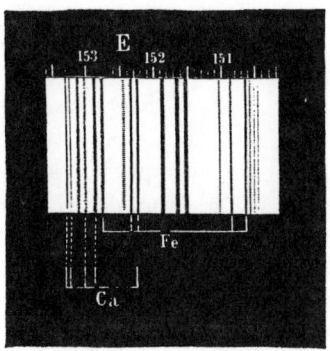
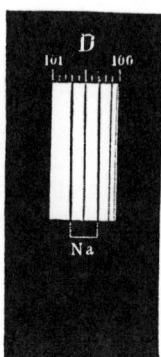

Figs. 3 and 4.

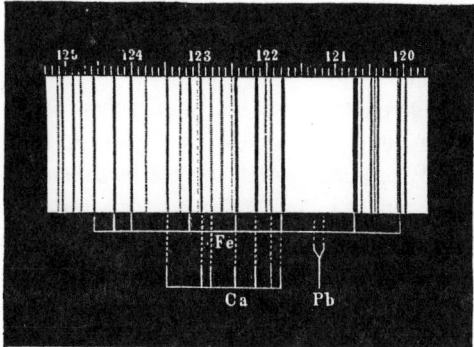

Fig. 5.

We cannot, indeed, say that we have explained all spectra; many fixed stars exhibit peculiarly banded spectra, probably belonging to gases whose molecules have not been completely resolved into their atoms by the high temperature. In the spectrum of the sun, also, are many lines which we cannot identify with those of terrestrial elements. It is possible that they may be due to substances unknown to us, it is also possible that they are produced by the excessively high temperature of the sun, far transcending anything we can produce. But this is certain, that the known terrestrial substances are widely diffused in space, and especially nitrogen, which constitutes the greater part of our atmosphere, and hydrogen, an element in water, which indeed is formed by its combustion. Both have been found in the irresolvable nebulae, and, from the inalterability of their shape, these must be masses of enormous dimensions and at an enormous distance. For this reason Sir W. Herschel considered that they did not belong to the system of our fixed stars, but were representatives of the manner in which other systems manifested themselves.

Spectrum analysis has further taught us more about the sun, by which he is brought nearer to us, as it were, than could formerly have seemed possible. You know that the sun is an enormous sphere, whose diameter is 112 times as great as that of the earth. We may consider what we see on its surface as a layer of incandescent vapour which, to judge from the appearances of the sun-spots, has a depth of about 500 miles. This layer of vapour which is continually radiating heat on the outside, and is certainly cooler than the inner masses of the sun, is, however, hotter than all our terrestrial flames—hotter even than the incandescent carbon points of the electrical arc, which represent the highest temperature attainable by terrestrial means. This can be deduced with certainty from Kirchhoff's law of the radiation of opaque bodies, from the greater luminous intensity of the sun. The older assumption, that the sun is a dark cool body, surrounded by a photosphere which only radiates heat and light externally, contains a physical impossibility.

Outside the opaque photosphere, the sun appears surrounded by a layer of transparent gases, which are hot enough to show in the spectrum bright coloured lines, and are hence called the *Chromosphere*. They show the bright lines of hydrogen, of sodium, of magnesium, and iron. In these layers of gas and of vapour about the sun enormous storms occur, which are as much greater than those of our earth in extent and in velocity as the sun is greater than the earth. Currents of ignited hydrogen burst out several thousands of miles high, like gigantic jets or tongues of flame, with clouds of smoke above them.[2] These structures could formerly only be viewed at the time of a total eclipse of the sun, forming what were called the rose-red protuberances. We now possess a method, devised by MM. Janssen and Lockyer, by which they may at any time be seen by the aid of the spectroscope.

On the other hand, there are individual darker parts on the sun's surface, what are called *sun-spots*, which were seen as long ago as by Galileo. They are funnel-shaped, the sides of the funnel are not so dark as the deepest part, the core. Fig. 6 represents such a spot according to Padre Secchi, as seen under powerful magnification. Their diameter is often more than many tens

---

2. According to H. C. Vogel's observations in Bothkamp to a height of 70,000 miles. The spectroscopic displacement of the lines showed velocities of 18 to 23 miles in a second; and, according to Lockyer, of even 37 to 42 miles.

of thousands of miles, so that two or three earths could lie in one of them. These spots may stand for weeks or months, slowly changing, before they are again resolved, and meanwhile several rotations of the sun may take

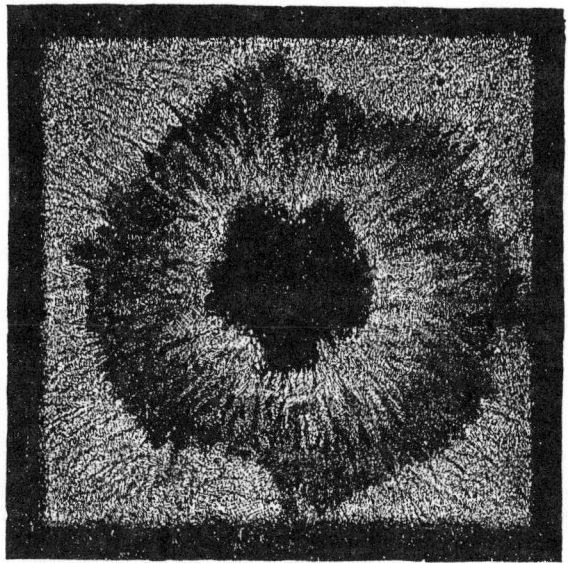

Fig. 6.

place. Sometimes, however, there are very rapid changes in them. That the core is deeper than the edge of the surrounding penumbra follows from their respective displacements as they come near the edge, and are therefore seen in a very oblique direction. Fig. 7 represents in A to E the different aspects of such a spot as it comes near the edge of the sun.

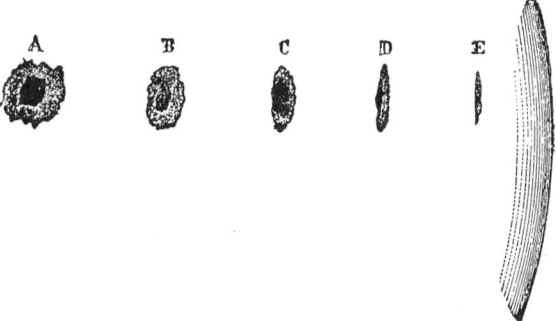

Fig. 7.

Just on the edge of these spots there are spectroscopic indications of the most violent motion, and in their vicinity there are often large protuberances; they show comparatively often a rotatory motion. They may be considered to be places where the cooler gases from the outer layers of the sun's atmosphere sink down, and perhaps produce local superficial coolings of the sun's mass. To understand the origin of these phenomena, it must be remembered that the gases, as they rise from the hot body of the sun, are charged with vapours of difficulty volatile metals, which expand as they ascend, and partly by their expansion, and partly by radiation into space, must become cooled. At the same time, they deposit their more difficulty volatile constituents as fog or cloud. This cooling can only, of course, be regarded as comparative; their temperature is probably, even then, higher than any temperature attainable on the earth. If now the upper layers, freed from the heavier vapours, sink down, there will be a space over the sun's body which is free from cloud. They appear then as depressions, because about them are layers of ignited vapours as much as 500 miles in height.

Violent storms cannot fail to occur in the sun's atmosphere, because it is cooled on the outside, and the coolest and comparatively densest and heaviest parts come to lie over the hotter and lighter ones. This is the reason why we have frequent, and at times sudden and violent, movements in the earth's atmosphere, because this is heated from the ground made hot by the sun and is cooled above. With the far more colossal magnitude and temperature of the sun, its meteorological processes are on a far larger scale, and are far more violent.

We will now pass to the question of the permanence of the present condition of our system. For a long time the view was pretty generally held that, in its chief features at any rate, it was unchangeable. This opinion was based mainly on the conclusions at which Laplace had arrived as the final results of his long and laborious investigations, of the influence of planetary disturbances. By disturbances of the planetary motion astronomers understand, as I have already mentioned, those deviations from the purely elliptical motion which are due to the attraction of various planets and satellites upon each other. The attraction of the sun, as by far the largest body of our system, is indeed the chief and preponderating force which produces the motion of the planets. If it alone were operative, each of the planets would move continuously in a constant ellipse whose axes would retain the same direction and the same magnitude, making the revolutions always in the same length of time. But, in point of fact, in addition to the attraction of the sun there are the attractions of all other planets, which, though small, yet, in long periods of time, do effect slow changes in the plane, the direction, and the magnitude of the axes of its elliptical orbit. It has been asked whether these attractions in the orbit of the planet could go so far as to cause two adjacent planets to encounter each other, so that individual ones fall into the sun. Laplace was able to reply that this could not be the case; that all alterations in the planetary orbits produced by this kind of disturbance must periodically increase and decrease, and again revert to a mean condition. But it must not be forgotten that this result of Laplace's investigations only applies to disturbances due to the reciprocal attraction of planets upon each other, and on the assumption that no forces of other kinds have any influence on their motions.

On our earth we cannot produce such an everlasting motion as that of the planets seems to be; for resisting forces are continually being opposed to all movements of terrestrial bodies. The best known of these are what we call friction, resistance of the air, and inelastic impact.

Hence the fundamental law of mechanics, according to which every motion of a body on which no force acts goes on in a straight line for ever with unchanged velocity, never holds fully.

Even if we eliminate the influence of gravity in a ball, for example, which rolls on a plane surface, we see it go on for a while, and the further the smoother is the path; but at the same time we hear the rolling ball make a clattering sound—that is, it produces waves of sound in the surrounding bodies; there is friction even on the smoothest surface; this sets the surrounding air in vibration, and imparts to it some of its own motion. Thus it happens that its velocity is continually less and less until it finally ceases. In like manner, even the most carefully constructed wheel which plays upon fine points, once made to turn, goes on for a quarter of an hour, or even more, but then stops. For there is always some friction on the axles, and in addition there is the resistance of the air, which resistance is mainly due to that of the particles of air against each other, due to their friction against the wheel.

If we could once set a body in rotation, and keep it from falling, without its being supported by another body, and if we could transfer the whole arrangement to an absolute vacuum, it would continue to move for ever with undiminished velocity. This case, which cannot be realized on terrestrial bodies, is apparently met with in the planets with their satellites. They appear to move in the perfectly vacuous cosmical space, without contact with any body which could produce friction, and hence their motion seems to be one which never diminishes.

You see, however, that the justification of this conclusion depends on the question whether cosmical space is really quite vacuous. Is there nowhere any friction in the motion of the planets?

From the progress which the knowledge of nature has made since the time of Laplace, we must now answer both questions in the negative.

Celestial space is not absolutely vacuous. In the first place, it is filled by that continuous medium the agitation of which constitutes light and radiant heat, and which physicists know as the luminiferous ether. In the second place, large and small fragments of heavy matter, from the size of huge stones to that of dust, are still everywhere scattered; at any rate, in those parts of space which our earth traverses. The existence of the luminiferous ether cannot be considered doubtful. That light and radiant heat are due to a motion which spreads in all directions has been sufficiently proved. For the transference of such a motion through space there must be something which can be moved. Indeed, from the magnitude of the action of this motion, or from that which the science of mechanics calls its *vis viva*, we may indeed assign certain limits for the density of this medium. Such a calculation has been made by Sir W. Thomson, the celebrated Glasgow physicist. He has found that the density may possibly be far less than that of the air in the most perfect exhaustion obtainable by a good air-pump; but that the mass of the ether cannot be absolutely equal to zero. A volume equal to that of the earth cannot contain less than 2,775 pounds of luminous ether.[3]

The phenomena in celestial space are in conformity with this. Just as a heavy stone flung through the air shows scarcely any influence of the resistance of the air, while a light feather is appreciably hindered; in like manner the medium which fills space is far too attenuated for any diminution to have been perceived in the motion of the planets since the time in which we possess astronomical observations of their path. It is different with the smaller bodies of our system. Encke in particular has shown, with reference to the well-known small comet which bears his name, that it circulates round the sun in ever-diminishing orbits and in ever shorter periods of revolution. Its motion is similar to that of the circular pendulum which we have mentioned, and which, having its velocity gradually delayed by the resistance of the air, describes circles about its centre of attraction, which continually become smaller and smaller. The reason for this phenomenon is the following: The force which offers a resistance to the attraction of the sun on all comets and planets, and which prevents them from getting continually nearer to the sun, is what is called the centrifugal force—that is, the tendency to continue their motion in a straight line in the direction of their path. As the force of their motion diminishes, they yield by a corresponding amount to the attraction of the sun, and get nearer to it. If the resistance continues, they will continue to get nearer the sun until they fall into it. Encke's comet is no doubt in this condition. But the resistance whose presence in space is hereby indicated, must act, and has long continued to act, in the same manner on the far larger masses of the planets.

The presence of partly fine and partly coarse heavy masses diffused in cosmical space is more distinctly revealed by the phenomena of asteroids

---

3. This calculation would, however, lose its bases if Maxwell's hypothesis were confirmed, according to which light depends on electrical and magnetical oscillations.

and of meteorites. We know now that these are bodies which ranged about in cosmical space, before they came within the region of our terrestrial atmosphere. In the more strongly resisting medium which this atmosphere offers they are delayed in their motion, and at the same time are heated by the corresponding friction. Many of them may still find an escape from the terrestrial atmosphere, and continue their path through space with an altered and retarded motion. Others fall to the earth; the larger ones as meteorites, while the smaller ones are probably resolved into dust by the heat, and as such fall without being seen. According to Alexander Herschel's estimate, we may figure shooting-stars as being on an average of the same size as paving-stones. Their incandescence mostly occurs in the higher and most attenuated regions of the atmosphere, eighteen miles and more above the surface of the earth. As they move in space under the influence of the same laws as the planets and comets, they possess a planetary velocity of from eighteen to forty miles in a second. By this, also, we observe that they are in fact *stelle cadente*, falling stars, as they have long been called by poets.

This enormous velocity with which they enter our atmosphere is undoubtedly the cause of their becoming heated. You all know that friction heats the bodies rubbed. Every match that we ignite, every badly greased coach-wheel, every auger which we work in hard wood, teaches this. The air, like solid bodies, not only becomes heated by friction, but also by the work consumed in its compression. One of the most important results of modern physics, the actual proof of which is mainly due to the Englishman Joule, is that, in such a case, the heat developed is exactly proportional to the work expended. If, like the mechanicians, we measure the work done by the weight which would be necessary to produce it, multiplied by the height from which it must fall, Joule has shown that the work, produced by a given weight of water falling through a height of 425 metres, would be just sufficient to raise the same weight of water through one degree Centigrade. The equivalent in work of a velocity of eighteen to twenty-four miles in a second may be easily calculated from known mechanical laws; and this, transformed into heat, would be sufficient to raise the temperature of a piece of meteoric iron to 900,000 to 2,500,000 degrees Centigrade, provided that all the heat were retained by the iron, and did not, as it undoubtedly does, mainly pass into the air. This calculation shows, at any rate, that the velocity of the shooting-stars is perfectly adequate to raise them to the most violent incandescence. The temperatures attainable by terrestrial means scarcely exceed 2,000 degrees. In fact, the outer crusts of meteoric stones generally show traces of incipient fusion; and in cases in which observers examined with sufficient promptitude the stones which had fallen they found them hot on the surface, while the interior of detached pieces seemed to show the intense cold of cosmical space.

To the individual observer who casually looks towards the starry sky the meteorites appear as a rare and exceptional phenomenon. If, however, they are continuously observed, they are seen with tolerable regularity, especially towards morning, when they usually fall. But a single observer only views but a small part of the atmosphere; and if they are calculated for the entire surface of the earth it results that about seven and a half millions fall every day. In our regions of space, they are somewhat sparse and distant from each other. According to Alexander Herschel's estimates, each stone is, on an average, at a distance of 450 miles from its neighbours. But the earth moves through 18 miles every second, and has a diameter of 7,820 miles, and therefore sweeps through 876 millions of cubic miles of space every second, and carries with it whatever stones are contained therein.

Many groups are irregularly distributed in space, being probably those which have already undergone disturbances by planets. There are also denser swarms which move in regular elliptical orbits, cutting the earth's orbit in definite places, and therefore always occur on particular days of the year. Thus the 10th of August of each year is remarkable, and every thirty-three years the splendid fireworks of the 12th to the 14th of November repeats itself for a few years. It is remarkable that certain comets accompany the paths of these swarms, and give rise to the supposition that the comets gradually split up into meteoric swarms.

This is an important process. What the earth does is done by the other planets, and in a far higher degree by the sun, towards which all the smaller bodies of our system must fall; those, therefore, that are more subject to the influence of the resisting medium, and which must fall the more rapidly, the smaller they are. The earth and the planets have for millions of years been sweeping together the loose masses in space, and they hold fast what they have once attracted. But it follows from this that the earth and the planets were once smaller than they are now, and that more mass was diffused in space; and if we follow out this consideration it takes us back to a state of things in which, perhaps, all the mass now accumulated in the sun and in the planets, wandered loosely diffused in space. If we consider, further, that the small masses of meteorites as they now fall, have perhaps been formed by the gradual aggregation of fine dust, we see ourselves led to a primitive condition of fine nebulous masses.

From this point of view, that the fall of shooting-stars and of meteorites is perhaps only a small survival of a process which once built up worlds, it assumes far greater significance.

This would be a supposition of which we might admit the possibility, but which could not perhaps claim any great degree of probability, if we did not find that our predecessors, starting from quite different considerations, had arrived at the same hypothesis.

You know that a considerable number of planets rotate around the sun besides the eight larger ones, Mercury, Venus, the Earth, Mars, Jupiter,

Saturn, Uranus, and Neptune; in the interval between Mars and Jupiter there circulate, as far as we know, 156 small planets or planetoids. Moons also rotate about the larger planets—that is, about the Earth and the four most distant ones, Jupiter, Saturn, Uranus, and Neptune; and lastly the Sun, and at any rate the larger planets, rotate about their own axes. Now, in the first place, it is remarkable that all the planes of rotation of the planets and of their satellites, as well as the equatorial planes of these planets, do not vary much from each other, and that in these planes all the rotation is in the same direction. The only considerable exceptions known are the moons of Uranus, whose plane is almost at right angles to the planes of the larger planets. It must at the same time be remarked that the coincidence, in the direction of these planes, is on the whole greater, the longer are the bodies and the larger the paths in question; while in the smaller bodies, and for the smaller paths, especially for the rotations of the planets about their own axes, considerable divergences occur. Thus the planes of all the planets, with the exception of Mercury and of the small ones between Mars and Jupiter, differ at most by three degrees from the path of the Earth. The equatorial plane of the Sun deviates by only seven and a half degrees, that of Jupiter only half as much. The equatorial plane of the Earth deviates, it is true, to the extent of twenty-three and a half degrees, and that of Mars by twenty-eight and a half degrees, and the separate paths of the small planet's satellites differ still more. But in these paths they all move direct, all in the same direction about the sun, and, as far as can be ascertained, also about their own axes, like the earth—that is, from west to east. If they had originated independently of each other, and had come together, any direction of the planes for each individual one would have been equally probable; a reverse direction of the orbit would have been just as probable as a direct one; decidedly elliptical paths would have been as probable as the almost circular ones which we meet with in all the bodies we have named. There is, in fact, a complete irregularity in the comets and meteoric swarms, which we have much reason for considering to be formations which have only accidentally come within the sphere of the sun's attraction.

The number of coincidences in the orbits of the planets and their satellites is too great to be ascribed to accident. We must inquire for the reason of this coincidence, and this can only be sought in a primitive connection of the entire mass. Now, we are acquainted with forces and processes which condense an originally diffused mass, but none which could drive into space such large masses, as the planets, in the condition we now find them. Moreover, if they had become detached from the common mass, at a place much nearer the sun, they ought to have a markedly elliptical orbit. We must assume, accordingly, that this mass in its primitive condition extended at least to the orbit of the outermost planets.

These were the essential features of the considerations which led Kant and Laplace to their hypothesis. In their view our system was originally a chaotic ball of nebulous matter, of which originally, when it extended to the path of the most distant planet, many billions of cubic miles could contain scarcely a gramme of mass. This ball, when it had become detached from the nebulous balls of the adjacent fixed stars, possessed a slow movement of rotation. It became condensed under the influence of the reciprocal attraction of its parts; and, in the degree in which it condensed, the rotatory motion increased, and formed it into a flat disk. From time to time masses at the circumference of this disk became detached under the influence of the increasing centrifugal force; that which became detached formed again into a rotating nebulous mass, which either simply condensed and formed a planet, or during this condensation again repelled masses from the periphery, which became satellites, or in one case, that of Saturn, remained as a coherent ring. In another case, the mass which separated from the outside of the chief ball, divided into many parts, detached from each other, and furnished the swarms of small planets between Mars and Jupiter.

Our more recent experience as to the nature of star showers teaches us that this process of the condensation of loosely diffused masses to form larger bodies is by no means complete, but still goes on, though the traces are slight. The form in which it now appears is altered by the fact that meanwhile the gaseous or dust-like mass diffused in space had united under the influence of the force of attraction, and of the force of crystallisation of their constituents to larger pieces than originally existed.

The showers of stars, as examples now taking place of the process which formed the heavenly bodies, are important from another point of view. They develop light and heat; and that directs us to a third series of considerations, which leads again to the same goal.

All life and all motion on our earth is, with few exceptions, kept up by a single force, that of the sun's rays, which bring to us light and heat. They warm the air of the hot zones, this becomes lighter and ascends, while the colder air flows towards the poles. Thus is formed the great circulation of the passage-winds. Local differences of temperature over land and sea, plains and mountains, disturb the uniformity of this great motion, and produce for us the capricious change of winds. Warm aqueous vapours ascend with the warm air, become condensed into clouds, and fall in the cooler zones, and upon the snowy tops of the mountains, as rain and as snow. The water collects in brooks, in rivers, moistens the plains, and makes life possible; crumbles the stones, carries their fragments along, and thus works at the geological transformation of the earth's surface. It is only under the influence of the sun's rays that the variegated covering of plants of the earth grows; and while they grow, they accumulate in their structure organic matter, which partly serves the whole animal kingdom as food, and serves man more particularly as fuel. Coals and lignites, the sources of power of our steam engines, are remains of primitive plants—the ancient production of the sun's rays.

Need we wonder if, to our forefathers of the Aryan race in India and Persia, the sun appeared as the fittest symbol of the Deity? They were right in regarding it as the giver of all life—as the ultimate source of almost all that has happened on earth.

But whence does the sun acquire this force? It radiates forth a more intense light than can be attained with any terrestrial means. It yields as much heat as if 1,500 pounds of coal were burned every hour upon each square foot of its surface. Of the heat which thus issues from it, the small fraction which enters our atmosphere furnishes a great mechanical force. Every steam-engine teaches us that heat can produce such force. The sun, in fact, drives on earth a kind of steam-engine whose performances are far greater than those of artificially constructed machines. The circulation of water in the atmosphere raises, as has been said, the water evaporated from the warm tropical seas to the mountain heights; it is, as it were, a water-raising engine of the most magnificent kind, with whose power no artificial machine can be even distantly compared. I have previously explained the mechanical equivalent of heat. Calculated by that standard, the work which the sun produces by its radiation is equal to the constant exertion of 7,000 horse-power for each square foot of the sun's surface.

For a long time experience had impressed on our mechanicians that a working force cannot be produced from nothing; that it can only be taken from the stores which nature possesses; which are strictly limited and which cannot be increased at pleasure—whether it be taken from the rushing water or from the wind; whether from the layers of coal, or from men and from animals, which cannot work without the consumption of food. Modern physics has attempted to prove the universality of this experience, to show that it applies to the great whole of all natural processes, and is independent of the special interests of man. These have been generalised and comprehended in the all-ruling natural law of the *Conservation of Force*. No natural process, and no series of natural processes, can be found, however manifold may be the changes which take place among them, by which a motive force can be continuously produced without a corresponding consumption. Just as the human race finds on earth but a limited supply of motive forces, capable of producing work, which it can utilise but not increase, so also must this be the case in the great whole of nature. The universe has its definite store of force, which works in it under ever varying forms; is indestructible, not to be increased, everlasting and unchangeable like matter itself. It seems as if Goethe had an idea of this when he makes the earth-spirit speak of himself as the representative of natural force.

In the currents of life, in the tempests of motion,

In the fervour of art, in the fire, in the storm,
    Hither and thither,
    Over and under,
    Wend I and wander.
    Birth and the grave,
    Limitless ocean,
    Where the restless wave
    Undulates ever
    Under and over,
    Their seething strife
    Heaving and weaving
    The changes of life.
At the whirling loom of time unawed,
I work the living mantle of God.

Let us return to the special question which concerns us here: Whence does the sun derive this enormous store of force which it sends out ?

On earth the processes of combustion are the most abundant source of heat. Does the sun's heat originate in a process of this kind? To this question we can reply with a complete and decided negative, for we now know that the sun contains the terrestrial elements with which we are acquainted. Let us select from among them the two, which, for the smallest mass, produce the greatest amount of heat when they combine; let us assume that the sun consists of hydrogen and oxygen, mixed in the proportion in which they would unite to form water. The mass of the sun is known, and also the quantity of heat produced by the union of known weights of oxygen and hydrogen. Calculation shows that under the above supposition, the heat resulting from their combustion would be sufficient to keep up the radiation of heat from the sun for 3,021 years. That, it is true, is a long time, but even profane history teaches that the sun has lighted and warmed us for 3,000 years, and geology puts it beyond doubt that this period must be extended to millions of years.

Known chemical forces are thus so completely inadequate, even on the most favourable assumption, to explain the production of heat which takes place in the sun, that we must quite drop this hypothesis.

We must seek for forces of far greater magnitude, and these we can only find in cosmical attraction. We have already seen that the comparatively small masses of shooting-stars and meteorites can produce extraordinarily large amounts of heat when their cosmical velocities are arrested by our atmosphere. Now the force which has produced these great velocities is gravitation. We know of this force as one acting on the surface of our planet when it appears as terrestrial gravity. We know that a weight *raised from the earth* can drive our clocks, and that in like manner the gravity of the water rushing down from the mountains works our mills.

If a weight falls from a height and strikes the ground its mass loses, indeed, the visible motion which it had as a whole—in fact, however, this motion is not lost; it is transferred to the smallest elementary particles of the mass, and this invisible vibration of the molecules is the motion of heat. Visible motion is transformed by impact, into the motion of heat.

That which holds in this respect for gravity, holds also for gravitation. A heavy mass, of whatever kind, which is suspended in space separated from another heavy mass, represents a force capable of work. For both masses attract each other, and, if unrestrained by centrifugal force, they move towards each other under the influence of this attraction; this takes place with ever-increasing velocity; and if this velocity is finally destroyed, whether this be suddenly, by collision, or gradually, by the friction of movable parts, it develops the corresponding quantity of the motion of heat, the amount of which can be calculated from the equivalence, previously established, between heat and mechanical work.

Now we may assume with great probability that very many more meteors fall upon the sun than upon the earth, and with greater velocity, too, and therefore give more heat. Yet the hypothesis, that the entire amount of the sun's heat which is continually lost by radiation, is made up by the fall of meteors, a hypothesis which was propounded by Mayer, and has been favourably adopted by several other physicists, is open, according to Sir W. Thomson's investigations, to objection; for, assuming it to hold, the mass of the sun should increase so rapidly that the consequences would have shown themselves in the accelerated motion of the planets. The entire loss of heat from the sun cannot at all events be produced in this way; at the most a portion, which, however, may not be inconsiderable.

If, now, there is no present manifestation of force sufficient to cover the expenditure of the sun's heat, the sun must originally have had a store of heat which it gradually gives out. But whence this store? We know that the cosmical forces alone could have produced it. And here the hypothesis, previously discussed as to the origin of the sun, comes to our aid. If the mass of the sun had been once diffused in cosmical space, and had then been condensed—that is, had fallen together under the influence of celestial gravity—if then the resultant motion had been destroyed by friction and impact, with the production of heat, the new world produced by such condensation must have acquired a store of heat not only of considerable, but even of colossal, magnitude.

Calculation shows that, assuming the thermal capacity of the sun to be the same as that of water, the temperature might be raised to 28,000,000 of degrees, if this quantity of heat could ever have been present in the sun at one time. This cannot be assumed, for such an increase of temperature would offer the greatest hindrance to condensation. It is probable rather that a great part of this heat, which was produced by condensation, began to radiate into space before this condensation was complete. But the heat which the sun could have previously developed by its condensation, would have been sufficient to cover its present expenditure for not less than 22,000,000 of years of the past.

And the sun is by no means so dense as it may become. Spectrum analysis demonstrates the presence of large masses of iron and of other known constituents of the rocks. The pressure which endeavours to condense the interior is about 800 times as great as that in the centre of the earth; and yet the density of the sun, owing probably to its enormous temperature, is less than a quarter of the mean density of the earth.

We may therefore assume with great possibility that the sun will still continue in its condensation, even if it only attained the density of the earth—though it will probably become far denser in the interior owing to the enormous pressure—this would develop fresh quantities of heat, which would be sufficient to maintain for an additional 17,000,000 of years the same intensity of sunshine as that which is now the source of all terrestrial life.

The smaller bodies of our system might become less hot than the sun, because the attraction of the fresh masses would be feebler. A body like the earth might, if even we put its thermal capacity as high as that of water, become heated to even 9,000 degrees, to more than our flames can produce. The smaller bodies must cool more rapidly as long as they are still liquid. The increase in temperature, with the depth, is shown in bore-holes and in mines. The existence of hot wells and of volcanic eruptions shows that in the interior of the earth there is a very high temperature, which can scarcely be anything than a remnant of the high temperature which prevailed at the time of its production. At any rate, the attempts to discover for the internal heat of the earth a more recent origin in chemical processes, have hitherto rested on very arbitrary assumptions; and, compared with the general uniform distribution of the internal heat, are somewhat insufficient.

On the other hand, considering the huge masses of Jupiter, of Saturn, of Uranus, and of Neptune, their small density, as well as that of the sun, is surprising, while the smaller planets and the moon approximate to the density of the earth. We are here reminded of the higher initial temperature, and the slower cooling, which characterises larger masses.[4] The moon, on the contrary, exhibits formations on its surface which are strikingly suggestive of volcanic craters, and point to a former state of ignition of our satellite. The mode of its rotation, moreover, that it always turns the same side towards the earth, is a peculiarity which might have been produced by the friction of a fluid. At present no trace of such a one can be perceived.

---

4. Mr. Zoellner concludes from photometric measurements, which, however, need confirmation, that Jupiter still possesses a light of its own.

You see, thus, by what various paths we are constantly led to the same primitive conditions. The hypothesis of Kant and Laplace is seen to be one of the happiest ideas in science, which at first astounds us, and then connects us in all directions with other discoveries, by which the conclusions are confirmed until we have confidence in them. In this case another circumstance has contributed—that is, the observation that this process of transformation, which the theory in question presupposes, goes on still, though on a smaller scale, seeing that all stages of that process can still be found to exist.

For as we have already seen, the larger bodies which are already formed go on increasing with the development of heat, by the attraction of the meteoric masses already diffused in space. Even now the smaller bodies are slowly drawn towards the sun by the resistance in space. We still find in the firmament of fixed stars, according to Sir J. Herschel's newest catalogue, over 5,000 nebulous spots, of which those whose light is sufficiently strong give for the most part a coloured spectrum of fine bright lines, as they appear in the spectra of the ignited gases. The nebulae are partly rounded structures, which are called *planetary nebulae* (fig. 8); sometimes wholly irregular in form, as the large nebula in Orion, represented in fig. 9; they are partly annular, as in the figures in fig. 10, from the Canes Venatici. They are for the most part feebly luminous over their whole surface, while the fixed stars only appear as luminous points.

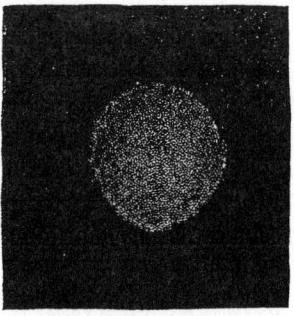

Fig. 8.

In many nebulae small stars can be seen, as in figs. 11 and 12, from Sagittarius and Aurigo. More stars are continually being discovered in them, the better are the telescopes used in their analysis. Thus, before the discovery of spectrum analysis, Sir W. Herschel's former view might be regarded as the most probable, that that which we see to be nebulae are only heaps of very fine stars, of other Milky Ways. Now, however, spectrum analysis has shown a gas spectrum in many nebulae which contains stars,

while actual heaps of stars show the continuous spectrum of ignited solid
bodies. Nebulae have in general three distinctly recognisable lines, one of
which, in the blue, belongs to hydrogen, a second in bluish-green to

Fig. 9.

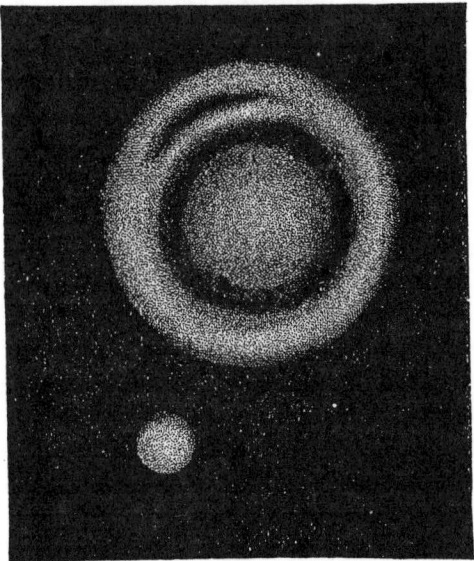

Fig. 10.

Figs. 11 and 12.

nitrogen,[5] while the third, between the two, is of unknown origin. Fig. 13 shows such a spectrum of a small but bright nebula in the Dragon. Traces of other bright lines are seen along with them, and sometimes also, as in fig. 13, traces of a continuous spectrum; all of which, however, are too feeble to admit of accurate investigation. It must be observed here that the light of

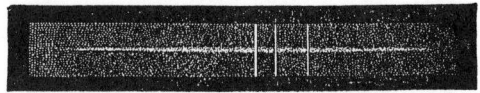

Fig. 13.

very feeble objects which give a continuous spectrum are distributed by the spectroscope over a large surface, and are therefore greatly enfeebled or even extinguished, while the undecomposable light of bright gas lines remains undecomposed, and hence can still be seen. In any case, the decomposition of the light of the nebulae shows that by far the greater part of their luminous surface is due to ignited gases, of which hydrogen forms a prominent constituent. In the planetary masses, the spherical or discoidal, it might be supposed that the gaseous mass had attained a condition of equilibrium; but most other nebulae exhibit highly irregular forms, which by no means correspond to such a condition. As, however, their shape has either not at all altered, or not appreciably, since they have been known and observed, they must either have very little mass, or they must be of colossal size and distance. The former does not appear very probable, because small masses very soon give out their heat, and hence we are left to the second

5. Or perhaps also to oxygen. The line occurs in the spectrum of atmospheric air, and according to H. C. Vogel's observation was wanting in the spectrum of pure oxygen.

alternative, that they are of huge dimensions and distances. The same conclusion had been originally drawn by Sir W. Herschel, on the assumption that the nebulae were heaps of stars.

With those nebulae which, besides the lines of gases, also show the continuous spectrum of ignited denser bodies, are connected spots which are partly irresolvable and partly resolvable into heaps of stars, which only show the light of the latter kind.

The countless luminous stars of the heavenly firmament, whose number increases with each newer and more perfect telescope, associate themselves with this primitive condition of the worlds as they are formed. They are like our sun in magnitude, in luminosity, and on the whole also in the chemical condition of their surface, although there may be differences in the quantity of individual elements.

But we find also in space a third stadium, that of extinct suns; and for this also there are actual evidences. In the first place, there are, in the course of history, pretty frequent examples of the appearance of new stars. In 1572 Tycho Brahe observed such a one, which, though gradually burning paler, was visible for two years, stood still like a fixed star, and finally reverted to the darkness from which it had so suddenly emerged. The largest of them all seems to have been that observed by Kepler in the year 1604, which was brighter than a star of the first magnitude, and was observed from September 27, 1604, until March 1606. The reason of its luminosity was probably the collision with a smaller world. In a more recent case, in which on May 12, 1866, a small star of the tenth magnitude in the Corona suddenly burst out to one of the second magnitude, spectrum analysis showed that it was an outburst of ignited hydrogen which produced the light. This was only luminous for twelve days.

In other cases obscure heavenly bodies have discovered themselves by their attraction on adjacent bright stars, and the motions of the latter thereby produced. Such an influence is observed in Sirius and Procyon. By means of a new refracting telescope Messrs. Alvan Clark and Bond, of Cambridge, U.S., have discovered in the case of Sirius a scarcely visible star, which has but little luminosity, but is almost seven times as heavy as the sun, has about half the mass of Sirius, and whose distance from Sirius is about equal to that of Neptune from the sun. The satellite of Procyon has not yet been seen; it appears to be quite dark.

Thus there are extinct suns. The fact that there are such lends new weight to the reasons which permit us to conclude that our sun also is a body which slowly gives out its store of heat, and thus will some time become extinct.

The term of 17,000,000 years which I have given may perhaps become considerably prolonged by the gradual abatement of radiation, by the new accretion of falling meteors, and by still greater condensation than that which I have assumed in that calculation. But we know of no natural process which could spare our sun the fate which has manifestly fallen upon other suns. This is a thought which we only reluctantly admit; it seems to us an insult to the beneficent Creative Power which we otherwise find at work in organisms and especially in living ones. But we must reconcile ourselves to the thought that, however we may consider ourselves to be the centre and final object of Creation, we are but as dust on the earth; which again is but a speck of dust in the immensity of space; and the previous duration of our race, even if we follow it far beyond our written history, into the era of the lake dwellings or of the mammoth, is but an instant compared with the primeval times of our planet; when living beings existed upon it, whose strange and unearthly remains still gaze at us from their ancient tombs; and far more does the duration of our race sink into insignificance compared with the enormous periods during which worlds have been in process of formation, and will still continue to form when our sun is extinguished, and our earth is either solidified in cold or is united with the ignited central body of our system.

But who knows whether the first living inhabitants of the warm sea on the young world, whom we ought perhaps to honour as our ancestors, would not have regarded our present cooler condition with as much horror as we look on a world without a sun? Considering the wonderful adaptability to the conditions of life which all organisms possess, who knows to what degree of perfection our posterity will have been developed in 17,000,000 of years, and whether our fossilised bones will not perhaps seem to them as monstrous as those of the Ichthyosaurus now do; and whether they, adjusted for a more sensitive state of equilibrium, will not consider the extremes of temperature, within which we now exist, to be just as violent and destructive as those of the older geological times appear to us? Yea, even if sun and earth should solidify and become motionless, who could say what new worlds would not be ready to develop life? Meteoric stones sometimes contain hydrocarbons; the light of the heads of comets exhibits a spectrum which is most like that of the electrical light in gases containing hydrogen and carbon. But carbon is the element, which is characteristic of organic compounds from which living bodies are built up. Who knows whether these bodies, which everywhere swarm through space, do not scatter germs of life wherever there is a new world, which has become capable of giving a dwelling-place to organic bodies? And this life we might perhaps consider as allied to ours in its primitive germ, however different might be the form which it would assume in adapting itself to its new dwelling-place.

However this may be, that which most arouses our moral feelings at the thought of a future, though possibly very remote, cessation of all living creation on the earth, is more particularly the question whether all this life is not an aimless sport, which will ultimately fall a prey to destruction by brute force? Under the light of Darwin's great thought we begin to see that not only pleasure and joy, but also pain, struggle, and death, are the powerful means by which nature has built up her finer and more perfect forms of life. And we men know more particularly that in our intelligence, our civic order, and our morality we are living on the inheritance which our forefathers have gained for us, and that which we acquire in the same way, will in like manner ennoble the life of our posterity. Thus the individual, who works for the ideal objects of humanity, even if in a modest position, and in a limited sphere of activity, may bear without fear the thought that the thread of his own consciousness will one day break. But even men of such free and large order of minds as Lessing and David Strauss could not reconcile themselves to the thought of a final destruction of the living race, and with it of all the fruits of all past generations.

As yet we know of no fact, which can be established by scientific observation, which would show that the finer and complex forms of vital motion could exist otherwise than in the dense material of organic life; that it can propagate itself as the sound-movement of a string can leave its originally narrow and fixed home and diffuse itself in the air, keeping all the time its pitch, and the most delicate shade of its colour-tint; and that, when it meets another string attuned to it, starts this again or excites a flame ready to sing to the same tone. The flame even, which, of all processes in inanimate nature, is the closest type of life, may become extinct, but the heat which it produces continues to exist—indestructible, imperishable, as an invisible motion, now agitating the molecules of ponderable matter, and then radiating into boundless space as the vibration of an ether. Even there it retains the characteristic peculiarities of its origin, and it reveals its history to the inquirer who questions it by the spectroscope. United afresh, these rays may ignite a new flame, and thus, as it were, acquire a new bodily existence.

Just as the flame remains the same in appearance, and continues to exist with the same form and structure, although it draws every minute fresh combustible vapour, and fresh oxygen from the air, into the vortex of its ascending current; and just as the wave goes on in unaltered form, and is yet being reconstructed every moment from fresh particles of water, so also in the living being, it is not the definite mass of substance, which now constitutes the body, to which the continuance of the individual is attached. For the material of the body, like that of the flame, is subject to continuous and comparatively rapid change—a change the more rapid, the livelier the activity of the organs in question. Some constituents are renewed from day to day, some from month to month, and others only after years. That which continues to exist as a particular individual is like the flame and the wave—only the form of motion which continually attracts fresh matter into its vortex and expels the old. The observer with a deaf ear only recognises the vibration of sound as long as it is visible and can be felt, bound up with heavy matter. Are our senses, in reference to life, like the deaf ear in this respect?

ADDENDUM

The sentences on pages 275–76 gave rise to a controversial attack by Mr. J. C. F. Zoellner, in his book 'On the Nature of the Comets,' on Sir W. Thomson, on which I took occasion to express myself briefly in the preface to the second part of the German translation of the 'Handbook of Theoretical Physics,' by Thomson and Tait. I give here the passage in question: — 'I will mention here a further objection. It refers to the question as to the possibility that organic germs may occur in meteoric stones, and be conveyed to the celestial bodies which have been cooled. In his opening Address at the Meeting of the British Association in Edinburgh, in August 1871, Sir W. Thomson had described this as "not unscientific." Here also, if there is an error, I must confess that I also am a culprit. I had mentioned the same view as a possible mode of explaining the transmission of organisms through space, even a little before Sir W. Thomson, in a lecture delivered in the spring of the same year at Heidelberg and Cologne, but not published. I cannot object if anyone considers this hypothesis to be in a high, or even in the highest, degree improbable. But to me it seems a perfectly correct scientific procedure, that when all our attempts fail in producing organisms from inanimate matter, we may inquire whether life has ever originated at all or not, and whether its germs have not been transported from one world to another, and have developed themselves wherever they found a favourable soil.

'Mr. Zoellner's so-called physical objections are but of very small weight. He recalls the history of meteoric stone, and adds (p. xxvi.): "If, therefore, that meteoric stones covered with organisms had escaped with a whole skin in the smashup of its mother-body, and had not shared the general rise of temperature, it must necessarily have first passed through the atmosphere of the earth, before it could deliver itself of its organisms for the purpose of peopling the earth."

'Now, in the first place, we know from repeated observations that the larger meteoric stones only become heated in their outside layer during their fall through the atmosphere, while the interior is cold, or even very cold. Hence all germs which there might be in the crevices would be safe from combustion in the earth's atmosphere. But even those germs which were collected on the surface when they reached the highest and most attenuated layer of the atmosphere would long before have been blown away by the powerful draught of air, before the stone reached the denser parts of the gaseous mass, where the compression would be sufficient to produce an appreciable heat. And, on the other hand, as far as the impact of two bodies is concerned, as Thomson assumes, the first consequences would be powerful mechanical motions, and only in the degree in which this would be destroyed by friction would heat be produced. We do not know whether that would last for hours, for days, or for weeks. The fragments, which at the first moment were scattered with planetary velocity, might escape without any disengagement of heat. I consider it even not improbable, that a stone, or shower of stones, flying through the higher regions of the atmosphere of a celestial body, carries with it a mass of air which contains unburned germs.

'As I have already remarked I am not inclined to suggest that all these possibilities are probabilities. They are questions the existence and signification of which we must remember, in order that if the case arise they may be solved by actual observations or by conclusions therefrom.'

Lecture delivered in Heidelberg and in Cologne, in 1871.

10

On the Relation of Optics to Painting

I FEAR that the announcement of my intention to address you on the subject of plastic art may have created no little surprise among some of my hearers. For I cannot doubt that many of you have had more frequent opportunities of viewing works of art, and have more thoroughly studied its historical aspects, than I can lay claim to have done; or indeed have had personal experience in the actual practice of art, in which I am entirely wanting. I have arrived at my artistic studies by a path which is but little trod, that is, by the physiology of the senses; and in reference to those who have a long acquaintance with, and who are quite at home in the beautiful fields of art, I may compare myself to a traveller who has entered upon them by a steep and stony mountain path, but who, in doing so, has passed many a stage from which a good point of view is obtained. If therefore I relate to you what I consider I have observed, it is with the understanding that I wish to regard myself as open to instruction by those more experienced than myself.

The physiological study of the manner in which the perceptions of our senses originate, how impressions from without pass into our nerves, and how the condition of the latter is thereby altered, presents many points of contact with the theory of the fine arts. On a former occasion I endeavoured to establish such a relation between the physiology of the sense of hearing, and the theory of music. Those relations in that case are particularly clear and distinct, because the elementary forms of music depend more closely on the nature and on the peculiarities of our perceptions than is the case in other arts, in which the nature of the material to be used and of the objects to be represented has a far greater influence. Yet even in those other branches of art, the especial mode of perception of that organ of sense by which the impression is taken up is not without importance; and a theoretical insight into its action, and into the principle of its methods, cannot be complete if this physiological element is not taken into account. Next to music this seems to predominate more particularly in painting, and this is the reason why I have chosen painting as the subject of my present lecture.

The more immediate object of the painter is to produce in us by his palette a lively visual impression of the objects which he has endeavoured to represent. The aim, in a certain sense, is to produce a kind of optical illusion; not indeed that, like the birds who pecked at the painted grapes of Apelles, we are to suppose we have present the real objects themselves, and not a picture; but in so far that the artistic representation produces in us a conception of their objects as vivid and as powerful as if we had them actually before us. The study of what are called illusions of the senses is however a very prominent and important part of the physiology of the senses; for just those cases in which external impressions evoke conceptions which are not in accordance with reality are particularly instructive for discovering the laws of those means and processes by which normal perceptions originate. We must look upon artists as persons whose observation of sensuous impressions is particularly vivid and accurate, and whose memory for these images is particularly true. That which long tradition has handed down to the men most gifted in this respect, and that which they have found by innumerable experiments in the most varied directions, as regards means and methods of representation, forms a series of important and significant facts, which the physiologist, who has here to learn from the artist, cannot afford to neglect. The study of works of art will throw great light on the question as to which elements and relations of our visual impressions are most predominant in determining our conception of what is seen, and what others are of less importance. As far as lies within his power, the artist will seek to foster the former at the cost of the latter.

In this sense then a careful observation of the works of the great masters will be serviceable, not only to physiological optics, but also because the investigation of the laws of the perceptions and of the observations of the senses will promote the theory of art, that is, the comprehension of its mode of action.

We have not here to do with a discussion of the ultimate objects and aims of art, but only with an examination of the action of the elementary means with which it works. The knowledge of the latter must, however, form an indispensable basis for the solution of the deeper questions, it we are to understand the problems which the artist has to solve, and the mode in which he attempts to attain his object.

I need scarcely lay stress on the fact, following as it does from what I have already said, that it is not my intention to furnish instructions according to which the artist is to work. I consider it a mistake to suppose that any kind of aesthetic lectures such as these can ever do so; but it is a mistake which those very frequently make who have only practical objects in view.

I. FORM.

The painter seeks to produce in his picture an image of external objects. The first aim of our investigation must be to ascertain what degree and what kind of similarity he can expect to attain, and what limits are assigned to him by the nature of his method. The uneducated observer usually requires nothing more than an illusive resemblance to nature: the more this is obtained, the more does he delight in the picture. An observer, on the contrary, whose taste in works of art has been more finely educated, will, consciously or unconsciously, require something more, and something different. A faithful copy of crude Nature he will at most regard as an artistic feat. To satisfy him, he will need artistic selection, grouping, and even idealisation of the objects represented. The human figures in a work of art must not be the everyday figures, such as we see in photographs; they must have expression, and a characteristic development, and if possible beautiful forms, which have perhaps belonged to no living individuals or indeed any individuals which ever have existed, but only to such a one as might exist, and as must exist, to produce a vivid perception of any particular aspect of human existence in its complete and unhindered development.

If however the artist is to produce an artistic arrangement of only idealised types, whether of man or of natural objects, must not the picture be an actual, complete, and directly true delineation of that which would appear if it anywhere came into being?

Since the picture is on a plane surface, this faithful representation can of course only give a faithful perspective view of the objects. Yet our eye, which in its optical properties is equivalent to a camera obscura, the well-known apparatus of the photographer, gives on the retina, which is its sensitive plate, only perspective views of the external world; these are stationary, like the drawing on a picture, as long as the standpoint of the eye is not altered. And, in fact, if we restrict ourselves in the first place to the form of the object viewed, and disregard for the present any consideration of colour, by a correct perspective drawing we can present to the eye of an observer, who views it from a correctly chosen point of view, the same forms of the visual image as the inspection of the objects themselves would present to the same eye, when viewed from the corresponding point of view.

But apart from the fact that any movement of the observer, whereby his eye changes its position, will produce displacements of the visual image, different when he stands before objects from those when he stands before the image, I could speak of only *one* eye for which equality of impression is to be established. We however see the world with *two* eyes, which occupy somewhat different positions in space, and which therefore show two different perspective views of objects before us. This difference of the images of the two eyes forms one of the most important means of estimating the distance of objects from our eye, and of estimating depth, and this is what is wanting to the painter, or even turns against him; since in binocular vision the picture distinctly forces itself on our perception as a plane surface.

You must all have observed the wonderful vividness which the solid form of objects acquires when good stereoscopic images are viewed in the stereoscope, a kind of vividness in which either of the pictures is wanting when viewed without the stereoscope. The illusion is most striking and instructive with figures in simple line; models of crystals and the like, in which there is no other element of illusion. The reason of this deception is, that looking with two eyes we view the world simultaneously from somewhat different points of view, and thereby acquire two different perspective images. With the right eye we see somewhat more of the right side of objects before us, and also somewhat more of those behind it, than we do with the left eye; and conversely we see with the left, more of the left side of an object, and of the background behind its left edges, and partially concealed by the edge. But a flat picture shows to the right eye absolutely the same picture, and all objects represented upon it, as to the left eye. If then we make for each eye such a picture as that eye would perceive if itself looked at the object, and if both pictures are combined in the stereoscope, so that each eye sees its corresponding picture, then as far as form is concerned the same impression is produced in the two eyes as the object itself produces. But if we look at a drawing or a picture with both eyes, we just as easily recognise that it is a representation on a plane surface, which is different from that which the actual object would show simultaneously to both eyes. Hence is due the well-known increase in the vividness of a picture if it is looked at with only one eye, and while quite stationary, through a dark tube; we thus exclude any comparison of its distance with that of adjacent objects in the room. For it must be observed that as we use different pictures seen with the two eyes for the perception of depth, in like manner as the body moves from one place to another, the pictures seen by the same eye serve for the same purpose. In moving, whether on foot or riding, the nearer objects are apparently displaced in comparison with the more distant ones; the former appear to recede, the latter appear to move with us. Hence arises a far stricter distinction between what is near and what is distant, than seeing with one eye from one and the same spot would ever afford us. If we move towards the picture, the sensuous impression that it is a flat picture hanging against the wall forces itself more strongly upon us than if we look at it while we are stationary. Compared with a large picture at a greater distance, all those elements which depend on binocular vision and on the movement of the body are less operative, because in very distant objects the differences between the images of the two eyes, or between the aspect from adjacent points of view, seem less. Hence large pictures furnish a less distorted aspect of their object than small ones, while the impression on a stationary eye, of a small picture close at hand, might be just the same as that of a large distant one. In a painting close at hand, the fact that it is a flat picture continually forces itself more powerfully and more distinctly on our perception.

The fact that perspective drawings, which are taken from too near a point of view, may easily produce a distorted impression, is, I think, connected with this. For here the want of the second representation for the other eye, which would be very different, is too marked. On the other hand, what are called geometrical projections, that is, perspective drawings which represent a view taken from an infinite distance, give in many cases a particularly favourable view of the object, although they correspond to a point of sight which does not in reality occur. Here the pictures of both eyes for such an object are the same.

You will notice that in these respects there is a primary incongruity, and one which cannot be got over, between the aspect of a picture and the aspect of reality. This incongruity may be lessened, but never entirely overcome. Owing to the imperfect action of binocular vision, the most important natural means is lost of enabling the observer to estimate the depth of objects represented in the picture. The painter possesses a series of subordinate means, partly of limited applicability, and partly of slight effect, of expressing various distances by depth. It is not unimportant to become acquainted with these elements, as arising out of theoretical considerations; for in the practice of the art of painting they have manifestly exercised great influence on the arrangement, selection, and mode of illumination of the objects represented. The distinctness of what is represented is indeed of subordinate importance when considered in reference to the ideal aims of art; it must not however be depreciated, for it is the first condition by which the observer attains an intelligibility of expression, which impresses itself without fatigue on the observer.

This direct intelligibility is again the preliminary condition for an undisturbed, and vivid action of the picture on the feelings and mood of the observer.

The subordinate methods of expressing depth which have been referred to, depend in the first place on perspective. Nearer objects partially conceal more distant ones, but can never themselves be concealed by the latter. If therefore the painter skilfully groups his objects, so that the feature in question comes into play, this gives at once a very certain gradation of far and near. This mutual concealment may even preponderate over the binocular perception of depth, if stereoscopic pictures are intentionally produced in which each counteracts the other. Moreover, in bodies of regular or of known form, the forms of perspective projection are for the most part characteristic for the depth of the object. If we look at houses, or other results of man's artistic activity, we know at the outset that the forms are for the most part plane surfaces at right angles to each other, with occasional circular or even spheroidal surfaces. And in fact, when we know so much, a correct perspective drawing is sufficient to produce the whole shape of the body. This is also the case with the figures of men and animals which are familiar to us, and whose forms moreover show two symmetrical halves. The best perspective drawing is however of but little avail in the case of irregular shapes, rough blocks of rock and ice, masses of foliage, and the like; that this is so, is best seen in photographs, where the perspective and shading may be absolutely correct, and yet the total impression is indistinct and confused.

When human habitations are seen in a picture, they represent to the observer the direction of the horizontal surfaces at the place at which they stand; and in comparison therewith the inclination of the ground, which without them would often be difficult to represent.

The apparent magnitude which objects, whose actual magnitude is known, present in different parts of the picture must also be taken into account. Men and animals, as well as familiar trees, are useful to the painter in this respect. In the more distant centre of the landscape they appear smaller than in the foreground, and thus their apparent magnitude furnishes a measure of the distance at which they are placed.

Shadows, and more especially double ones, are of great importance. You all know how much more distinct is the impression which a well-shaded drawing gives as distinguished from an outline; the shading is hence one of the most difficult, but at the same time most effective, elements in the productions of the draughtsman and painter. It is his task to imitate the fine gradation and transitions of light and shade on rounded surfaces, which are his chief means of expressing their modelling, with all their fine changes of curvature; he must take into account the extension or restriction of the sources of light, and the mutual reflection of the surfaces on each other. While the modifications of the lighting on the surface of bodies themselves is often dubious—for instance, an intaglio of a medal may, with a particular illumination, produce the impression of reliefs which are only illuminated from the other side—double shadows, on the contrary, are undoubted indications that the body which throws the shadow is nearer the source of light than that which receives the shadow. This rule is so completely without exception, that even in stereoscopic views a falsely placed double shadow may destroy or confuse the entire illusion.

The various kinds of illumination are not all equally favourable for obtaining the full effect of shadows. When the observer looks at the objects in the same direction as that in which light falls upon them, he sees only their illuminated sides and nothing of the shadow; the whole relief which the shadows could give then disappears. If the object is between the source of light and the observer he only sees the shadows. Hence we need lateral illumination for a picturesque shading; and over surfaces which like those of plane or hilly land only present slightly moving figures, we require light which is almost in the direction of the surface itself, for only such a one gives shadows. This is one of the reasons which makes illumination by the rising or the setting sun so effective. The forms of the landscape become more distinct. To this must also be added the influence of colour, and of aerial light, which we shall subsequently discuss.

Direct illumination from the sun, or from a flame, makes the shadows sharply defined, and hard. Illumination from a very wide luminous surface, such as a cloudy sky, makes them confused, or destroys them altogether. Between these two extremes there are transitions; illumination by a portion of the sky, defined by a window, or by trees, &c., allows the shadows to be more or less prominent according to the nature of the object. You must have seen of what importance this is to photographers, who have to modify their light by all manner of screens and curtains in order to obtain well-modelled portraits.

Of more importance for the representation of depth than the elements hitherto enumerated, and which are more or less of local and accidental significance, is what is called *aerial perspective*. By this we understand the optical action of the light, which the illuminated masses of air, between the observer and distant objects, give. This arises from a fine opacity in the atmosphere, which never entirely disappears. If, in a transparent medium, there are fine transparent particles of varying density and varying refrangibility, in so far as they are struck by it, they deflect the light passing through such a medium, partly by reflection and partly by refraction; to use an optical expression, they *scatter* it in all directions. If the opaque particles are sparsely distributed, so that a great part of the light can pass through them without being deflected, distant objects are seen in sharp, well-defined outlines through such a medium, while at the same time a portion of the light which is deflected is distributed in the transparent medium as an opaque halo. Water rendered turbid by a few drops of milk shows this dispersion of the light and cloudiness very distinctly. The light in this case is deflected by the microscopic globules of butter which are suspended in the milk.

In the ordinary air of our rooms, this turbidity is very apparent when the room is closed, and a ray of sunlight is admitted through a narrow aperture. We see then some of these solar particles, large enough to be distinguished by the naked eye, while others form a fine homogeneous turbidity. But even the latter must consist mainly of suspended particles of organic substances, for, according to an observation of Tyndall, they can be burnt. If the flame of a spirit lamp is placed directly below the path of these rays, the air rising from the flame stands out quite dark in the surrounding bright turbidity; that is to say, the air rising from the flame has been quite freed from dust. In the open air, besides dust and occasional smoke, we must often also take into account the turbidity arising from incipient aqueous deposits, where the temperature of moist air sinks so far that the water retained in it can no longer exist as invisible vapour. Part of the water settles then in the form of fine drops, as a kind of the very finest aqueous dust, and forms a finer or denser fog; that is to say, cloud. The turbidity which forms in hot sunshine and dry air may arise, partly from dust which the ascending currents of warm air whirl about; and partly from the irregular mixture of cold and warm layers of air of different density, as is seen in the tremulous motion of the lower layers of air over surfaces irradiated by the sun. But science can as yet give no explanation of the turbidity in the higher regions of the atmosphere which produces the blue of the sky; we do not know whether it arises from suspended particles of foreign substances, or whether the molecules of air themselves may not act as turbid particles in the luminous ether.

The colour of the light reflected by the opaque particles mainly depends on their magnitude. When a block of wood floats on water, and by a succession of falling drops we produce small wave-rings near it, these are repelled by the floating wood as if it were a solid wall. But in the long waves of the sea, a block of wood would be rocked about without the waves being thereby materially disturbed in their progress. Now light is well known to be an undulatory motion of the ether which fills all space. The red and yellow rays have the longest waves, the blue and violet the shortest. Very fine particles, therefore, which disturb the uniformity of the ether, will accordingly reflect the latter rays more markedly than the red and yellow rays. The light of turbid media is bluer, the finer are the opaque particles; while the larger particles of uniform light reflect all colours, and therefore give a whitish turbidity. Of this kind is the celestial blue, that is, the colour of the turbid atmosphere as seen against dark cosmical space. The purer and the more transparent the air, the bluer is the sky. In like manner it is bluer and darker when we ascend high mountains, partly because the air at great heights is freer from turbidity, and partly because there is less air above us. But the same blue, which is seen against the dark celestial space, also occurs against dark terrestrial objects; for instance, when a thick layer of illuminated air is between us and masses of deeply shaded or wooded hills. The same aerial light makes the sky blue, as well as the mountains; excepting that in the former case it is pure, while in the latter it is mixed with the light from objects behind; and moreover it belongs to the coarser turbidity of the lower regions of the atmosphere, so that it is whiter. In hot countries, and with dry air, the aerial turbidity is also finer in the lower regions of the air, and therefore the blue in front of distant terrestrial objects is more like that of the sky. The clearness and the pure colours of Italian landscapes depend mainly on this fact. On high mountains, particularly in the morning, the aerial turbidity is often so slight that the colours of the most distant objects can scarcely be distinguished from those of the nearest. The sky may then appear almost bluish-black.

Conversely, the denser turbidity consists mainly of coarser particles, and is therefore whitish. As a rule, this is the case in the lower layers of air, and in states of weather in which the aqueous vapour in the air is near its point of condensation.

On the other hand, the light which reaches the eye of the observer after having passed through a long layer of air, has been robbed of part of its violet and blue by scattered reflections; it therefore appears yellowish to reddish-yellow or red, the former when the turbidity is fine, the latter when it is coarse. Thus the sun and the moon at their rising and setting, and also distant brightly illuminated mountain-tops, especially snow-mountains, appear coloured.

These colourations are moreover not peculiar to the air, but occur in all cases in which a transparent substance is made turbid by the admixture of another transparent substance. We see it, as we have observed, in diluted milk, and in water to which a few drops of eau de Cologne have been added, whereby the ethereal oils and resins dissolved by the latter, separate out and produce the turbidity. Excessively fine blue clouds, bluer even than the air, may be produced, as Tyndall has observed, when the sun's light is allowed to exert its decomposing action on the vapours of certain carbon compounds. Goethe called attention to the universality of this phenomenon, and endeavoured to base upon it his theory of colour.

By aerial perspective we understand the artistic representation of aerial turbidity; for the greater or less predominance of the aerial colour above the colour of the objects, shows their varying distance very definitely; and landscapes more especially acquire the appearance of depth. According to the weather, the turbidity of the air may be greater or less, more white or more blue. Very clear air, as sometimes met with after continued rain, makes the distant mountains appear small and near; whereas, when the air contains more vapour, they appear large and distant.

This latter is decidedly better for the landscape painter, and the high transparent landscapes of mountainous regions, which so often lead the Alpine climber to under-estimate the distance and the magnitude of the mountain-tops before him, are also difficult to turn to account in a pictur-esque manner. Views from the valleys, and from seas and plains in which the aerial light is faintly but markedly developed, are far better; not only do they allow the various distances and magnitudes of what is seen to stand out, but they are on the other hand favourable to the artistic unity of colouration.

Although aerial colour is most distinct in the greater depths of landscape, it is not entirely wanting in front of the near objects of a room. What is seen to be isolated and well defined, when sunlight passes into a dark room through a hole in the shutter, is also not quite wanting when the whole room is lighted. Here, also, the aerial lighting must stand out against the background, and must somewhat deaden the colours in comparison with those of nearer objects; and these differences, also, although far more delicate than against the background of a landscape, are important for the historical, genre, or portrait painter; and when they are carefully observed and imitated, they greatly heighten the distinctness of his representation.

II. Shade.

The circumstances which we have hitherto discussed indicate a profound difference, and one which is exceedingly important for the perception of solid form, between the visual image which our eyes give, when we stand before objects, and that which the picture gives. The choice of the objects to be represented in pictures is thereby at once much restricted. Artists are well aware that there is much which cannot be represented by the means at their disposal. Part of their artistic skill consists in the fact that by a suitable grouping, position, and turn of the objects, by a suitable choice of the point of view, and by the mode of lighting, they learn to overcome the unfavourable conditions which are imposed on them in this respect.

It might at first sight appear that of the requisite truth to nature of a picture, so much would remain that, seen from the proper point of view, it would at least produce the same distribution of light, colour, and shadow in its field of view, and would produce in the interior of the eye exactly the same image on the retina as the object represented would do if we had it actually before us, and looked at it from a definite, fixed point of view. It might seem to be an object of pictorial skill to aim at producing, under the given limitations, the *same* effect as is produced by the object itself.

If we proceed to examine whether, and how far, painting can satisfy such a condition, we come upon difficulties before which we should perhaps shrink, if we did not know that they had been already overcome.

Let us begin with the simplest case; with the quantitative relations between luminous intensities. If the artist is to imitate exactly the impression which the object produces on our eye, he ought to be able to dispose of brightness and darkness equal to that which nature offers. But of this there can be no idea. Let me give a case in point. Let there be, in a picture-gallery, a desert-scene, in which a procession of Bedouins, shrouded in white, and of dark negroes, marches under the burning sunshine; close to it a bluish moonlight scene, where the moon is reflected in the water, and groups of trees, and human forms, are seen to be faintly indicated in the darkness. You know from experience that both pictures, if they are well done, can produce with surprising vividness the representation of their objects; and yet, in both pictures, the brightest parts are produced with the same white-lead, which is but slightly altered by admixtures; while the darkest parts are produced with the same black. Both, being hung on the same wall, share the same light, and the brightest as well as the darkest parts of the two scarcely differ as concerns the degree of their brightness.

How is it, however, with the actual degrees of brightness represented? The relation between the brightness of the sun's light, and that of the moon, was measured by Wollaston, who compared their intensities with that of the light of candles of the same material. He thus found that the luminosity of the sun is 800,000 times that of the brightest light of a full moon.

An opaque body, which is lighted from any source whatever, can, even in the most favourable case, only emit as much light as falls upon it. Yet, from Lambert's observations, even the whitest bodies only reflect about two fifths of the incident light. The sun's rays, which proceed parallel from the sun, whose diameter is 85,000 miles, when they reach us, are distributed uniformly over a sphere 195 millions of miles in diameter. Its density and illuminating power is here only the one forty-thousandth of that with which it left the sun's surface; and Lambert's number leads to the conclusion that even the brightest white surface on which the sun's rays fall vertically, has only the one hundred-thousandth part of the brightness of the sun's disk. The moon however is a gray body, whose mean brightness is only about one fifth of that of the purest white.

And when the moon irradiates a body of the purest white on the earth, its brightness is only the hundred-thousandth part of the brightness of the moon itself; hence the sun's disk is 80,000 million times brighter than a white which is irradiated by the full moon.

Now pictures which hang in a room are not lighted by the direct light of the sun, but by that which is reflected from the sky and clouds. I do not know of any direct measurements of the ordinary brightness of the light in a picture gallery, but estimates may be made from known data. With strong upper light and bright light from the clouds, the brightest white on a picture has probably 1-20th of the brightness of white directly lighted by the sun; it will generally be only 1-40th, or even less.

Hence the painter of the desert, even if he gives up the representation of the sun's disk, which is always very imperfect, will have to represent the glaringly lighted garments of his Bedouins with a white which, in the most favourable case, shows only the 1-20th part of the brightness which corresponds to actual fact. If he could bring it, with its lighting unchanged, into the desert near the white there, it would seem like a dark grey. I found in fact, by an experiment, that lamp-black, lighted by the sun, is not less than half as bright, as shaded white in the brighter part of a room.

On the picture of the moon, the same white which has been used for depicting the Bedouins' garments must be used for representing the moon's disk, and its reflection in the water; although the real moon has only one fifth of this brightness, and its reflection in water still less. Hence white garments in moonlight, or marble surfaces, even when the artist gives them a grey shade, will always be ten to twenty times as bright in his picture as they are in reality.

On the other hand, the darkest black which the artist could apply would be scarcely sufficient to represent the real illumination of a white object on which the moon shone. For even the deadest black coatings of lamp-black, black velvet, when powerfully lighted appear grey, as we often enough know to our cost, when we wish to shut off superfluous light. I investigated a coating of lamp-black, and found its brightness to be about $\frac{1}{100}$ that of white paper. The brightest colours of a painter are only about one hundred times as bright as his darkest shades.

The statements I have made may perhaps appear exaggerated. But they depend upon measurements, and you can control them by well-known observations. According to Wollaston, the light of the full moon is equal to that of a candle burning at a distance of 12 feet. You know that we cannot read by the light of the full moon, though we can read at a distance of three or four feet from a candle. Now assume that you suddenly passed from a room in daylight to a vault perfectly dark, with the exception of the light of a single candle. You would at first think you were in absolute darkness, and at most you would only recognise the candle itself. In any case, you would not recognise the slightest trace of any objects at a distance of 12 feet from the candle. These however are the objects whose illumination is the same as that which the moonlight gives. You would only become accustomed to the darkness after some time, and you would then find your way about without difficulty.

If, now, you return to the daylight, which before was perfectly comfortable, it will appear so dazzling that you will perhaps have to close the eyes, and only be able to gaze round with a painful glare. You see thus that we are concerned here not with minute, but with colossal, differences. How now is it possible that, under such circumstances, we can imagine there is any similarity between the picture and reality?

Our discussion of what we did not see at first, but could afterwards see in the vault, points to the most important element in the solution; it is the varying extent to which our senses are deadened by light; a process to which we can attach the same name, fatigue, as that for the corresponding one in the muscle. Any activity of our nervous system diminishes its power for the time being. The muscle is tired by work, the brain is tired by thinking, and by mental operations; the eye is tired by light, and the more so the more powerful the light. Fatigue makes it dull and insensitive to new impressions, so that it appreciates strong ones only moderately, and weak ones not at all.

But now you see how different is the aim of the artist when these circumstances are taken into account. The eye of the traveller in the desert, who is looking at the caravan, has been dulled to the last degree by the dazzling sunshine; while that of the wanderer by moonlight has been raised to the extreme of sensitiveness. The condition of one who is looking at a picture differs from both the above cases by possessing a certain mean degree of sensitiveness. Accordingly, the painter must endeavour to produce by his colours, on the moderately sensitive eye of the spectator, the same impression as that which the desert, on the one hand, produces on the deadened, and the moonlight, on the other hand, creates on the untired eye of its observer. Hence, along with the actual luminous phenomena of the outer world, the different physiological conditions of the eye play a most important part in the work of the artist. What he has to give is not a mere transcript of the object, but a translation of his impression into another scale of sensitiveness, which belongs to a different degree of impressibility of the observing eye, in which the organ speaks a very different dialect in responding to the impressions of the outer world.

In order to understand to what conclusions this leads, I must first of all explain the law which Fechner discovered for the scale of sensitiveness of the eye, which is a particular case of the more general *psychophysical law* of the relations of the various sensuous impressions to the irritations which produce them. This law may be expressed as follows: *Within very wide limits of brightness, differences in the strength of light are equally distinct or appear equal in sensation, if they form an equal fraction of the total quantity of light compared.* Thus, for instance, differences in intensity of one hundredth of the total amount can be recognised without great trouble with very different strengths of light, without exhibiting material differences in the certainty and facility of the estimate, whether the brightest daylight or the light of a good candle be used.

The easiest method of producing accurately measurable differences in the brightness of two white surfaces, depends on the use of rapidly rotating disks. If a disk, like the adjacent one in Fig. 1, is made to rotate very rapidly (that is, 20 to 30 times in a second), it appears to the eye to be covered with three grey rings as in Fig. 2. The reader must, however, figure to himself the grey of these rings, as it appears on the rotating disk of Fig. 1, as a scarcely perceptible shade of the ground. When the rotation is rapid each ring of the disk appears illuminated, as if all the light which fell upon it had been uniformly distributed over its entire surface. Those rings, in which are the black bands, have somewhat less light than the quite white ones, and if the breadth of the marks is compared with the length of half the circumference of the corresponding ring, we get the fraction by which the intensity of the

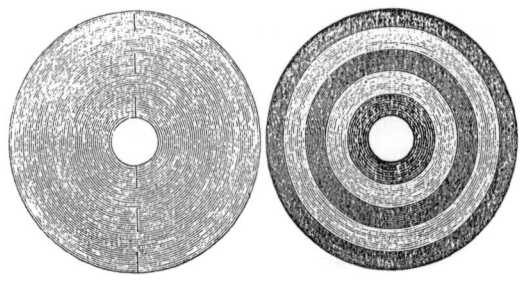

Figs. 1 and 2.

light in the white ground of the disk is diminished in the ring in question. If the bands are all equally broad, as in Fig. 1, the inner rings appear darker than the outer ones, for in this latter case the same loss of light is distributed over a larger area than in the former. In this way extremely delicate shades of brightness may be obtained, and by this method, when the strength of the illumination varies, the brightness always diminishes by the *same proportion* of its total value. Now it is found, in accordance with Fechner's law, that the distinctness of the rings is nearly constant for very different strengths of light. We exclude, of course, the cases of too dazzling or of too dim a light. In both cases the finer distinctions can no longer be perceived by the eye.

The case is quite different when for different strengths of illumination we produce differences which always correspond to the same quantity of light. If, for instance, we close the shutter of a room at daytime, so that it is quite dark, and now light it by a candle, we can discriminate without difficulty the shadows, such as that of the hand, thrown by the candle on a sheet of white paper. If, however, the shutters are again opened, so that daylight enters the room, for the same position of the hand we can no longer recognise the shadow, although there falls on that part of the white sheet, which is not struck by this shadow, the same excess of candle-light as upon the parts shaded by the hand. But this small quantity of light disappears in comparison with the newly added daylight, provided that this strikes all parts of the white sheet uniformly. You see then that, while the difference between candle-light and darkness can be easily perceived, the equally great difference between daylight, on the one hand, and daylight plus candle-light on the other, can be no longer recognised.

This law is of great importance in discriminating between various degrees of brightness of natural objects. A white body appears white because it reflects a large fraction, and a grey body appears grey because it reflects a small fraction, of incident light. For different intensities of illuminations the difference of brightness between the two will always correspond to the same fraction of their total brightness, and hence will be equally perceptible to our eyes, provided we do not approach too near to the upper or the lower limit of the brightness, for which Fechner's law no longer holds. Hence, on the whole, the painter can produce what appears an equal difference for the spectator of his picture, notwithstanding the varying strength of light in the gallery, provided he gives to his colours the same *ratio* of brightness as that which actually exists.

For, in fact, in looking at natural objects, the absolute brightness in which they appear to the eye varies within very wide limits, according to the intensity of the light, and the sensitiveness of the eye. That which is constant is only the ratio of the brightness in which surfaces of various depth of colour appear to us when lighted to the same amount. But this ratio of brightness is for us the perception, from which we form our judgment as to the lighter or darker colour of the bodies we see. Now this ratio can be imitated by the painter without restraint, and in conformity with nature, to evoke in us the same conception as to the nature of the bodies seen. A truthful imitation in this respect would be attained within the limits in which Fechner's law holds, if the artist reproduced the fully lighted parts of the objects which he has to represent with pigments, which, with the same light, were equal to the colours to be represented. This is approximately the case. On the whole, the painter chooses coloured pigments which almost exactly reproduce the colours of the bodies represented, especially for objects of no great depth, such as portraits, and which are only darker in the shaded parts. Children begin to paint on this principle; they imitate one colour by another; and, in like manner also, nations in which painting has remained in a childish stage. Perfect artistic painting is only reached when we have succeeded in imitating the action of light upon the eye, and not merely the pigments; and only when we look at the object of pictorial representation from this point of view, will it be possible to understand the variations from nature which artists have to make in the choice of their scale of colour and of shade.

These are, in the first case, due to the circumstance that Fechner's law only holds for mean degrees of brightness; while, for a brightness which is too high or too low, appreciable divergences are met with.

At both extremes of luminous intensity the eye is less sensitive for differences in light than is required by that law. With a very strong light it is dazzled; that is, its internal activity cannot keep pace with the external excitations; the nerves are too soon tired. Very bright objects appear almost always to be equally bright, even when there are, in fact, material differences in their luminous intensity. The light at the edge of the sun is only about half as bright as that at the centre, yet none of you will have noticed that, if you have not looked through coloured glasses, which reduce the brightness to a convenient extent. With a weak light the eye is also less sensitive, but from the opposite reason. If a body is so feebly illuminated that we scarcely perceive it, we shall not be able to perceive that its brightness is lessened by a shadow by the one hundredth or even by a tenth. It follows from this, that, with moderate illumination, darker objects become more like the darkest objects, while with greater illumination brighter objects become more like the brightest than should be the case in accordance with Fechner's law, which holds for mean degrees of illumination. From this results, what, for painting, is an extremely characteristic difference between the impression of very powerful and very feeble illumination.

When painters wish to represent glowing sunshine, they make all objects almost equally bright, and thus produce with their moderately bright colours the impression which the sun's glow makes upon the dazzled eye of the observer. If, on the contrary, they wish to represent moonshine, they only indicate the very brightest objects, particularly the reflection of moonlight on shining surfaces, and keep everything so dark as to be almost unrecognisable; that is to say, they make all dark objects more like the deepest dark which they can produce with their colours, than should be the case in accordance with the true ratio of the luminosities. In both cases they express, by their gradation of the lights, the *insensitiveness* of the eye for differences of too bright or too feeble lights. If they could employ the colour of the dazzling brightness of full sunshine, or of the actual dimness of moonlight, they would not need to represent the gradation of light in their picture other than it is in nature; the picture would then make the same impression on the eye as is produced by equal degrees of brightness of actual objects. The alteration in the scale of shade which has been described is necessary because the colours of the picture are seen in the mean brightness of a moderately lighted room, for which Fechner's law holds; and therewith objects are to be represented whose brightness is beyond the limits of this law.

We find that the older masters, and pre-eminently Rembrandt, employ the same deviation, which corresponds to that actually seen in moonlight landscapes; and this in cases in which it is by no means wished to produce the impression of moonshine, or of a similar feeble light. The brightest parts of the objects are given in these pictures in bright, luminous yellowish colours; but the shades towards the black are made very marked, so that the darker objects are almost lost in an impermeable darkness. But this darkness is covered with the yellowish haze of powerfully lighted aërial masses, so that, notwithstanding their darkness, these pictures give the impression of sunlight, and the very marked gradation of the shadows, the contours of the faces and figures, are made extremely prominent. The deviation from strict truth to nature is very remarkable in this shading, and yet these pictures give particularly bright and vivid aspects of the objects. Hence they are of particular interest for understanding the principles of pictorial illumination.

In order to explain these actions we must, I think, consider that while Fechner's law is approximately correct for those mean lights which are agreeable to the eye, the deviations which are so marked, for too high or too low lights, are not without some influence in the region of the middle lights. We have to observe more closely in order to perceive this influence. It is found, in fact, that when the very finest differences of shade are produced on a rotating disk, they are only visible by a light which about corresponds to the illumination of a white paper on a bright day, which is lighted by the light of the sky, but is not directly struck by the sun. With such a light, shades of $\frac{1}{150}$ or $\frac{1}{180}$ of the total intensity can be recognised. The light in which pictures are looked at is, on the contrary, much feebler; and if we are to retain the same distinctness of the finest shadows and of the modelling of the contours which it produces, the gradations of shade in the picture must be somewhat stronger than corresponds to the exact luminous intensities. The darkest objects of the picture thereby become unnaturally dark, which is however not detrimental to the object of the artist if the attention of the observer is to be directed to the brighter parts. The great artistic effectiveness of this manner shows us that the chief emphasis is to be laid on imitating difference of brightness and not on absolute brightness; and that the greatest differences in this latter respect can be borne without perceptible incongruity, if only their gradations are imitated with expression.

III. Colour.

With these divergences in brightness are connected certain divergences in colour, which, physiologically, are caused by the fact that the scale of sensitiveness is different for different colours. The strength of the sensation produced by light of a particular colour, and for a given intensity of light, depends altogether on the special reaction of that complex of nerves which are set in operation by the action of the light in question. Now all our sensations of colour are admixtures of three simple sensations; namely, of red, green, and violet,[1] which, by a not improbable supposition of Thomas Young, can be apprehended quite independently of each other by three different systems of nerve-fibres. To this independence of the different sensations of colour corresponds their independence in the gradation of

---

1. Helmholtz's "The Recent Progress of the Theory of Vision," pp. 152–65.

intensity. Recent measurements[2] have shown that the sensitiveness of our eye for feeble shadows is greatest in the blue and least in the red. A difference of $\frac{1}{205}$ to $\frac{1}{268}$ of the intensity can be observed in the blue, and with an untired eye of $\frac{1}{16}$ in the red; or when the colour is dimmed by being looked at for a long time, a difference of $\frac{1}{50}$ to $\frac{1}{70}$.

Red therefore acts as a colour towards whose shades the eye is relatively less sensitive than towards that of blue. In agreement with this, the impression of glare, as the intensity increases, is feebler in red than in blue. According to an observation of Dove, if a blue and a red paper be chosen which appear of equal brightness under a mean degree of white light, as the light is made much dimmer the blue appears brighter, and as the light is much strengthened, the red. I myself have found that the same differences are seen, and even in a more striking manner, in the red and violet spectral colours, and, when their intensity is increased only moderately, by the same fraction for both.

Now the impression of white is made up of the impressions which the individual spectral colours make on our eye. If we increase the brightness of white, the strength of the sensation for the red and yellow rays will relatively be more increased than that for the blue and violet. In bright white, therefore, the former will produce a relatively stronger impression than the latter; in dull white the blue and bluish colours will have this effect. Very bright white appears therefore yellowish, and dull white appears bluish. In our ordinary way of looking at the objects about us, we are not so readily conscious of this; for the direct comparison of colours of very different shade is difficult, and we are accustomed to see in this alteration in the white the result of different illumination of one and the same white object, so that in judging pigment-colours we have learnt to eliminate the influence of brightness.

If however to the painter is put the problem of imitating, with faint colours, white irradiated by the sun, he can attain a high degree of resemblance; for by an admixture of yellow in his white he makes this colour preponderate just as it would preponderate in actual bright light, owing to the impression on the nerves. It is the same impression as that produced if we look at a clouded landscape through a yellow glass, and thereby give it the appearance of a sunny light. The artist will, on the contrary, give a bluish tint to moonlight, that is, a faint white; for the colours on the picture must, as we have seen, be far brighter than the colour to be represented. In moonshine scarcely any other colour can be recognised than blue; the blue starry sky or blue colours may still appear distinctly coloured, while yellow and red can only be seen as obscurations of the general bluish white or grey.

I will again remind you that these changes of colour would not be

---

2. Dobrowolsky in *Graefe's Archiv für Ophthalmologie*, vol. xviii. part i. pp. 24–92.

necessary if the artist had at his disposal colours of the same brightness, or the same faintness, as are actually shown by the bodies irradiated by the sun or by the moon.

The change of colour, like the scale of shade, previously discussed, is a subjective action which the artist must represent objectively on his canvas, since moderately bright colours cannot produce them.

We observe something quite similar in regard to the phenomena of *Contrast.* By this term we understand cases in which the colour or brightness of a surface appears changed by the proximity of a mass of another colour or shade, and, in such a manner, that the original colour appears darker by the proximity of a brighter shade, and brighter by that of a darker shade; while by a colour of a different kind it tends towards the complementary tint.

The phenomena of contrast are very various, and depend on different causes. One class, *Chevreul's simultaneous Contrast,* is independent of the motions of the eyes, and occurs with surfaces where there are very slight differences in colour and shade. This contrast appears both on the picture and in actual objects, and is well known to painters. Their mixtures of colours on the palette often appear quite different to what they are on the picture. The changes of colour which are here met with are often very striking; I will not, however, enter upon them, for they produce no divergence between the picture and reality.

The second class of phenomena of contrast, and one which, for us, is more important, is met with in changes of direction of the glance, and more especially between surfaces in which there are great differences of shade and of colour. As the eye glides over bright and dark, or coloured objects and surfaces, the impression of each colour changes, for it is depicted on portions of the retina which directly before were struck by other colours and lights, and were therefore changed in their sensitiveness to an impression. This kind of contrast is therefore essentially dependent on movements of the eye, and has been called by Chevreul, '*successive Contrast.*'

We have already seen that the retina is more sensitive in the dark to feeble light than it was before. By strong light, on the contrary, it is dulled, and is less sensitive to feeble lights which it had before perceived. This latter process is designated as 'Fatigue' of the retina; an exhaustion of the capability of the retina by its own activity, just as the muscles by their activity become tired.

I must here remark that the fatigue of the retina by light does not necessarily extend to the whole surface; but when only a small portion of this membrane is struck by a minute, defined picture it can also be locally developed in this part only.

You must all have observed the dark spots which move about in the field of vision, when we have been looking for only a short time towards the setting sun, and which physiologists call *negative after-images* of the sun. They are due to the fact that only those parts of the retina which are actually struck by the image of the sun in the eye, have become insensitive to a new impression of light. If, with an eye which is thus locally tired, we look towards a uniformly bright surface, such as the sky, the tired parts of the retina are more feebly and more darkly affected than the other portions, so that the observer thinks he sees dark spots in the sky, which move about with his sight. We have then in juxtaposition, in the bright parts of the sky, the impression which these make upon the untired parts of the retina, and in the dark spots their action on the tired portions. Objects, bright like the sun, produce negative after-images in the most striking manner; but with a little attention they may be seen even after much more moderate impressions of light. A longer time is required in order to develop such an impression, so that it may be distinctly recognised, and a definite point of the bright object must be fixed, without moving the eye, so that its image may be distinctly formed on the retina, and only a limited portion of the retina be excited and tired, just as in producing sharp photographic portraits, the object must be stationary during the time of exposure in order that its image may not be displaced on the sensitive plate. The after-image in the eye is, as it were, a photograph on the retina, which becomes visible owing to the altered sensitiveness towards fresh light, but only remains stationary for a short time; it is longer, the more powerful and durable was the action of light.

If the object viewed was coloured, for instance red paper, the after-image is of the complementary colour on a grey ground; in this case of a bluish green.[3] Rose-red paper, on the contrary, gives a pure green after-image, green a rose-red, blue a yellow, and yellow a blue. These phenomena show that in the retina partial fatigue is possible for the several colours. According to Thomas Young's hypothesis of the existence of three systems of fibres in the visual nerves,[4] of which one set perceives red whatever the kind of irritation, the second green, and the third violet, with green light, only those fibres of the retina which are sensitive to green are powerfully excited and tired. If this same part of the retina is afterwards illuminated with white light, the sensation of green is enfeebled, while that of red and violet is vivid and predominant; their sum gives the sensation of purple,

---

3. In order to see this kind of image as distinctly as possible, it is desirable to avoid all movements of the eye. On a large sheet of dark grey paper a small black cross is drawn, the centre of which is steadily viewed, and a quadrangular sheet of paper of that colour whose after-image is to be observed is slid from the side, so that one of its corners touches the cross. The sheet is allowed to remain for a minute or two, the cross being steadily viewed, and it is then drawn suddenly away, without relaxing the view. In place of the sheet removed the after-image appears then on the dark ground.

4. See Helmholtz's "The Recent Progress of the Theory of Vision," p. 161.

which mixed with the unchanged white ground forms rose-red.

In the ordinary way of looking at light and coloured objects, we are not accustomed to fix continuously one and the same point; for following with the gaze the play of our attentiveness, we are always turning it to new parts of the object as they happen to interest us. This way of looking, in which the eye is continually moving, and therefore the retinal image is also shifting about on the retina, has moreover the advantage of avoiding disturbances of sight, which powerful and continuous after-images would bring with them. Yet here also, after-images are not wanting; only they are shadowy in their contours, and of very short duration.

If a red surface be laid upon a grey ground, and if we look from the red over the edge towards the grey, the edges of the grey will seem as if struck by such an after-image of red, and will seem to be of a faint bluish green. But as the after-image rapidly disappears, it is mostly only those parts of the grey, which are nearest the red, which show the change in a marked degree.

This also is a phenomenon which is produced more strongly by bright light and brilliant saturated colours than by fainter light and duller colours. The artist, however, works for the most part with the latter. He produces most of his tints by mixture; each mixed pigment is, however, greyer and duller than the pure colour of which it is mixed, and even the few pigments of a highly saturated shade, which oil-painting can employ, are comparatively dark. The pigments employed in water-colours and coloured chalks are again comparatively white. Hence such bright contrasts, as are observed in strongly coloured and strongly lighted objects in nature, cannot be expected from their representation in the picture. If, therefore, with the pigments at his command, the artist wishes to reproduce the impression which objects give, as strikingly as possible, he must paint the contrasts which they produce. If the colours on the picture are as brilliant and luminous as in the actual objects, the contrasts in the former case would produce themselves as spontaneously as in the latter. Here, also, subjective phenomena of the eye must be objectively introduced into the picture, because the scale of colour and of brightness is different upon the latter.

With a little attention you will see that painters and draughtsmen generally make a plain uniformly lighted surface brighter, where it is close to a dark object, and darker, where it is near a light object. You will find that uniform grey surfaces are given a yellowish tint at the edge where there is a background of blue, and a rose-red tint where they impinge on green, provided that none of the light collected from the blue or green can fall upon the grey. Where the sun's rays passing through the green leafy shade of trees strike against the ground, they appear to the eye, tired with looking at the predominant green, of a rose-red tint; the whole daylight, entering through a slit, appears blue, compared with reddish yellow candle-light. In this way they are represented by the painter, since the colours of his pictures are not bright enough to produce the contrast without such help.

To the series of subjective phenomena, which artists are compelled to represent objectively in their pictures, must be associated certain phenomena of *irradiation*. By this is understood cases in which any bright object in the field spreads its light or colour over the neighbourhood. The phenomena are the more marked the brighter is the radiating object, and the halo is brightest in the immediate neighbourhood of the bright object, but diminishes at a greater distance. These phenomena of irradiation are most striking around a very bright light on a dark ground. If the view of the flame itself is closed by a narrow dark object such as the finger, a bright misty halo disappears, which covers the whole neighbourhood, and, at the same time, any objects there may be in the dark part of the field of view are seen more distinctly. If the flame is partly screened by a ruler, this appears jagged where the flame projects beyond it. The luminosity in the neighbourhood of the flame is so intense, that its brightness can scarcely be distinguished from that of the flame itself; as is the case with all bright objects, the flame appears magnified, and as if spreading over towards the adjacent dark objects.

The cause of this phenomenon is quite similar to that of aërial perspective. It is due to a diffusion of light which arises from the passage of light through dull media, excepting that for the phenomena of aërial perspective the turbidity is to be sought in the air in front of the eye, while for true phenomena of irradiation it is to be sought in the transparent media of the eye. When even the healthiest human eye is examined by powerful light, the best being a pencil of sunlight concentrated on the side by a condensing lens, it is seen that the sclerotica and crystalline lens are not perfectly clear. If strongly illuminated, they both appear whitish and as if rendered turbid by a fine mist. Both are, in fact, tissues of fibrous structure, and are not therefore so homogeneous as a pure liquid or a pure crystal. Every inequality, however small, in the structure of a transparent body can, however, reflect some of the incident light—that is, can diffuse it in all directions.[5]

The phenomena of irradiation also occur with moderate degrees of brightness. A dark aperture in a sheet of paper illuminated by the sun, or a small dark object on a coloured glass plate which is held against the clear sky, appear as if the colour of the adjacent surface were diffused over them.

Hence the phenomena of irradiation are very similar to those which produce the opacity of the air. The only essential difference lies in this, that the opacity by luminous air is stronger before distant objects which have a greater mass of air in front of them than before near ones; while irradiation

---

5. I disregard here the view that irradiation in the eye depends on a diffusion of the excitation in the substance of the nerves, as this appears to me too hypothetical. Moreover, we are here concerned with the phenomena and not with their cause.

in the eye sheds its halo uniformly over near and over distant objects. Irradiation also belongs to the subjective phenomena of the eye which the artist represents objectively, because painted lights and painted sunlight are not bright enough to produce a distinct irradiation in the eye of the observer. The representation which the painter has to give of the lights and colours of his object I have described as a translation, and I have urged that, as a general rule, it cannot give a copy true in all its details. The altered scale of brightness which the artist must apply in many cases is opposed to this. It is not the colours of the objects, but the impression which they have given, or would give, which is to be imitated, so as to produce as distinct and vivid a conception as possible of those objects. As the painter must change the scale of light and colour in which he executes his picture, he only alters something which is subject to manifold change according to the lighting, and the degree of fatigue of the eye. He retains the more essential, that is, the *gradations* of brightness and tint. Here present themselves a series of phenomena which are occasioned by the manner in which the eye replies to an external irritation; and since they depend upon the intensity of this irritation they are not directly produced by the varied luminous intensity and colours of the picture. These objective phenomena, which occur on looking at the object, would be wanting if the painter did not represent them objectively on his canvas. The fact that they are represented is particularly significant for the kind of problem which is to be solved by a pictorial representation.

Now, in all translations, the individuality of the translator plays a part. In artistic productions many important points are left to the choice of the artist, which he can decide according to his individual taste, or according to the requirements of his subject. Within certain limits he can freely select the absolute brightness of his colours, as well as the strength of the shadows. Like Rembrandt, he may exaggerate them in order to obtain strong relief; or he may diminish them, with Fra Angelico and his modern imitators, in order to soften earthly shadows in the representation of sacred objects. Like the Dutch school, he may represent the varying light of the atmosphere, now bright and sunny, and now pale, or warm and cold, and thereby evoke in the observer moods which depend on the illumination and on the state of the weather; or by means of undisturbed air he may cause his figures to stand out objectively clear as it were, and uninfluenced by subjective impressions. By this means, great variety is attained in what artists call 'style' or 'treatment,' and indeed in their purely pictorial elements.

IV. HARMONY OF COLOUR.

We here naturally raise the question: If, owing to the small quantity of light and saturation of his colours, the artist seeks, in all kinds of indirect ways, by imitating subjective impressions to attain resemblance to nature, as close as possible, but still imperfect, would it not be more convenient to seek for means of obviating these evils? Such there are indeed. Frescoes are sometimes viewed in direct sunlight; transparencies and paintings on glass can utilise far higher degrees of brightness, and far more saturated colours; in dioramas and in theatrical decorations we may employ powerful artificial light, and, if need be, the electric light. But when I enumerate these branches of art, it will at once strike you that those works which we admire as the greatest masterpieces of painting, do not belong to this class; but by far the larger number of the great works of art are executed with the comparatively dull water or oil-colours, or at any rate for rooms with softened light. If higher artistic effects could be attained with colours lighted by the sun, we should undoubtedly have pictures which took advantage of this. Fresco painting would have led to this; or the experiments of Münich's celebrated optician Steinheil, which he made as a matter of science, that is, to produce oil paintings which should be looked at in bright sunshine, would not be isolated.

Experiment seems therefore to teach, that moderation of light and of colours in pictures is ever advantageous, and we need only look at frescoes in direct sunlight, such as those of the new Pinakothek in Münich, to learn in what this advantage consists. Their brightness is so great that we cannot look at them steadily for any length of time. And what in this case is so painful and so tiring to the eye, would also operate in a smaller degree if, in a picture, brilliant colours were used, even locally and to a moderate extent, which were intended to represent bright sunlight, and a mass of light shed over the picture. It is much easier to produce an accurate imitation of the feeble light of moonshine with artificial light in dioramas and theater decorations.

We may therefore designate truth to Nature of a beautiful picture as an ennobled fidelity to Nature. Such a picture reproduces all that is essential in the impression, and attains full vividness of conception, but without injury or tiring the eye by the nude lights of reality. The differences between Art and Nature are chiefly confined, as we have already seen, to those matters which we can in reality only estimate in an uncertain manner, such as the absolute intensities of light.

That which is pleasant to the senses, the beneficial but not exhausting fatigue of our nerves, the feeling of comfort, corresponds in this case, as in others, to those conditions which are most favourable for perceiving the outer world, and which admit of the finest discrimination and observation.

It has been mentioned above that the discrimination of the finest shadows, and of the modelling which they express, is the most delicate under a certain mean brightness. I should like to direct your attention to another point which has great importance in painting: I refer to our natural delight in colours, which has undoubtedly a great influence upon our pleasure in the works of the painter. In its simplest expression, as pleasure in gaudy flowers, feathers, stones, in fireworks, and Bengal lights, this inclination has but little to do with man's sense of art; it only appears as the natural pleasure of the perceptive organism in the varying and multifarious excitation of its various nerves, which is necessary for its healthy continuance and productivity. But the thorough fitness in the construction of living organisms, whatever their origin, excludes the possibility that in the majority of healthy individuals an instinct should be developed or maintain itself which did not serve some definite purpose.

We have not far to seek for the delight in light and in colours, and for the dread of darkness; this coincides with the endeavour to see and to recognise surrounding objects. Darkness owes the greater part of the terror which it inspires to the fright of what is unknown and cannot be recognised. A coloured picture gives a far more accurate, richer, and easier conception than a similarly executed drawing, which only retains the contrasts of light and shade. A picture retains the latter, but has in addition the material for discrimination which colours afford; by which surfaces which appear equally bright in the drawing, owing to their different colour, are now assigned to various objects, or again as alike in colour are seen to be parts of the same, or of similar objects. In utilising the relations thus naturally given, the artist, by means of prominent colours, can direct and enchain the attention of the observer upon the chief objects of the picture; and by the variety of the garments he can discriminate the figures from each other, but complete each individual one in itself. Even the natural pleasure in pure, strongly saturated colours, finds its justification in this direction. The case is analogous to that in music, with the full, pure, well-sounding tones of a beautiful voice. Such a one is more expressive; that is, even the smallest change of its pitch, or its quality—any slight interruption, any tremulousness, any rising or falling in it—is at once more distinctly recognised by the hearer than could be the case with a less regular sound; and it seems also that the powerful excitation which it produces in the ear of the listener, arouses trains of ideas and passions more strongly than does a feebler excitation of the same kind. A pure, fundamental colour bears to small admixtures the same relation as a dark ground on which the slightest shade of light is visible. Any of the ladies present will have known how sensitive clothes of uniform saturated shades are to dirt, in comparison with grey or greyish-brown materials. This also corresponds to the conclusions from Young's theory of colours. According to this theory, the perception of each of the three fundamental colours arises from the excitation of only one kind of sensitive fibres, while the two others are at rest; or at any rate are but feebly excited. A brilliant, pure colour produces a powerful stimulus, and yet, at the same time, a great degree of sensitiveness to the admixture of other colours, in those systems of nerve-fibres which are at rest. The modelling of a coloured surface mainly depends upon the reflection of light of other colours which falls upon them from without. It is more particularly when the material glistens that the reflections of the bright places are preferably of the colour of the incident light. In the depth of the folds, on the contrary, the coloured surface reflects against itself, and thereby makes its own colour more saturated. A white surface, on the contrary, of great brightness, produces a dazzling effect, and is thereby insensitive to slight degrees of shade. Strong colours thus, by the powerful irritation which they produce, can enchain the eye of the observer, and yet be expressive for the slightest change of modelling or of illumination; that is, they are expressive in the artistic sense.

If, on the other hand, we coat too large surfaces, they produce fatigue for the prominent colour, and a diminution in sensitiveness towards it. This colour then becomes more grey, and on all surfaces of a different colour the complementary tint appears, especially on grey or black surfaces. Hence therefore clothes, and more particularly curtains, which are of too bright a single colour, produce an unsatisfactory and fatiguing effect; the clothes have moreover the disadvantage for the wearer that they cover face and hands with the complementary colour. Blue produces yellow, violet gives greenish yellow, bright purple gives green, scarlet gives blue, and, conversely, yellow gives blue, etc. There is another circumstance which the artist has to consider, that colour is for him an important means of attracting the attention of the observer. To be able to do this he must be sparing in the use of the pure colours, otherwise they distract the attention, and the picture becomes glaring. It is necessary, on the other hand, to avoid a one-sided fatigue of the eye by too prominent a colour. This is effected either by introducing the prominent colour to a moderate extent upon a dull, slightly coloured ground, or by the juxtaposition of variously saturated colours, which produce a certain equilibrium of irritation in the eye, and, by the contrast in their after-images, strengthen and increase each other. A green surface on which the green after-image of a purple one falls, appears to be a far purer green than without such an after-image. By fatigue towards purple, that is towards red and violet, any admixture of these two colours in the green is enfeebled, while this itself produces its full effect. In this way the sensation of green is purified from any foreign admixture. Even the purest and most saturated green, which Nature shows in the prismatic spectrum, may thus acquire a higher degree of saturation. We find thus that the other pairs of complementary colours, which we have mentioned, make each other more brilliant by their contrast, while colours which are very similar are detrimental to each other, and acquire a grey tint.

These relations of the colours to each other have manifestly a great influence on the degree of pleasure which different combinations of colours afford. Two colours may, without injury, be juxtaposed, which indeed are so similar as to look like varieties of the same colour, produced by varying degrees of light and shade. Thus, upon scarlet the more shaded parts appear of a carmine, or on a straw-colour they appear of a golden yellow.

If we pass beyond these limits, we arrive at unpleasant combinations, such as carmine and orange, or orange and straw-yellow. The distance of the colours must then be increased, so as to create pleasing combinations once more. The complementary colours are those which are most distant from each other. When these are combined, such, for instance, as straw-colour and ultramarine, or verdigris and purple, they have something insipid but crude; perhaps because we are prepared to expect the second colour to appear as an after-image of the first, and it does not sufficiently appear to be a new and independent element in the compound. Hence, on the whole, combinations of those pairs are most pleasing in which the second colour of the complementary tint is near the first, though with a distinct difference. Thus, scarlet and greenish blue are complementary. The combination produced when the greenish blue is allowed to glide either into ultramarine, or yellowish green (sap green), is still more pleasing. In the latter case, the combination tends towards yellow, and in the former, towards rose-red. Still more satisfactory combinations are those of three tints which bring about equilibrium in the impression of colour, and, notwithstanding the great body of colour, avoid a one-sided fatigue of the eye, without falling into the baldness of complementary tints. To this belongs the combination which the Venetian masters used so much—red, green, and violet; as well as Paul Veronese's purple, greenish blue, and yellow. The former triad corresponds approximately to the three fundamental colours, in so far as these can be produced by pigments; the latter gives the mixtures of each pair of fundamental colours. It is however to be observed, that it has not yet been possible to establish rules for the harmony of colours with the same precision and certainty as for the consonance of tones. On the contrary, a consideration of the facts shows that a number of accessory influences come into play,[6] when once the coloured surface is also to produce, either wholly or in part, a representation of natural objects or of solid forms, or even if it only offers a resemblance with the representation of a relief, of shaded and of non-shaded surfaces. It is moreover often difficult to establish, as a matter of fact, what are the colours which produce the harmonic impression. This is pre-eminently the case with pictures in which the aërial colour, the coloured reflection and

6. Conf. E. Brücke, *Die Physiologie der Farben für die Zwecke der Kunstgewerbe.* Leipzig, 1866. W. v. Bezold, *Die Farbenlehre in Hinblick auf Kunst und Kunstgewerbe.* Braunschweig, 1874.

shade, so variously alter the tint of each single coloured surface when it is not perfectly smooth, that it is hardly possible to give an indisputable determination of its tint. In such cases, moreover, the direct action of the colour upon the eye is only subordinate means; for, on the other hand, the prominent colours and lights must also serve for directing the attention to the more important points of the representation. Compared with these more poetical and psychological elements of the representation, considerations as to the pleasing effect of the colours are thrown into the background. Only in the pure ornamentation on carpets, draperies, ribbons, or architectonic surfaces is there free scope for pure pleasure in the colours, and only there can it develop itself according to its own laws.

In pictures, too, there is not, as a general rule, perfect equilibrium between the various colours, but one of them preponderates to an extent which corresponds to the dominant light. This is occasioned, in the first case, by the truthful imitation of physical circumstances. If the illumination is rich in yellow light, yellow colours will appear brighter and more brilliant than blue ones; for yellow bodies are those which preferably reflect yellow light; while that of blue is only feebly reflected, and is mainly absorbed. Before the shaded parts of blue bodies, the yellow aërial light produces its effect, and imparts to the blue more or less of a grey tint. The same thing happens in front of red and green, though to a less extent, so that, in their shadows, these colours merge into yellow. This also is closely in accordance with the aesthetic requirements of artistic unity of composition in colour. This is caused by the fact that the divergent colours show a relation to the predominant colour, and point to it most distinctly in their shades. Where this is wanting, the various colours are hard and crude; and, since each one calls attention to itself, they make a motley and disturbing impression; and, on the other hand, a cold one, for the appearance of a flood of light thrown over the objects is wanting.

We have a natural type of the harmony which a well-executed illumination of masses of air can produce in a picture, in the light of the setting sun, which throws over the poorest regions a flood of light and colour, and harmoniously brightens them. The natural reason for this increase of aërial illumination lies in the fact, that the lower and more opaque layers of air are in the direction of the sun, and therefore reflect more powerfully; while at the same time the yellowish red colour of the light which has passed through the atmosphere becomes more distinct as the length of path increases which it has to traverse, and that further, this colouration is more pronounced as the background falls into shadow.

In summing up once more these considerations, we have first seen what limitations are imposed on truth to Nature in artistic representation; how the painter links the principal means which nature furnishes of recognising depths in the field of view, namely binocular vision, which indeed is even turned against him, as it shows unmistakably the flatness of the picture; how therefore the painter must carefully select, partly the perspective arrangement of his subject, its position and its aspect, and partly the lighting and shading, in order to give us a directly intelligible image of its magnitude, its shape, and distance, and how a truthful representation of aërial light is one of the most important means of attaining the object.

We then saw that even the scale of luminous intensity, as met with in the objects, must be transformed in the picture to one differing sometimes by a hundredfold; how here, the colour of the object cannot be simply represented by the pigment; that indeed it is necessary to introduce important changes in the distribution of light and dark, of yellowish and of bluish tints.

The artist cannot transcribe Nature; he must translate her; yet this translation may give us an impression in the highest degree distinct and forcible, not merely of the objects themselves, but even of the greatly altered intensities of light under which we view them. The altered scale is indeed in many cases advantageous, as it gets rid of everything which, in the actual objects, is too dazzling, and too fatiguing for the eye. Thus the imitation of Nature in the picture is at the same time an ennobling of the impression on the senses. In this respect we can often give ourselves up more calmly and continuously, to the consideration of a work of art, than to that of a real object. The work of art can produce those gradations of light, and those tints in which the modelling of the forms is most distinct and therefore most expressive. It can bring forward a fulness of vivid fervent colours, and by skilful contrast can retain the sensitiveness of the eye in advantageous equilibrium. It can fearlessly apply the entire energy of powerful sensuous impressions, and the feeling of delight associated therewith, to direct and enchain the attention; it can use their variety to heighten the direct understanding of what is represented, and yet keep the eye in a condition of excitation most favourable and agreeable for delicate sensuous impressions.

If, in these considerations, my having continually laid much weight on the lightest, finest, and most accurate sensuous intelligibility of artistic representation, may seem to many of you as a very subordinate point—a point which, if mentioned at all by writers on aesthetics, is treated as quite accessory—I think this is unjustly so. The sensuous distinctness is by no means a low or subordinate element in the action of works of art; its importance has forced itself the more strongly upon me the more I have sought to discover the physiological elements in their action.

What effect is to be produced by a work of art, using this word in its highest sense? It should excite and enchain our attention, arouse in us, in easy play, a host of slumbering conceptions and their corresponding feelings, and direct them towards a common object, so as to give a vivid perception of all the features of an ideal type, whose separate fragments lie scattered in our imagination and overgrown by the wild chaos of accident. It seems as if we can only refer the frequent preponderance, in the mind, of art over reality, to the fact that the latter mixes something foreign, disturbing, and even injurious; while art can collect all the elements for the desired impression, and allow them to act without restraint. The power of this impression will no doubt be greater the deeper, the finer, and the truer to nature is the sensuous impression which is to arouse the series of images and the effects connected therewith. It must act certainly, rapidly, unequivocably, and with accuracy if it is to produce a vivid and powerful impression. These essentially are the points which I have sought to comprehend under the name of intelligibility of the work of art.

Then the peculiarities of the painters' technique (*Technik*), to which physiological optical investigation have led us, are often closely connected with the highest problems of art. We may perhaps think that even the last secret of artistic beauty—that is, the wondrous pleasure which we feel in its presence—is essentially based on the feeling of an easy, harmonic, vivid stream of our conceptions, which, in spite of manifold changes, flow towards a common object, bring to light laws hitherto concealed, and allow us to gaze in the deepest depths of sensation of our own minds.

Being the substance of a series of Lectures delivered in Cologne, Berlin, and Bonn.

11

On Thought in Medicine

IT is now thirty-five years since, on the 2nd August, I stood on the rostrum in the Hall of this Institute, before another such audience as this, and read a paper on the operation of Venal Tumours. I was then a pupil of this Institution, and was just at the end of my studies. I had never seen a tumour cut, and the subject-matter of my lecture was merely compiled from books; but book knowledge played at that time a far wider and a far more influential part in medicine than we are at present disposed to assign to it. It was a period of fermentation, of the fight between learned tradition and the new spirit of natural science, which would have no more of tradition, but wished to depend upon individual experience. The authorities at that time judged more favourably of my Essay than I did myself, and I still possess the books which were awarded to me as the prize.

The recollections which crowd in upon me on this occasion have brought vividly before my mind a picture of the then condition of our science, of our endeavours and of our hopes, and have led me to compare the past state of things with that into which it has developed. Much indeed has been accomplished.

Although all that we hoped for has not been fulfilled, and many things have turned out differently from what we wished, yet we have gained much for which we could not have dared to hope. Just as the history of the world has made one of its few giant steps before our eyes, so also has our science; hence an old student, like myself, scarcely recognises the somewhat matronly aspect of Dame Medicine, when he accidentally comes again in relation to her, so vigorous and so capable of growth has she become in the fountain of youth of the Natural Sciences.

I may, perhaps, retain the impression of this antagonism, more freshly than those of my contemporaries whom I have the honour to see assembled before me; and who, having remained permanently connected with science and practice, have been less struck and less surprised by great changes, taking place as they do by slow steps. This must be my excuse for speaking to you about the metamorphosis which has taken place in medicine during this period, and with the results of whose development you are better acquainted than I am. I should like the impression of this development and of its causes not to be quite lost on the younger of my hearers. They have no special incentive for consulting the literature of that period; they would meet with principles which appear as if written in a lost tongue, so that it is by no means easy for us to transfer ourselves into the mode of thought of a period which is so far behind us. The course of development of medicine is an instructive lesson on the true principles of scientific inquiry, and the positive part of this lesson has, perhaps, in no previous time been so impressively taught as in the last generation.

The task falls to me, of teaching that branch of the natural sciences which has to make the widest generalisations, and has to discuss the meaning of fundamental ideas; and which has, on that account, been not unfitly termed Natural Philosophy by the English-speaking peoples. Hence it does not fall too far out of the range of my official duties and of my own studies, if I attempt to discourse here of the principles of scientific method, in reference to the sciences of experience.

As regards my acquaintance with the tone of thought of the older medicine, independently of the general obligation, incumbent on every educated physician, of understanding the literature of his science and the direction as well as the conditions of its progress, there was in my case a special incentive. In my first professorship at Königsberg, from the year 1849 to 1856, I had to lecture each winter on general pathology—that is, on that part of the subject which contains the general theoretical conceptions of the nature of disease, and of the principles of its treatment.

General pathology was regarded by our elders as the fairest blossom of medical science. But in fact, that which formed its essence possesses only historical interest for the disciples of modern natural science.

Many of my predecessors have broken a lance for the scientific defence of this essence, and more especially Henle and Lotze. The latter, whose starting-point was also medicine, had, in his general pathology and therapeutics, arranged it very thoroughly and methodically and with great critical acumen.

My own original inclination was towards physics; external circumstances compelled me to commence the study of medicine, which was made possible to me by the liberal arrangements of this Institution. It had, however, been the custom of a former time to combine the study of medicine with that of the Natural Sciences, and whatever in this was compulsory I must consider fortunate; not merely that I entered medicine at a time in which any one who was even moderately at home in physical considerations found a fruitful virgin soil for cultivation; but I consider the study of medicine to have been that training which preached more impressively and more convincingly than any other could have done, the everlasting principles of all scientific work; principles which are so simple and yet are ever forgotten again; so clear and yet always hidden by a deceptive veil.

Perhaps only he can appreciate the immense importance and the fearful practical scope of the problems of medical theory, who has watched the fading eye of approaching death, and witnessed the distracted grief of affection, and who has asked himself the solemn questions, Has all been done which could be done to ward off the dread event? Have all the resources and all the means which Science has accumulated become exhausted?

Provided that he remains undisturbed in his study, the purely theoretical inquirer may smile with calm contempt when, for a time, vanity and conceit seek to swell themselves in science and stir up a commotion. Or he may consider ancient prejudices to be interesting and pardonable, as remains of poetic romance, or of youthful enthusiasm. To one who has to contend with the hostile forces of fact, indifference and romance disappear; that which he knows and can do, is exposed to severe tests; he can only use the hard and clear light of facts, and must give up the notion of lulling himself in agreeable illusions.

I rejoice, therefore, that I can once more address an assembly consisting almost exclusively of medical men who have gone through the same school. Medicine was once the intellectual home in which I grew up, and even the emigrant best understands and is best understood by his native land.

If I am called upon to designate in one word the fundamental error of that former time, I should be inclined to say that it pursued a false ideal of science in a one-sided and erroneous reverence for the deductive method. Medicine, it is true, was not the only science which was involved in this error, but in no other science have the consequences been so glaring, or have so hindered progress, as in medicine. The history of this science claims, therefore, a special interest in the history of the development of the human mind. None other is, perhaps, more fitted to show that a true criticism of the sources of cognition is also practically an exceedingly important object of true philosophy.

The proud word of Hippokrates,

ἰητρὸς φιλόσοφος ἰσόθεος,

'Godlike is the physician who is a philosopher,' served, as it were, as a banner of the old deductive medicine.

We may admit this if only we once agree what we are to understand as a philosopher. For the ancients, philosophy embraced all theoretical knowledge; their philosophers pursued Mathematics, Physics, Astronomy, Natural History, in close connection with true philosophical or metaphysical considerations. If, therefore, we are to understand the medical philosopher of Hippokrates to be a man who has a *perfected* insight into causal connection of natural processes, we shall in be able to say with Hippokrates, Such a one can give help like a god.

Understood in this sense, the aphorism describes in three words the ideal which our science has to strive after. But who can allege that it will ever attain this ideal?

But those disciples of medicine who thought themselves divine even in their own lifetime, and who wished to impose themselves upon others as such, were not inclined to postpone their hopes for so long a period. The requirements for the φιλόσοφος were considerably moderated. Every adherent of any given cosmological system, in which, for well or ill, facts must be made to correspond with reality, felt himself to be a philosopher. The philosophers of that time knew little more of the laws of Nature than the unlearned layman; but the stress of their endeavours was laid upon thinking, upon the logical consequence and completeness of the system. It is not difficult to understand how in periods of youthful development, such a one-sided over-estimate of thought could be arrived at. The superiority of man over animals, of the scholar over the barbarian, depends upon thinking; sensation, feeling, perception, on the contrary, he shares with his lower fellow-creatures, and in acuteness of the senses many of these are even superior to him. That man strives to develop his thinking faculty to the utmost is a problem on the solution of which the feeling of his own dignity, as well as of his own practical power, depends; and it is a natural error to have considered unimportant the dowry of mental capacities which Nature had given to animals, and to have believed that thought could be liberated from its natural basis, observation and perception, to begin its Icarian flight of metaphysical speculation.

It is, in fact, no easy problem to ascertain completely the origins of our knowledge. An enormous amount is transmitted by speech and writing. This power which man possesses of gathering together the stores of knowledge of generations, is the chief reason of his superiority over the animal, who is restricted to an inherited blind instinct and to its individual experience. But all transmitted knowledge is handed on already formed; whence the reporter has derived it, or how much criticism he has bestowed upon it, can seldom be made out, especially if the tradition has been handed down through several generations. We must admit it all upon good faith; we cannot arrive at the source; and when many generations have contented themselves with such knowledge, have brought no criticism to bear upon it; have, indeed, gradually added all kinds of small alterations, which ultimately grew up to large ones—after all this, strange things are often reported and believed under the authority of primeval wisdom. A curious case of this kind is the of the circulation of the blood, of which we shall still have to speak.

But another kind of tradition by speech, which long remained undetected, is even still more confusing for one who reflects upon the origin of knowledge. Speech cannot readily develop names for classes of objects or for classes of processes, if we have not been accustomed very often to mention together the corresponding individuals, things, and separate cases, and to assert what there is in common about them. They must, therefore, possess many points in common. Or if we, reflecting scientifically upon this, select some of these characteristics, and collate them to form a definition, the common possession of these selected characteristics must necessitate that in the given cases a great number of other characteristics are to be regularly met with; there must be a natural connection between the first and the last-named characteristics. If, for instance, we assign the name of mammals to those animals which, when young, are suckled by their mothers, we can assert further, in reference to them, that they are all warm-blooded animals, born alive, that they have a spinal column but no quadrate bone, breathe through lungs, have separate divisions of the heart, &c. Hence the fact, that in the speech of an intelligent observing people a certain class of things are included in one name, indicates that these things or cases fall under a common natural relationship; by this alone a host of experiences are transmitted from preceding generations without this appearing to be the case.

The adult, moreover, when he begins to reflect upon the origin of his knowledge, is in possession of a huge mass of every-day experiences, which in great part reach back to the obscurity of his first childhood. Everything individual has long been forgotten, but the similar traces which the daily repetition of similar cases has left in his memory have deeply engraved themselves. And since only that which is in conformity with law is always repeated with regularity, these deeply impressed remains of all previous conceptions are just the conceptions of what is conformable to law in the things and processes.

Thus man, when he begins to reflect, finds that he possesses a wide range of acquirements of which he knows not whence they came, which he has possessed as long as he can remember. We need not refer even to the possibility of inheritance by procreation.

The conceptions which he has formed, which his mother tongue has transmitted, assert themselves as regulative powers, even in the objective world of fact, and as he does not know that he or his forefathers have developed these conceptions from the things themselves, the world of facts seems to him, like his conceptions, to be governed by intellectual forces. We recognise this psychological anthropomorphism, from the *Ideas* of Plato, to the immanent dialectic of the cosmical process of Hegel, and to the unconscious will of Schopenhauer.

Natural science, which in former times was virtually identical with medicine, followed the path of philosophy; the deductive method seemed to be capable of doing everything. Socrates, it is true, had developed the inductive conception in the most instructive manner. But the best which he accomplished remained virtually misunderstood.

I will not lead you through the motley confusion of pathological theories which, according to the varying inclination of their authors, sprouted up in consequence of this or the other increase of natural knowledge, and were mostly put forth by physicians, who obtained fame and renown as great observers and empirics, independently of their theories. Then came the less gifted pupils, who copied their master, exaggerated his theory, made it more one-sided and more logical, without regard to any discordance with Nature. The more rigid the system, the fewer and the more thorough were the methods to which the healing art was restricted. The more the schools were driven into a corner by the increase in actual knowledge, the more did they depend upon the ancient authorities, and the more intolerant were they against innovation. The great reformer of anatomy, Vesalius, was cited before the Theological faculty of Salamanca; Servetus was burned at Geneva along with his book, in which he described the circulation of the lungs; and the Paris faculty prohibited the teaching of Harvey's doctrine of the circulation of the blood in its lecture rooms.

At the same time the bases of the systems from which these schools started were mostly views on natural science which it would have been quite right to utilise within a narrow circle. What was not right was the delusion that it was more scientific to refer all diseases to one kind of explanation, than to several. What was called the solidar pathology wanted to deduce everything from the altered mechanism of the solid parts, especially from their altered tension; from the *strictum* and *laxum*, from tone and want of tone, and afterwards from strained or relaxed nerves and from obstructions in the vessels. Humoral pathology was only acquainted with alterations in mixture. The four cardinal fluids, representatives of the classical four elements, blood, phlegm, black and yellow gall; with others, the acrimonies or dyscrasies, which had to be expelled by sweating and purging; in the beginning of our modern epoch, the acids and alkalies or the alchymistic spirits, and the occult qualities of the substances assimilated—all these were the elements of this chemistry.

Along with these were found all kinds of physiological conceptions, some of which contained remarkable foreshadowings, such as the ἔμφυτον θέρμον, the inherent vital force of Hippokrates, which is kept up by nutritive substances, this again boils in the stomach and is the source of all motion; here the thread is begun to be spun which subsequently led a physician to the law of the conservation of force. On the other hand, the πνεῦμα, which is half spirit and half air, which can be driven from the lungs into the arteries and fills them, has produced much confusion. The fact that air is generally found in the arteries of dead bodies, which indeed only penetrates in the moment in which the vessels are cut, led the ancients to the belief that air is also present in the arteries during life. The veins only remained then in which blood could circulate. It was believed to be formed in the liver, to move from there to the heart, and through the veins to the organs. Any careful observation of the operation of blood-letting must have taught that, in the veins, it comes from the periphery, and flows towards the heart. But this false theory had become so mixed up with the explanation of fever and of inflammation, that it acquired the authority of a dogma, which it was dangerous to attack.

Yet the essential and fundamental error of this system was, and still continued to be, the false kind of logical conclusion to which it was supposed to lead; the conception that it must be possible to build a complete system which would embrace all forms of disease, and their cure, upon any one such simple explanation. Complete knowledge of the causal connection of one class of phenomena gives finally a logical coherent system. There is no prouder edifice of the most exact thought than modern astronomy, deduced even to the minutest of its small disturbances, from Newton's law of gravitation. But Newton had been preceded by Kepler, who had by induction collated all the facts; and the astronomers have never believed that Newton's force excluded the simultaneous action of other forces. They have been continually on the watch to see whether friction, resisting media, and swarms of meteors have not also some influence. The older philosophers and physicians believed they could deduce, before they had settled their general principles by induction. They forgot that a deduction can have no more certainty than the principle from which it is deduced; and that each new induction must in the first place be a new test, by experience, of its own bases. That a conclusion is deduced by the strictest logical method from an uncertain premise does not give it a hair's breadth of certainty or of value.

One characteristic of the schools which built up their system on such hypotheses, which they assumed as dogmas, is the intolerance of expression which I have already partially mentioned. One who works upon a well-ascertained foundation may readily admit an error; he loses, by so doing, nothing more than that in which he erred. If, however, the starting-point has been placed upon a hypothesis, which either appears guaranteed by authority, or is only chosen because it agrees with that which it is *wished* to believe true, any crack may then hopelessly destroy the whole fabric of conviction. The convinced disciples must therefore claim for each individual part of such a fabric the same degree of infallibility; for the anatomy of Hippokrates just as much as for fever crises; every opponent must only appear then as stupid or depraved, and the dispute will thus, according to old precedent, be so much the more passionate and personal, the more uncertain is the basis which is defended. We have frequent opportunities of confirming these general rules in the schools of dogmatic deductive medicine. They turned their intolerance partly against each other, and partly against the eclectics who found various explanations for various forms of disease. This method, which in its essence is completely justified, had, in the eyes of systematists, the defect of being illogical. And yet the greatest physicians and observers, Hippokrates at the head, Aretaeus, Galen, Sydenham, and Boerhaave, had become eclectics, or at any rate very lax systematists.

About the time when we seniors commenced the study of medicine, it was still under the influence of the important discoveries which Albrecht von Haller had made on the excitability of nerves; and which he had placed in connection with the vitalistic theory of the nature of life. Haller had observed the excitability in the nerves and muscles of amputated members. The most surprising thing to him was, that the most varied external actions, mechanical, chemical, thermal, to which electrical ones were subsequently added, had always the same result; namely, that they produced muscular contraction. They were only quantitatively distinguished as regards their action on the organism, that is, only by the strength of the excitation; he designated them by the common name of *stimulus*; he called the altered condition of the nerve the *excitation*, and its capacity of responding to a stimulus the *excitability*, which was lost at death. This entire condition of things, which physically speaking asserts no more than that the nerves, as concerns the changes which take place in them after excitation, are in an exceedingly unstable state of equilibrium; this was looked upon as the fundamental property of animal life, and was unhesitatingly transferred to the other organs and tissues of the body, for which there was no similar justification. It was believed that none of them were active of themselves, but must receive an impulse by a stimulus from without; air and nourishment were considered to be the normal stimuli. The *kind* of activity seemed, on the contrary, to be conditioned by the specific energy of the organ, under the influence of the vital force. Increase or diminution of the excitability was the category under which the whole of the acute diseases were referred, and from which indications were taken as to whether the treatment should be lowering or stimulating. The rigid one-sidedness and the unrelenting logic with which Robert Brown had once worked out this system was broken, but it always furnished the leading points of view.

The vital force had formerly lodged as ethereal spirit, as a Pneuma in the arteries; it had then with Paracelsus acquired the form of an Archeus, a kind of useful Kobold, or indwelling alchymist, and had acquired its clearest scientific position as 'soul of life,' *anima inscia*, in Georg Ernst Stahl, who, in the first half of the last century, was professor of chemistry and pathology in Halle. Stahl had a clear and acute mind, which is informing and stimulating, from the way in which he states the proper question, even in those cases in which he decides against our present views. He it is who established the first comprehensive system of chemistry, that of phlogiston. If we translate his phlogiston into latent heat, the theoretical bases of his system passed essentially into the system of Lavoisier; Stahl did not then know oxygen, which occasioned some false hypotheses; for instance, on the negative gravity of phlogiston. Stahl's 'soul of life' is, on the whole, constructed on the pattern on which the pietistic communities of that period represented to themselves the sinful human soul; it is subject to errors and passions, to sloth, fear, impatience, sorrow, indiscretion, despair. The physician must first appease it, or then incite it, or punish it, and compel it to repent. And the way in which, at the same time, he established the necessity of the physical and vital actions was well thought out. The soul of life governs the body, and only acts by means of the physico-chemical forces of the substances assimilated. But it has the power to bind and to loose these forces, to allow them full play or to restrain them. After death the restrained forces become free, and evoke putrefaction and decomposition. For the refutation of this hypothesis of binding and loosing, it was necessary to discover the law of the conservation of force.

The second half of the previous century was too much possessed by the principles of rationalism to recognise openly Stahl's 'soul of life.' It was presented more scientifically as vital force, *Vis vitalis*, while in the main it retained its functions, and under the name of 'Nature's healing power' it played a prominent part in the treatment of diseases.

The doctrine of vital force entered into the pathological system of changes in irritability. The attempt was made to separate the direct actions of the virus which produce disease, in so far as they depended on the play of blind natural forces, the *symptomata morbi*, from those which brought on the reaction of vital force, the *symptomata reactionis*. The latter were principally seen in inflammation and in fever. It was the function of the physician to observe the strength of this reaction, and to stimulate or moderate it according to circumstances.

The treatment of fever seemed at that time to be the chief point; to be that part of medicine which had a real scientific foundation, and in which the local treatment fell comparatively into the background. The therapeutics of febrile diseases had thereby become very monotonous, although the means indicated by theory were still abundantly used, and especially blood-letting, which since that time has almost been entirely abandoned. Therapeutics became still more impoverished as the younger and more critical generation grew up, and tested the assumptions of that which was considered to be scientific. Among the younger generation were many who, in despair as to their science, had almost entirely given up therapeutics, or on principle had grasped at an empiricism such as Rademacher then taught, which regarded any expectation of a scientific explanation as a vain hope.

What we learned at that time were only the ruins of the older dogmatism, but their doubtful features soon manifested themselves.

The vitalistic physician considered that the essential part of the vital processes did not depend upon natural forces, which, doing their work with blind necessity and according to a fixed law, determined the result. What these forces could do appeared quite subordinate, and scarcely worthy of a minute study. He thought that he had to deal with a soul-like being, to which a thinker, a philosopher, and an intelligent man must be opposed. May I elucidate this by a few outlines?

At this time auscultation and percussion of the organs of the chest were being regularly practiced in the clinical wards. But I have often heard it maintained that they were a coarse mechanical means of investigation which a physician with a clear mental vision did not need; and it indeed lowered and debased the patient, who was anyhow a human being, by treating him as a machine. To feel the pulse seemed the most direct method of learning the mode of action of the vital force, and it was practised, therefore, as by far the most important means of investigation. To count with a repeater was quite usual, but seemed to the old gentlemen as a method not quite in good taste. There was, as yet, no idea of measuring temperature in cases of disease. In reference to the ophthalmoscope, a celebrated surgical colleague said to me that he would never use the instrument, it was too dangerous to admit crude light into diseased eyes; another said the mirror might be useful for physicians with bad eyes, his, however, were good, and he did not need it.

A professor of physiology of that time, celebrated for his literary activity, and noted as an orator and intelligent man, had a dispute on the images in the eye with his colleague the physicist. The latter challenged the physiologist to visit him and witness the experiment. The physiologist, however, refused his request with indignation; alleging that a physiologist had nothing to do with experiments; they were of no good but for the physicist. Another aged and learned professor of therapeutics, who occupied himself much with the reorganisation of the Universities, was urgent with me to divide physiology, in order to restore the good old time; that I myself should lecture on the really intellectual part, and should hand over the lower experimental part to a colleague whom he regarded as good enough for the purpose. He quite gave me up when I said that I myself considered experiments to be the true basis of science.

I mention these points, which I myself have experienced, to elucidate the feeling of the older schools, and indeed of the most illustrious representatives of medical science, in reference to the progressive set of ideas of the natural sciences; in literature these ideas naturally found feebler expression, for the old gentlemen were cautious and worldly wise.

You will understand how great a hindrance to progress such a feeling on the part of influential and respected men must have been. The medical education of that time was based mainly on the study of books; there were still lectures, which were restricted to mere dictation; for experiments and demonstrations in the laboratory the provision made was sometimes good and sometimes the reverse; there were no physiological and physical laboratories in which the student himself might go to work. Liebig's great deed, the foundation of the chemical laboratory, was complete, as far as chemistry was concerned, but his example had not been imitated elsewhere. Yet medicine possessed in anatomical dissections a great means of education for independent observation, which is wanting in the other faculties, and to which I am disposed to attach great weight. Microscopic demonstrations were isolated and infrequent in the lectures. Microscopic instruments were costly and scarce. I came into possession of one by having spent my autumn vacation in 1841 in the Charité, prostrated by typhoid fever; as pupil, I was nursed without expense, and on my recovery I found myself in possession of the savings of my small resources. The instrument was not beautiful, yet I was able to recognise by its means the prolongations of the ganglionic cells in the invertebrata, which I described in my dissertation, and to investigate the vibrions in my research on putrefaction and fermentation.

Any of my fellow-students who wished to make experiments had to do so at the cost of his pocket-money. One thing we learned thereby, which the younger generation does not, perhaps, learn so well in the laboratories—that is, to consider in all directions the ways and means of attaining the end, and to exhaust all possibilities, in the consideration, until a practicable path was found. We had, it is true, an almost uncultivated field before us, in which almost every stroke of the spade might produce remunerative results.

It was one man more especially who aroused our enthusiasm for work in the right direction—that is, Johannes Müller, the physiologist. In his theoretical views he favoured the vitalistic hypothesis, but in the most essential points he was a natural philosopher, firm and immovable; for him, all theories were but hypotheses, which had to be tested by facts and about which facts could alone decide. Even the views upon those points which most easily crystallise into dogmas, on the mode of activity of the vital force and the activity of the conscious soul, he tried continually to define more precisely, to prove or to refute by means of facts.

And, although the art of anatomical investigation was most familiar to him, and he therefore recurred most willingly to this, yet he worked himself into the chemical and physical methods which were more foreign to him. He furnished the proof that fibrine is dissolved in blood; he experimented on the propagation of sound in such mechanisms as are found in the drum of the ear; he treated the action of the eye as an optician. His most important performance for the physiology of the nervous system, as well as for the theory of cognition, was the actual definite establishment of the doctrine of the specific energies of the nerves. In reference to the separation of the nerves of motor and sensible energy, he showed how to make the experimental proof of Bell's law of the roots of the spinal cord so as to be free from errors; and in regard to the sensible energies he not only established the general law, but carried out a great number of separate investigations, to eliminate objections, and to refute false indications and evasions. That which hitherto had been imagined from the data of everyday experience, and which had been sought to be expressed in a vague manner, in which the true was mixed up with the false; or which had just been established for individual branches, such as by Dr. Young for the theory of colours, or by Sir Charles Bell for the motor nerves, that emerged from Müller's hands in a state of classical perfection—a scientific achievement whose value I am inclined to consider as equal to that of the discovery of the law of gravitation.

His scientific tendency, and more especially his example, were continued in his pupils. We had been preceded by Schwann, Henle, Reichert, Peters, Remak; I met as fellow-students E. Du Bois-Reymond, Virchow, Brücke, Ludwig, Traube, J. Meyer, Lieberkühn, Hallmann; we were succeeded by A. von Graefe, W. Busch, Max Schultze, A. Schneider.

Microscopic and pathological anatomy, the study of organic types, physiology, experimental pathology and therapeutics, ophthalmology, developed themselves in Germany under the influence of this powerful impulse far beyond the standard of rival adjacent countries. This was helped by the labours of those of similar tendencies among Müller's contemporaries, among whom the three brothers Weber of Leipzig must first of all be mentioned, who have built solid foundations in the mechanism of the circulation, of the muscles, of the joints, and of the ear. The attack was made wherever a way could be perceived of understanding one of the vital processes; it was assumed that they could be understood, and success justified this assumption. A delicate and copious technical apparatus has been developed in the methods of microscopy, of physiological chemistry, and of vivisection; the latter greatly facilitated more particularly by the use of anaesthetic ether and of the paralyzing curara, by which a number of deep problems became open to attack, which to our generation seemed hopeless. The thermometer, the ophthalmoscope, the auricular speculum, the laryngoscope, nervous irritation on the living body opened out to the physician possibilities of delicate and yet certain diagnosis where there seemed to be absolute darkness. The continually increasing number of proved parasitical organisms substitute tangible objects for mystical entities, and teach the surgeon to forestall the fearfully subtle diseases of decomposition.

But do not think, gentlemen, that the struggle is at an end. As long as there are people of such astounding conceit as to imagine that they can effect by a few clever strokes, that which man can otherwise only hope to achieve by toilsome labour, hypotheses will be started which, propounded as dogmas, at once promise to solve all riddles. And as long as there are people who believe implicitly in that which they wish to be true, so long will the hypotheses of the former find credence. Both classes will certainly not die out, and to the latter the majority will always belong.

There are two characteristics more particularly which metaphysical systems have always possessed. In the first place man is always desirous of feeling himself to be a being of a higher order, far beyond the standard of the rest of nature; this wish is satisfied by the spiritualists. On the other hand, he would like to believe that by his thought he was unrestrained lord of the world, and of course by his thinking with those conceptions, to the development of which he has attained; this is attempted to be satisfied by the materialists.

But one who, like the physician, has actively to face natural forces which bring about weal or woe, is also under the obligation of seeking for a knowledge of the truth, and of the truth only; without considering whether, what he finds, is pleasant in one way or the other. His aim is one which is firmly settled; for him the success of facts is alone finally decisive. He must endeavour to ascertain beforehand, what will be the result of his attack if he pursues this or that course. In order to acquire this foreknowledge of what is coming, but of what has not been settled by observations, no other method is possible than that of endeavouring to arrive at the laws of facts by observations; and we can only learn them by induction, by the careful selection, collation, and observation of those cases which fall under the law. When we fancy that we have arrived at a law, the business of deduction commences. It is then our duty to develop the consequences of our law as completely as may be, but in the first place only to apply to them the test of experience, so far as they can be tested, and then to decide by this test whether the law holds, and to what extent. This is a test which really never ceases. The true natural philosopher reflects at each new phenomenon, whether the best established laws of the best known forces may not experience a change; it can of course only be a question of a change which does not contradict the whole store of our previously collected experiences. It never thus attains unconditional truth, but such a high degree of probability that it is practically equal to certainty. The metaphysicians may amuse themselves at this; we will take their mocking to heart when they are in a position to do better, or even as well. The old words of Socrates, the prime master of inductive definitions, in reference to them are just as fresh as they were 2,000 years ago: 'They imagined they knew what they did not know, and he at any rate had the advantage of not pretending to know what he did not know.' And again, he was surprised at its not being clear to them that it is not possible for men to discover such things; since even those who most prided themselves on the speeches made on the matter, did not agree among themselves, but behaved to each other like madmen (τοῖς μαινομένοις ὁμοίως).[1] Socrates calls them τοὺς μέγιστον φρονοῦντας. Schopenhauer[2] calls himself a Mont Blanc, by the side of a mole-heap, when he compares himself with a natural philosopher. The pupils admire these big words and try to imitate the master.

In speaking against the empty manufacture of hypotheses, do not by any means suppose that I wish to diminish the real value of original thoughts. The first discovery of a new law, is the discovery of a similarity which has hitherto been concealed in the course of natural processes. It is a manifestation of that which our forefathers in a serious sense described as 'wit'; it is of the same quality as the highest performances of artistic perception in the discovery of new types of expression. It is something which cannot be forced, and which cannot be acquired by any known method. Hence all those aspire after it who wish to pass as the favoured children of genius. It seems, too, so easy, so free from trouble, to get by so sudden mental flashes an unattainable advantage over our contemporaries. The true artist and the true inquirer knows that great works can only be produced by hard work. The proof that the ideas

---

1. Xenophon, *Memorabil.* I.i.11.
2. Arthur Schopenhauer, *Von ihm, über ihn von Frauenstadt und Lindner*. Berlin, 1863, p. 653.

formed do not merely scrape together superficial resemblances, but are produced by a quick glance into the connection of the whole, can only be acquired when these ideas are completely developed—that is, for a newly discovered natural law, only by its agreement with facts. This estimate must by no means be regarded as depending on external success, but the success is here closely connected with the depth and completeness of the preliminary perceptions.

To find superficial resemblances is easy; it is amusing in society, and witty thoughts soon procure for their author the name of a clever man. Among the great number of such ideas, there must be some which are ultimately found to be partially or wholly correct; it would be a stroke of skill *always* to guess falsely. In such a happy chance a man can loudly claim his priority for the discovery; if otherwise, a lucky oblivion conceals the false conclusions. The adherents of such a process are glad to certify the value of a first thought. Conscientious workers who are shy at bringing their thoughts before the public before they have tested them in all directions, solved all doubts, and have firmly established the proof, these are at a decided disadvantage. To settle the present kind of questions of priority, only by the date of their first publication, and without considering the ripeness of the research, has seriously favoured this mischief.

In the 'type case' of the printer all the wisdom of the world is contained which has been or can be discovered; it is only requisite to know how the letters are to be arranged. So also, in the hundreds of books and pamphlets which are every year published about ether, the structure of atoms, the theory of perception, as well as on the nature of the asthenic fever and carcinoma, all the most refined shades of possible hypotheses are exhausted, and among these there must necessarily be many fragments of the correct theory. But who knows how to find them?

I insist upon this in order to make clear to you that all this literature, of untried and unconfirmed hypotheses, has no value in the progress of science. On the contrary, the few sound ideas which they may contain are concealed by the rubbish of the rest; and one who wants to publish something really new—facts—sees himself open to the danger of countless claims of priority, unless he is prepared to waste time and power in reading beforehand a quantity of absolutely useless books, and to destroy his readers' patience by a multitude of useless quotations.

Our generation has had to suffer under the tyranny of spiritualistic metaphysics; the newer generation will probably have to guard against that of the materialistic hypotheses. Kant's rejection of the claims of pure thought has gradually made some impression, but Kant allowed one way of escape. It was as clear to him as to Socrates that all metaphysical systems which up to that time had been propounded were tissues of false conclusions. His *Kritik der reinen Vernunft* is a continual sermon against the use of the category of thought beyond the limits of possible experience. But geometry seemed to him to do something which metaphysics was striving after; and hence geometrical axioms, which he looked upon as *a priori* principles antecedent to all experience, he held to be given by transcendental intuition, or as the inherent form of all external intuition. Since that time, pure *a priori* intuition has been the anchoring-ground of metaphysicians. It is even more convenient than pure thought, because everything can be heaped on it without going into chains of reasoning, which might be capable of proof or of refutation. The nativistic theory of perception of the senses is the expression of this theory in physiology. All mathematicians united to fight against any attempt to resolve the intuitions into their natural elements; whether the so-called pure or the empirical, the axioms of geometry, the principles of mechanics, or the perceptions of vision. For this reason, therefore, the mathematical investigations of Lobaschewsky, Gauss, and Riemann on the alterations which are logically possible in the axioms of geometry; and the proof that the axioms are principles which are to be confirmed or perhaps even refuted by experience, and can accordingly be acquired from experience—these I consider to be very important steps. That all metaphysical sects get into a rage about this must not lead you astray, for these investigations lay the axe at the bases of apparently the firmest supports which their claims still possess. Against those investigators who endeavour to eliminate from among the perceptions of the senses, whatever there may be of the actions of memory, and of the repetition of similar impressions, which occur in memory; whatever, in short, is a matter of experience, against them it is attempted to raise a party cry that they are spiritualists. As if memory, experience, and custom were not also facts, whose laws are to be sought, and which are not to be explained away because they cannot be glibly referred to reflex actions, and to the complex of the prolongation of ganglionic cells, and of the connection of nerve-fibres in the brain.

Indeed, however self-evident, and however important the principle may appear to be, that natural science has to seek for the laws of facts, this principle is nevertheless often forgotten. In recognising the law found, as a force which rules the processes in nature, we conceive it objectively as a *force* and such a reference of individual cases to a force which under given conditions produces a definite result, that we designate as a causal explanation of phenomena. We cannot always refer to the forces of atoms; we speak of a refractive force, of electromotive and of electrodynamic force. But do not forget the *given conditions* and the *given result*. If these cannot be given, the explanation attempted is merely a modest confession of ignorance, and then it is decidedly better to confess this openly.

If any process in vegetation is referred to forces in the cells, without a closer definition of the conditions among which, and of the direction in which, they work, this can at most assert that the more remote parts of the organism are without influence; but it would be difficult to confirm this with certainty in more than a few cases. In like manner, the originally definite sense which Johannes Müller gave to the idea of reflex action, is gradually evaporated into this, that when an impression has been made on any part of the nervous system, and an action occurs in any other part, this is supposed to have been explained by saying that it is a reflex action. Much may be imposed upon the irresolvable complexity of the nerve-fibres of the brain. But the resemblance to the *qualitates occultae* of ancient medicine is very suspicious.

From the entire chain of my argument it follows that what I have said against metaphysics is not intended against philosophy. But metaphysicians have always tried to plume themselves on being philosophers, and philosophical amateurs have mostly taken an interest in the high-flying speculations of the metaphysicians, by which they hope in a short time, and at no great trouble, to learn the whole of what is worth knowing. On another occasion[3] I compared the relationship of metaphysics to philosophy with that of astrology to astronomy. The former had the most exciting interest for the public at large, and especially for the fashionable world, and turned its alleged connoisseurs into influential persons. Astronomy, on the contrary, although it had become the ideal of scientific research, had to be content with a small number of quietly working disciples.

In like manner, philosophy, if it gives up metaphysics, still possesses a wide and important field, the knowledge of mental and spiritual processes and their laws. Just as the anatomist, when he has reached the limits of microscopic vision, must try to gain an insight into the action of his optical instrument, in like manner every scientific enquirer must study minutely the chief instrument of his research as to its capabilities. The groping of the medical schools for the last two thousand years is, among other things, an illustration of the harm of erroneous views in this respect. And the physician, the statesman, the jurist, the clergyman, and the teacher, ought to be able to build upon a knowledge of physical processes if they wish to acquire a true scientific basis for their practical activity. But the true science of philosophy has had, perhaps, to suffer more from the evil mental habits and the false ideals of metaphysics than even medicine itself.

One word of warning. I should not like you to think that my statements are influenced by personal irritation. I need not explain that one who has

---

3. Preface to the German translation of Tyndall's *Scientific Fragments*, p. xxii.

such opinions as I have laid before you, who impresses on his pupils, whenever he can, the principle that 'a metaphysical conclusion is either a false conclusion or a concealed experimental conclusion,' that he is not exactly beloved by the votaries of metaphysics or of intuitive conceptions. Metaphysicians, like all those who cannot give any decisive reasons to their opponents, are usually not very polite in their controversy; one's own success may approximately be estimated from the increasing want of politeness in the replies.

My own researches have led me more than other disciples of the school of natural science into controversial regions; and the expressions of metaphysical discontent have perhaps concerned me even more than my friends, as many of you are doubtless aware.

In order, therefore, to leave my own personal opinions quite on one side, I have allowed two unsuspected warrantors to speak for me —Socrates and Kant—both of whom were certain that all metaphysical systems established up to their time were full of empty false conclusions, and who guarded themselves against adding any new ones. In order to show that the matter has not changed, either in the last 2,000 years or in the last 100 years, let me conclude with a sentence of one who was unfortunately too soon taken away from us, Frederick Albert Lange, the author of the 'History of Materialism.' In this posthumous 'Logical Studies,' which he wrote in anticipation of his approaching end, he gives the following picture, which struck me because it would hold just as well in reference to solidar or humoral pathologists, or any other of the old dogmatic schools of medicine.

Lange says: The Hegelian ascribes to the Herbartian a less perfect knowledge than to himself, and conversely; but neither hesitates to consider the knowledge of the other to be higher compared with that of the empiricist, and to recognise in it at any rate an approximation to the only true knowledge. It is seen, also, that here no regard is paid to the validity of the proof, and that a mere statement in the form of a deduction from the entirety of a system is recognised as 'apodictic knowledge.'

Let us, then, throw no stones at our old medical predecessors, who in dark ages, and with but slight preliminary knowledge, fell into precisely the same errors as the great intelligences of what wishes to be thought the illuminated nineteenth century. They did no worse than their predecessors except that the nonsense of their method was more prominent in the matter of natural science. Let us work on. In this work of true intelligence physicians are called upon to play a prominent part. Among those who are continually called upon actively to preserve and apply their knowledge of nature, you are those who begin with the best mental preparation, and are acquainted with the most varied regions of natural phenomena.

In order, finally, to conclude our consultation on the condition of Dame Medicine correctly with the epikrisis, I think we have every reason to be content with the success of the treatment which the school of natural science has applied, and we can only recommend the younger generation to continue the same therapeutics.

An Address delivered August 2, 1877, on the Anniversary of the Foundation of the Institute for the Education of Army Surgeons.

12

On Academic Freedom in German Universities

IN entering upon the honourable office to which the confidence of my colleagues has called me, my first duty is once more openly to express my thanks to those who have thus honoured me by their confidence. I have the more reason to appreciate it highly, as it was conferred upon me, notwithstanding that I have been but few years among you, and notwithstanding that I belong to a branch of natural science which has come within the circle of University instruction in some sense as a foreign element; which has necessitated many changes in the old order of University teaching, and which will, perhaps, necessitate other changes. It is indeed just in that branch (Physics) which I represent, and which forms the theoretical basis of all other branches of Natural Science, that the particular characteristics of their methods are most definitely pronounced. I have already been several times in the position of having to propose alterations in the previous regulations of the University, and I have always had the pleasure of meeting with the ready assistance of my colleagues in the faculty, and of the Senate. That you have made me the Director of the business of this University for this year, is a proof that you regard me as no thoughtless innovator. For, in fact, however the objects, the methods, the more immediate aims of investigations in the natural sciences may differ externally from those of the mental sciences, and however foreign their results and however remote their interest may often appear, to those who are accustomed only to the direct manifestations and products of mental activity, there is in reality, as I have endeavoured to show in my discourse as Rector at Heidelberg, the closest connection in the essentials of scientific methods, as well as in the ultimate aims of both classes of the sciences. Even if most of the objects of investigation of the natural sciences are not directly connected with the interests of the mind, it cannot, on the other hand, be forgotten that the power of true scientific method stands out in the natural sciences far more prominently—that the real is far more sharply separated from the unreal, by the incorruptible criticism of facts, than is the case with the more complex problems of mental science.

And not merely the development of this new side of scientific activity, which was almost unknown to antiquity, but also the influence of many political, social, and even international relationships make themselves felt, and require to be taken into account. The circle of our students has had to be increased; a changed national life makes other demands upon those who are leaving; the sciences become more and more specialised and divided; exclusive of the libraries, larger and more varied appliances for study are required. We can scarcely foresee what fresh demands and what new problems we may have to meet in the more immediate future.

On the other hand, the German Universities have conquered a position of honour not confined to their fatherland; the eyes of the civilised world are upon them. Scholars speaking the most different languages crowd towards them, even from the farthest parts of the earth. Such a position would be easily lost by a false step, but would be difficult to regain.

Under these circumstances it is our duty to get a clear understanding of the reason for the previous prosperity of our Universities; we must try to find what is the feature in their arrangements which we must seek to retain as a precious jewel, and where, on the contrary, we may give way when changes are required. I consider myself by no means entitled to give a final opinion on this matter. The point of view of any single individual is restricted; representatives of other sciences will be able to contribute something. But I think that a final result can only be arrived at when each one becomes clear as to the state of things as seen from his point of view.

The European Universities of the Middle Age had their origin as free private unions of their students, who came together under the influence of celebrated teachers, and themselves arranged their own affairs. In recognition of the public advantage of these unions they soon obtained from the State, privileges and honourable rights, especially that of an independent jurisdiction, and the right of granting academic degrees. The students of that time were mostly men of mature years, who frequented the University more immediately for their own instruction and without any direct practical object; but younger men soon began to be sent, who, for the most part, were placed under the superintendence of the older members. The separate Universities split again into closer economic unions, under the name of 'Nations,' 'Bursaries,' 'Colleges,' whose older members, the seniors, governed the common affairs of each such union, and also met together for regulating the common affairs of the University. In the courtyard of the University of Bologna are still to be seen the coats-of-arms, and lists of members and seniors, of many such Nations in ancient times. The older graduated members were regarded as permanent life members of such Unions, and they retained the right of voting, as is still the case in the College of Doctors in the University of Vienna, and in the Colleges of Oxford and of Cambridge, or was until recently.

Such a free confederation of independent men, in which teachers as well as taught were brought together by no other interest than that of love of science; some by the desire of discovering the treasure of mental culture which antiquity had bequeathed, others endeavouring to kindle in a new generation the ideal enthusiasm which had animated their lives. Such was the origin of Universities, based, in the conception, and in the plan of their organisation, upon the most perfect freedom. But we must not think here of freedom of teaching in the modern sense. The majority was usually very intolerant of divergent opinions. Not unfrequently the adherents of the minority were compelled to quit the University in a body. This was not restricted to those cases in which the Church intermeddled, and where political or metaphysical propositions were in question. Even the medical faculties—that of Paris, the most celebrated of all at the head—allowed no divergence from that which they regarded as the teaching of Hippocrates. Anyone who used the medicines of the Arabians or who believed in the circulation of the blood was expelled.

The change, in the Universities, to their present constitution, was caused mainly by the fact that the State granted to them material help, but required, on the other hand, the right of co-operating in their management. The course of this development was different in different European countries, partly owing to divergent political conditions and partly to that of national character.

Until lately, it might have been said that the least change has taken place in the old English Universities, Oxford and Cambridge. Their great endowments, the political feeling of the English for the retention of existing rights, had excluded almost all change, even in directions in which such change was urgently required. Until of late both Universities had in great measure retained their character as schools for the clergy, formerly of the Roman and now of the Anglican Church, whose instruction laymen might also share in so far as it could serve the general education of the mind; they were subjected to such a control and mode of life, as was formerly considered to be good for young priests. They lived, as they still live, in colleges, under the superintendence of a number of older graduate members (Fellows) of the College; in other respects in the style and habits of the well-to-do classes in England.

The range and the method of the instruction is a more highly developed gymnasial instruction; though in its limitation to what is afterwards required in the examination, and in the minute study of the contents of prescribed text-books, it is more like the Repetitoria which are here and there held in our Universities. The acquirements of the students are controlled by searching examinations for academical degrees, in which very special knowledge is required, though only for limited regions. By such examinations the academical degrees are acquired.

While the English Universities give but little for the endowment of the positions of approved scientific teachers, and do not logically apply even that little for this object, they have another arrangement which is apparently of great promise for scientific study, but which has hitherto not effected much; that is the institution of Fellowships. Those who have passed the best examinations are elected as Fellows of their college, where they have a home, and along with this, a respectable income, so that they can devote the whole of their leisure to scientific pursuits. Both Oxford and Cambridge have each more than 500 such fellowships. The Fellows *may*, but *need* not act as tutors for the students. They need not even live in the University Town, but may spend their stipends where they like, and in many cases may retain the fellowships for an indefinite period. With some exceptions, they only lose it in case they marry, or are elected to certain offices. They are the real successors of the old corporation of students, by and for which the University was founded and endowed. But however beautiful this plan may seem, and notwithstanding the enormous sums devoted to it, in the opinion of all unprejudiced Englishmen it does but little for science; manifestly because most of these young men, although they are the pick of the students, and in the most favourable conditions possible for scientific work, have in their student-career not come sufficiently in contact with the living spirit of inquiry, to work on afterwards on their own account, and with their own enthusiasm.

In certain respects the English Universities do a great deal. They bring up their students as cultivated men, who are expected not to break through the restrictions of their political and ecclesiastical party, and, in fact, do not thus break through. In two respects we might well endeavour to imitate them. In the first place, together with a lively feeling for the beauty and youthful freshness of antiquity, they develop in a high degree a sense for delicacy and precision in writing which shows itself in the way in which they handle their mother-tongue. I fear that one of the weakest sides in the instruction of German youth is in this direction. In the second place the English Universities, like their schools, take greater care of the bodily health of their students. They live and work in airy, spacious buildings, surrounded by lawns and groves of trees; they find much of their pleasure in games which excite a passionate rivalry in the development of bodily energy and skill, and which in this respect are far more efficacious than our gymnastic and fencing exercises. It must not be forgotten that the more young men are cut off from fresh air and from the opportunity of vigorous exercise, the more induced will they be to seek an apparent refreshment in the misuse of tobacco and of intoxicating drinks. It must also be admitted that the English Universities accustom their students to energetic and accurate work, and keep them up to the habits of educated society. The *moral* effect of the more rigorous control is said to be rather illusory.

The Scotch Universities and some smaller English foundations of more recent origin—University College and King's College in London, and Owens College in Manchester—are constituted more on the German and Dutch model. The development of French Universities has been quite different, and indeed almost in the opposite direction. In accordance with the tendency of the French to throw overboard everything of historic development to suit some rationalistic theory, their faculties have logically become purely institutes for instruction—special schools, with definite regulations for the course of instruction, developed and quite distinct from those institutions which are to further the progress of science, such as the *Collège de France*, the *Jardin des Plantes*, and the *École des Études Supérieures*. The faculties are entirely separated from one another, even when they are in the same town. The course of study is definitely prescribed, and is controlled by frequent examinations. French teaching is confined to that which is clearly established, and transmits this in a well-arranged, well worked-out manner, which is easily intelligible, and does not excite doubt nor the necessity for deeper inquiry. The teachers need only possess good receptive talents. Thus in France it is looked upon as a false step when a young man of promising talent takes a professorship in a faculty in the provinces. The method of instruction in France is well adapted to give pupils, of even moderate capacity, sufficient knowledge for the routine of their calling. They have no choice between different teachers, and they swear *in verba magistri*; this gives a happy self-satisfaction and freedom from doubts. If the teacher has been well chosen, this is sufficient in ordinary cases, in which the pupil does what he has seen his teacher do. It is only unusual cases that test how much actual insight and judgment the pupil has acquired. The French people are moreover gifted, vivacious, and ambitious, and this corrects many defects in their system of teaching.

A special feature in the organisation of French Universities consists in the fact that the position of the teacher is quite independent of the favour of his hearers; the pupils who belong to his faculty are generally compelled to attend his lectures, and the far from inconsiderable fees which they pay flow into the chest of the Minister of Education; the regular salaries of the University professors are defrayed from this source; the State gives but an insignificant contribution towards the maintenance of the University. When, therefore, the teacher has no real pleasure in teaching, or is not ambitious of having a number of pupils, he very soon becomes indifferent to the success of his teaching, and is inclined to take things easily.

Outside the lecture-rooms, the French students live without control, and associate with young men of other callings, without any special *esprit de corps* or common feeling.

The development of the German Universities differs characteristically from these two extremes. Too poor in their own possessions not to be compelled, with increasing demands for the means of instruction, eagerly to accept the help of the State, and too weak to resist encroachments upon their ancient rights in times in which modern States attempt to consolidate themselves, the German Universities have had to submit themselves to the controlling influence of the State. Owing to this latter circumstance the decision in all important University matters has in principle been transferred to the State, and in times of religious or political excitement this supreme power has occasionally been unscrupulously exerted. But in most cases the States which were working out their own independence were favourably disposed towards the Universities; they required intelligent officials, and the fame of their country's University conferred a certain lustre upon the Government. The ruling officials were, moreover, for the most part students of the University; they remained attached to it. It is very remarkable how among wars and political changes in the States fighting with the decaying Empire for the consolidation of their young sovereignties, while almost all other privileged orders were destroyed, the Universities of Germany saved a far greater nucleus of their internal freedom and of the most valuable side of this freedom, than in conscientious Conservative England, and than in France with its wild chase after freedom.

We have retained the old conception of students, as that of young men responsible to themselves, striving after science of their own free will, and to whom it is left to arrange their own plan of studies as they think best. If attendance on particular lectures was enjoined for certain callings—what are called 'compulsory lectures'—these regulations were not made by the University, but by the State, which was afterwards to admit candidates to these callings. At the same time the students had, and still have, perfect freedom to migrate from one German University to another, from Dorpat to Zurich, from Vienna to Gratz; and in each University they had free choice among the teachers of the same subject, without reference to their position as ordinary or extraordinary professors or as private docents. The students are, in fact, free to acquire any part of their instruction from books; it is highly desirable that the works of great men of past times should form an essential part of study.

Outside the University there is no control over the proceedings of the students, so long as they do not come in collision with the guardians of public order. Beyond these cases the only control to which they are subject is that of their colleagues, which prevents them from doing anything which is repugnant to the feeling of honour of their own body. The Universities of the Middle Ages formed definite close corporations, with their own jurisdiction, which extended to the right over life and death of their own members. As they lived for the most part on foreign soil, it was necessary to have their own jurisdiction, partly to protect the members from the caprices of foreign judges, partly to keep up that degree of respect and order, within the society, which was necessary to secure the continuation of the rights of hospitality on a foreign soil; and partly, again, to settle disputes among the members. In modern times the remains of this academic jurisdiction have by degrees been completely transferred to the ordinary courts, or will be so transferred; but it is still necessary to maintain certain restrictions on a union of strong and spirited young men, which guarantee the peace of their fellow-students and that of the citizens. In cases of collision this is the object of the disciplinary power of the University authorities. This object, however, must be mainly attained by the sense of honour of the students; and it must be considered fortunate that German students have retained a vivid sense of corporate union, and of what is intimately connected therewith, a requirement of honourable behaviour in the individual. I am by no means prepared to defend every individual regulation in the Codex of Students' Honour; there are many Middle Age remains among them which were better swept away; but that can only be done by the students themselves.

For most foreigners the uncontrolled freedom of German students is a subject of astonishment; the more so as it is usually some obvious excrescences of this freedom which first meet their eyes; they are unable to understand how young men can be so left to themselves without the greatest detriment. The German looks back to his student life as to his golden age; our literature and our poetry are full of expressions of this feeling. Nothing of this kind is but even faintly suggested in the literature of other European peoples. The German student alone has this perfect joy in the time, in which, in the first delight in youthful responsibility, and freed more immediately from having to work for extraneous interests, he can devote himself to the task of striving after the best and noblest which the human race has hitherto been able to attain in knowledge and in speculation, closely joined in friendly rivalry with a large body of associates of similar aspirations, and in daily mental intercourse with teachers from whom he learns something of the workings of the thoughts of independent minds.

When I think of my own student life, and of the impression which a man like Johannes Müller, the physiologist, made upon us, I must place a very high value upon this latter point. Anyone who has once come in contact with one or more men of the first rank must have had his whole mental standard altered for the rest of his life. Such intercourse is, moreover, the most interesting that life can offer.

You, my younger friends, have received in this freedom of the German students a costly and valuable inheritance of preceding generations. Keep it—and hand it on to coming races, purified and ennobled if possible. You have to maintain it, by each, in his place, taking care that the body of German students is worthy of the confidence which has hitherto accorded such a measure of freedom. But freedom necessarily implies responsibility. It is as injurious a present for weak, as it is valuable for strong characters. Do not wonder if parents and statesmen sometimes urge that a more rigid system of supervision and control, like that of the English, shall be introduced even among us. There is no doubt that, by such a system, many a one would be saved who is ruined by freedom. But the State and the Nation is best served by those who can bear freedom, and have shown that they know how to work and to struggle, from their own force and insight and from their own interest in science.

My having previously dwelt on the influence of mental intercourse with distinguished men, leads me to discuss another point in which German Universities are distinguished from the English and French ones. It is that we start with the object of having instruction given, if possible, only by teachers who have proved their own power of advancing science. This also is a point in respect to which the English and French often express their surprise. They lay more weight than the Germans on what is called the 'talent for teaching'—that is, the power of explaining the subjects of instruction in a well-arranged and clear manner, and, if possible, with eloquence, and so as to entertain and to fix the attention. Lectures of eloquent orators at the Collège de France, Jardin des Plantes, as well as in Oxford and Cambridge, are often the centres of the elegant and the educated world. In Germany we are not only indifferent to, but even distrustful of, oratorical ornament, and often enough are more negligent than we should be of the outer forms of the lecture. There can be no doubt that a good lecture can be followed with far less exertion than a bad one; that the matter of the first can be more certainly and completely apprehended; that a well arranged explanation, which develops the salient points and the divisions of the subject, and which brings it, as it were, almost intuitively before us, can impart more information in the same time than one which has the opposite qualities. I am by no means prepared to defend what is, frequently, our too great contempt for form in speech and in writing. It cannot also be doubted that many original men, who have done considerable scientific work, have often an uncouth, heavy, and hesitating delivery. Yet I have not infrequently seen that such teachers had crowded lecture-rooms, while empty-headed orators excited astonishment in the first lecture, fatigue in the second, and were deserted in the third. Anyone who desires to give his hearers a perfect conviction of the truth of his principles must, first of all, know from his own experience how conviction is acquired and how not. He must have known how to acquire conviction where no predecessor had been before him—that is, he must have worked at the confines of human knowledge, and have conquered for it new regions. A teacher who retails convictions which are foreign to him, is sufficient for those pupils who depend upon authority as the source of their knowledge, but not for such as

require bases for their conviction which extend to the very bottom.
You will see that this is an honourable confidence which the nation
reposes in you. Definite courses and specified teachers are not prescribed
to you. You are regarded as men whose unfettered conviction is to be
gained; who know how to distinguish what is essential from what is only
apparent; who can no longer be appeased by an appeal to any authority, and
who no longer let themselves be so appeased. Care is also always taken that
you yourselves should penetrate to the sources of knowledge in so far as
these consist in books and monuments, or in experiments, and in the
observation of natural objects and processes.

Even the smaller German Universities have their own libraries,
collections of casts, and the like. And in the establishment of laboratories
for chemistry, microscopy, physiology, and physics, Germany has preceded
all other European countries, who are now beginning to emulate her. In our
own University we may in the next few weeks expect the opening of two
new institutions devoted to instruction in natural science.

The free conviction of the student can only be acquired when freedom of
expression is guaranteed to the teacher's own conviction—the *liberty of
teaching*. This has not always been ensured, either in Germany or in the
adjacent countries. In times of political and ecclesiastical struggle the ruling
parties have often enough allowed themselves to encroach; this has always
been regarded by the German nation as an attack upon their sanctuary. The
advanced political freedom of the new German Empire has brought a cure
for this. At this moment, the most extreme consequences of materialistic
metaphysics, the boldest speculations upon the basis of Darwin's theory of
evolution, may be taught in German Universities with as little restraint as
the most extreme deification of Papal Infallibility. As in the tribune of
European Parliaments it is forbidden to suspect motives or indulge in abuse
of the personal qualities of our opponents, so also is any incitement to such
acts as are legally forbidden. But there is no obstacle to the discussion of a
scientific question in a scientific spirit. In English and French Universities
there is less idea of liberty of teaching in this sense. Even in the Collège de
France the lectures of a man of Renan's scientific importance and earnestness are forbidden.

I have to speak of another aspect of our liberty of teaching. That is, the
extended sense in which German Universities have admitted teachers. In the
original meaning of the word, a doctor is a 'teacher,' or one whose capacity
as teacher is recognised. In the Universities of the Middle Ages any doctor
who found pupils could set up as teacher. In course of time the practical
signification of the title was changed. Most of those who sought the title did
not intend to act as teachers, but only needed it as an official recognition of
their scientific training. Only in Germany are there any remains of this
ancient right. In accordance with the altered meaning of the title of doctor,

and the minuter specialisation of the subjects of instruction, a special proof of more profound scientific proficiency, in the particular branch in which they wish to habilitate, is required from those doctors who desire to exercise the right of teaching. In most German Universities, moreover, the legal status of these habilitated doctors as teachers is exactly the same as that of the ordinary professors. In a few places they are subject to some slight restrictions which, however, have scarcely any practical effect. The senior teachers of the University, especially the ordinary professors, have this amount of favour, that, on the one hand, in those branches in which special apparatus is needed for instruction, they can more freely dispose of the means belonging to the State; while on the other it falls to them to hold the examinations in the faculty, and, as a matter of fact, often also the State examination. This naturally exerts a certain pressure on the weaker minds among the students. The influence of examinations is, however, often exaggerated. In the frequent migrations of our students, a great number of examinations are held in which the candidates have never attended the lectures of the examiners.

On no feature of our University arrangements do foreigners express their astonishment so much as about the position of private docents. They are surprised, and even envious, that we have such a number of young men who, without salary, for the most part with insignificant incomes from fees, and with very uncertain prospects for the future, devote themselves to strenuous scientific work. And, judging us from the point of view of basely practical interests, they are equally surprised that the faculties so readily admit young men who at any moment may change from assistants to competitors; and further, that only in the most exceptional cases is anything ever heard of unworthy means of competition in what is a matter of some delicacy.

The appointment to vacant professorships, like the admission of private docents, rests, though not unconditionally, and not in the last resort, with the faculty, that is with the body of ordinary professors. These form, in German Universities, that residuum of former colleges of doctors to which the rights of the old corporations have been transferred. They form as it were a select committee of the graduates of a former epoch, but established with the co-operation of the Government. The usual form for the nomination of new ordinary professors is that the faculty proposes three candidates to Government for its choice; where the Government, however, does not consider itself restricted to the candidates proposed. Excepting in times of heated party conflict it is very unusual for the proposals of the faculty to be passed over. If there is not a very obvious reason for hesitation it is always a serious personal responsibility for the executive officials to elect, in opposition to the proposals of competent judges, a teacher who has publicly to prove his capacity before large circles.

The professors have, however, the strongest motives for securing to the faculty the best teachers. The most essential condition for being able to work with pleasure at the preparation of lectures is the consciousness of having not too small a number of intelligent listeners; moreover, a considerable fraction of the income of many teachers depends upon the number of their hearers. Each one must wish that his faculty, as a whole, shall attract as numerous and as intelligent a body of students as possible. That, however, can only be attained by choosing as many able teachers, whether professors or docents, as possible. On the other hand, a professor's attempt to stimulate his hearers to vigorous and independent research can only be successful when it is supported by his colleagues; besides this, working with distinguished colleagues makes life in University circles interesting, instructive, and stimulating. A faculty must have greatly sunk, it must not only have lost its sense of dignity, but also even the most ordinary worldly prudence, if other motives could preponderate over these; and such a faculty would soon ruin itself.

With regard to the spectre of rivalry among University teachers with which it is sometimes attempted to frighten public opinion, there can be none such if the students and their teachers are of the right kind. In the first place, it is only in large Universities that there are two to teach one and the same branch; and even if there is no difference in the official definition of the subject, there will be a difference in the scientific tendencies of the teachers; they will be able to divide the work in such a manner that each has that side which he most completely masters. Two distinguished teachers who are thus complementary to each other, form then so strong a centre of attraction for the students that both suffer no loss of hearers, though they may have to share among themselves a number of the less zealous ones.

The disagreeable effects of rivalry will be feared by a teacher who does not feel quite certain in his scientific position. This can have no considerable influence on the official decisions of the faculty when it is only a question of one, or of a small number, of the voters.

The predominance of a distinct scientific school in a faculty may become more injurious than such personal interests. When the school has scientifically outlived itself, students will probably migrate by degrees to other Universities. This may extend over a long period, and the faculty in question will suffer during that time.

We see best how strenuously the Universities under this system have sought to attract the scientific ability of Germany when we consider how many pioneers have remained outside the Universities. The answer to such an inquiry is given in the not infrequent jest or sneer that all wisdom in Germany is professorial wisdom. If we look at England, we see men like Humphry Davy, Faraday, Mill, Grote, who have had no connection with English Universities. If, on the other hand, we deduct from the list of

German men of science those who, like David Strauss, have been driven away by Government for ecclesiastical or for political reasons, and those who, as members of learned Academies, had the right to deliver lectures in the Universities, as Alexander and Wilhelm von Humboldt, Leopold von Buch, and others, the rest will only form a small fraction of the number of the men of equal scientific standing who have been at work in the Universities; while the same calculation made for England would give exactly the opposite result. I have often wondered that the Royal Institution of London, a private Society, which provides for its members and others short courses of lectures on the Progress of Natural Science, should have been able to retain permanently the services of men of such scientific importance as Humphry Davy and Faraday. It was no question of great emoluments; these men were manifestly attracted by a select public consisting of men and women of independent mental culture. In Germany the Universities are unmistakably the institutions which exert the most powerful attraction on the taught. But it is clear that this attraction depends on the teacher's hope that he will not only find in the University a body of pupils enthusiastic and accustomed to work, but such also as devote themselves to the formation of an independent conviction. It is only with such students that the intelligence of the teacher bears any further fruit.

The entire organisation of our Universities is thus permeated by this respect for a free independent conviction, which is more strongly impressed on the Germans than on their Aryan kindred of the Celtic and Romanic branches, in whom practical political motives have greater weight. They are able, and as it would seem with perfect conscientiousness, to restrain the inquiring mind from the investigation of those principles which appear to them to be beyond the range of discussion, as forming the foundation of their political, social, and religious organisation; they think themselves quite justified in not allowing their youth to look beyond the boundary which they themselves are not disposed to overstep.

If, therefore, any region of questions is to be considered as outside the range of discussion, however remote and restricted it may be, and however good may be the intention, the pupils must be kept in the prescribed path, and teachers must be appointed who do not rebel against authority. We can then, however, only speak of free conviction in a very limited sense.

You see how different was the plan of our forefathers. However violently they may at times have interfered with individual results of scientific inquiry, they never wished to pull it up by the roots. An opinion which was not based upon independent conviction appeared to them of no value. In their hearts they never lost faith that freedom alone could cure the errors of freedom, and a riper knowledge the errors of what is unripe. The same spirit which overthrew the yoke of the Church of Rome, also organised the German Universities.

But any institution based upon freedom must also be able to calculate on the judgment and reasonableness of those to whom freedom is granted. Apart from the points which have been previously discussed, where the students themselves are left to decide on the course of their studies and to select their teachers, the above considerations show how the students react upon their teachers. To produce a good course of lectures is a labour which is renewed every term. New matter is continually being added which necessitates a reconsideration and a rearrangement of the old from fresh points of view. The teacher would soon be dispirited in his work if he could not count upon the zeal and the interest of his hearers. The estimate which he places on his task will depend on how far he is followed by the appreciation of a sufficient number of, at any rate, his more intelligent hearers. The influx of hearers to the lectures of a teacher has no slight influence upon his fame and promotion, and, therefore, upon the composition of the body of teachers. In all these respects, it is assumed that the general public opinion among the students cannot go permanently wrong. The majority of them—who are, as it were, the representatives of the general opinion—must come to us with a sufficiently logically trained judgment, with a sufficient habit of mental exertion, with a tact sufficiently developed on the best models, to be able to discriminate truth from the babbling appearance of truth. Among the students are to be found those intelligent heads who will be the mental leaders of the next generation, and who, perhaps, in a few years, will direct to themselves the eyes of the world. Occasional errors in youthful and excitable spirits naturally occur; but, on the whole, we may be pretty sure that they will soon set themselves right.

Thus prepared, they have hitherto been sent to us by the Gymnasiums. It would be very dangerous for the Universities if large numbers of students frequented them, who were less developed in the above respects. The general self-respect of the students must not be allowed to sink. If that were the case, the dangers of academic freedom would choke its blessings. It must therefore not be looked upon as pedantry, or arrogance, if the Universities are scrupulous in the admission of students of a different style of education. It would be still more dangerous if, for any extraneous reasons, teachers were introduced into the faculty, who have not the complete qualifications of an independent academical teacher.

Do not forget, my dear colleagues, that you are in a responsible position. You have to preserve the noble inheritance of which I have spoken, not only for your own people, but also as a model to the widest circles of humanity. You will show that youth also is enthusiastic, and will work for independence of conviction. I say work; for independence of conviction is not the facile assumption of untested hypotheses, but can only be acquired as the fruit of conscientious inquiry and strenuous labour. You must show that a conviction which you yourselves have worked out is a more fruitful germ of fresh insight, and a better guide for action, than the best-intentioned guidance by authority. Germany—which in the sixteenth century first revolted for the right of such conviction, and gave its witness in blood—is still in the van of this fight. To Germany has fallen an exalted historical task, and in it you are called upon to co-operate.

Inaugural Address as Rector of the Frederick William University of Berlin. Delivered October 15, 1877.

13

The Facts in Perception

HONORED ASSEMBLY!

Today we celebrate our university's commemoration holiday on the day of our founder's birth, the much-tested King Frederick William III. The year of this founding, 1811, came at the time of our State's greatest foreign troubles: a considerable amount of territory was lost, and the country was fully exhausted by the preceding war and hostile occupation. The warlike pride that had remained from the days of the great Electors and the Great King was deeply humiliated. And yet, when we look back, that very same time now seems to us so rich in goods of a spiritual kind—in inspiration, energy, ideal hopes, and creative thoughts—that, notwithstanding the relatively brilliant foreign situation in which the State and Nation now find themselves, we want to look back on that period almost with envy. That the King thought of founding the University during this distressing situation before attending to other material demands, and that he risked the throne and his life in order to entrust himself to the Nation's resolute inspiration in the struggle against the conqueror—this shows how deeply he too, a man who was disinclined to simple, spirited expressions of feeling, won the trust of his people's spiritual forces.

At the time, Germany had a stately series of praiseworthy names for display in both art and science, names whose bearers are partly to be counted among the first rank of all times and peoples in the history of human cultural education.

Goethe and Beethoven lived then; Schiller, Kant, Herder, and Haydn had yet to experience the century's first years. Wilhelm von Humboldt outlined the new science of comparative linguistics; Niebuhr, Fr. Aug. Wolf, and Savigny taught ancient history, poetry, and law with a penetrating, lively understanding; Schleiermacher sought seriously to comprehend the intellectual content of religion; and Joh. Gottlieb Fichte, our University's second rector and the powerful, fearless speaker, swept his listeners away with the stream of his moral inspiration and the bold intellectual flight of his Idealism.

Even the aberrations of this way of thinking, which expressed itself in the weaknesses of Romanticism that are so readily seen, have something attractive compared to dry, calculating egoism. One admired oneself in the beautiful feelings in which one knew to revel; one sought to cultivate the art of having such feelings; one believed that the more fantasy freed itself from the rules of the understanding, the more one had to admire it as a creative force. Therein resided much vanity—yet it was a vanity that reveled in high ideals.

The older ones among us still knew the men of that period: men who had entered into the army as the first volunteers; who were always ready to immerse themselves in the discussion of metaphysical problems; who were well read in the works of Germany's great poets; and who still burned with anger when Napoleon I, inspiration and pride, and the acts of the War of Liberation were discussed.

How things have changed! We cry out with astonishment in an age in which cynical contempt for all the human race's ideal goods has spread into the streets and the press, and has culminated in two abominable crimes which chose our Emperor's head as their object simply because his person united all that humanity to date had considered worthy of honor and gratitude.

It almost requires an effort for us to remember that only eight years have passed since the great moment when all classes of our people, full of joyful sacrifice and inspired by love of the Fatherland, unhesitatingly rose to the call of that same monarch and engaged in a dangerous war against an opponent whose power and bravery was not unknown to us. And it almost requires an effort for us to recall the broad scope that political and humane efforts have taken in the activities and thoughts of the educated classes, not least to give the poorer classes of our people a more carefree and humane existence, to think of how very much their lot has really improved in material and legal relations.

It seems to be humanity's nature that, along with much light, there is also found much shadow: political freedom at first gives the baser motives greater license to manifest themselves and mutually to encourage one another—so long as an armed public opinion does not confront them in energetic opposition. Even in the years prior to the War of Liberation, when Fichte held penitential sermons, these elements were not lacking. He describes governing conditions and convictions that remind us of the worst of our time. "In its fundamental principle, the present age arrogantly looks down upon those who, by some dream of virtue, allow themselves to be extricated from pleasures; and it is glad that it may be beyond such things, and in this way allows nothing to be imposed upon itself."[1] He refers to the only pleasure—beyond the purely sensual—with which the era's representatives are acquainted as "comfort in their own cunning." Nonetheless, a powerful impetus belonging to our history's most glorious events prepared itself in this very same era.

If we need not, therefore, think of our era as hopelessly lost, we should also not too easily calm ourselves with the consolation that it was not better in other times than it is now. It is, nonetheless, advisable that in such serious circumstances each person be watchful in the circle in which he works and which he knows, as to how the work of humanity's immortal goals is proceeding, that they are being kept in view, whether we may have drawn nearer to them. During our University's youth, science was also youthfully bold and filled with hope: its eye was turned chiefly to the highest goals. If these goals were not as easily reached as that generation had hoped, and if it also became clear that detailed individual investigations had to prepare the way—and thus, by the very nature of the tasks themselves, at first another type of work, one less enthusiastic and less directly, aimed at the ideal goals was needed—it would, for all that, doubtless be damaging if our generation should, thanks to secondary and practically useful tasks, lose sight of humanity's eternal ideals.

Epistemology was at that time the fundamental problem posited at the start of all science: "What is truth in our intuitions and thought? and in what sense do our ideas correspond to reality?" Philosophy and natural science approached this problem from two opposite sides; it is a common task of both. The first, which considers the intellectual side, seeks to exclude from our knowledge and ideas that which originates from the influences of the corporeal world in order to be able to state that which belongs to the mind's own activity. Natural science, by contrast, seeks to divide off that which is definition, designation, form of representation, and hypothesis in order to retain as pure residue that which belongs to the world of reality, whose laws it seeks. Both seek to accomplish the same division, even if each is interested in another part of the divide. Even the natural scientist cannot avoid these questions in the theory of sense perceptions and in the investigations on the fundamental principles of geometry, mechanics, and physics. Since my own work has often concerned both fields, I shall try to give you an overview (from the side of natural science) of what has been done in this direction. The laws of thought in men who pursue natural science are, of course, ultimately no different than in those who do philosophy. In all cases where the facts of daily experience—whose abundance is indeed already very large—suffice to give a rigorous thinker

---

1. Fichte's *Werke*, VII, p. 40.

with an unprejudiced feel for the truth enough material to reach a correct judgment, the natural researcher must be content to acknowledge that the methodical and complete collection of the facts of experience simply confirms results previously gained. However, opposite cases also occur. This is an apology—if apology it must be—that in what follows not all answers are new, but are, for the most part, long-known answers to the relevant questions. Often enough, even an older concept, when compared with new facts, takes on a livelier coloration and new authority.

Shortly before the start of the new century, Kant had developed the theory of what is given prior to all experience, or, as he called it, "transcendental" forms of intuitions and thought, wherein all the content of our representations must necessarily be received if it should become an idea. For the qualities of sensation, Locke had already considered the contribution which our corporeal and intellectual organization make to how things appear to us. In this way, the investigations in sensory physiology—that (above all) Johannes Müller completed, critically envisioned, and then summarized in the law of specific energies of sensory nerves—have now brought the fullest confirmation—one can almost say to a degree unexpected—and thereby at once presented and made evident the nature and meaning of one such *a priori*, subjective form of sensation in a very decisive and tangible way. This theme has already often been reviewed; hence I need discuss it only briefly today.

Among the different kinds of sense perceptions there occur two different types of distinction. The most fundamental is the distinction among sensations that belong to the different senses, as among blue, sweet, warm, and high-pitched. I have permitted myself to call this a distinction in the modality of sensation. It is so fundamental that it excludes any transition from one to the other, from any relation of greater or lesser similarity. For example, it simply cannot be an issue as to whether sweet may be similar to blue or red. By contrast, the second type of distinction, which is less fundamental, is that among different sensations of the same sense. To it I restrict the designation of a difference in quality. Fichte summarized these qualities of each sense as a circle of qualities, and calls what I called difference in modality a difference in the circle of qualities. A transition and comparison is possible within each such circle. We can move from blue through violet and carmine red into scarlet red; and, for example, assert that yellow may be more similar to orange red than to blue. Now physiological investigations teach that that fundamental distinction does not completely depend on the type of external impression by which the sensation is stimulated; rather, it is determined completely, solely, and exclusively by the sensory nerve that has been affected by the impression. Stimulation of the optic nerve produces only sensations of light, no matter whether it is struck by objective light—that is, by vibrations of the aether—or by electric currents which are conducted through the eye, or by pressure on the eyeball, or by stretching the nerve endings during a rapid change of view. The sensation that originates with the latter effects is so similar to that of objective light that one had long believed in an actual development of light in the eye. Johannes Müller showed that such a development definitely does not occur and that the sensation of light is most certainly only there because the optic nerve is stimulated.

On the one hand, just as each sensory nerve, stimulated by the most manifold influences, always gives sensations only from its particular circle of qualities, so, on the other hand, do the same external influences produce, when they meet different sensory nerves, the most different types of sensations, and these latter are always taken from the circle of qualities of the nerves in question. The same aether vibrations which the eye feels as light, the skin feels as heat. The same aerial vibrations which the skin feels as whirring motions, the ear feels as sound. Here, once again, the impression's heterogeneity is so great that physicists were first assuaged with the idea that agents like light and radiating heat, which appear so different, may be of the same kind and partly identical, after the complete homogeneity of their physical behavior was determined in every way by laborious experimental investigations.

Yet the most unexpected incongruencies also occur within the circle of qualities of each individual sense, where the type of effective object at least co-determines the quality of the sensation produced. In this regard, the comparison of eye and ear is instructive, since the objects of both—light and sound—are vibrating motions, which, depending on the speed of their vibrations, stimulate different sensations of different colors in the eye, of different pitches in the ear. If, for greater clarity, we may designate the light's vibrational relations with the names of musical intervals formed by the corresponding sound vibrations, then the following occurs: The ear senses about 10 octaves of different tones, the eye only a sixth, although the vibrations lying beyond these limits occur in sound, as in light, and can be demonstrated physically. The eye (in its short scale) has only three sensations fundamentally different from one another, by which all its qualities are composed through addition: namely, red, green, and blue-violet. These mix together in the sensation without disturbing one another. The ear, by contrast, distinguishes an enormous number of sounds of different levels. No chord sounds like any other chord which is composed of different sounds, while in the eye precisely the analogue is the case: for a similar shade of white can be produced by red and green-blue of the spectrum; by yellow and ultramarine blue; by green-yellow and violet; by green, red, and violet; or, in any case, by two, three or all of these mixtures together. Were the relations in the ear similar, then the chords C and F would be equal-sounding with D and G, with E and A, or with C, D, E, F,

G, A, etc. Moreover, concerning the objective meaning of color it is noteworthy that, apart from the effect on the eye, no single physical relation has been found in which similarly appearing light may be regularly equivalent. Finally, the entire foundation of the musical effect of consonance and dissonance depends on the characteristic phenomenon of beats. These are based on a rapid change in the intensity of sound from which thereby arises that two nearly equal high sounds cooperate in exchanging the same and opposite phases, and, accordingly, stimulate first strong, then weak vibrations of co-vibrating bodies. The physical phenomenon could occur quite similarly in the cooperation of two light-wave impulses, as it recurs in the cooperation of two sound-wave impulses. However, the nerve must first be able to be affected by both wave impulses; and second, it must be able to follow rapidly enough the change of stronger and weaker intensity. In the latter relation, the auditory nerve is far superior to the optic nerve. At the same time, each fibre of the auditory nerve is sensitive only to sounds from a narrow interval of the scale, so that only sounds situated very closely to one another can, in general, work together in it, while those situated far apart from one another cannot or cannot directly do so. If they do so, then this arises from accompanying upper harmonics or combination tones. Hence, the distinction of humming and non-humming intervals occurs in the ear, that is, of consonance and dissonance. By contrast, each optic nerve fibre senses throughout the entire spectrum, although it does so with different strengths in different parts. If the optic nerve could in general follow the tremendously rapid beats of light oscillations in the sensation, then each color mixture would operate as dissonance.

You see how all these distinctions as to the manner in which light and sound produce effects are conditioned by the type, by how the nerve apparatus reacts to them.

Our sensations are precisely effects produced by external causes in our organs, and the manner in which one such effect expresses itself depends, of course, essentially on the type of apparatus which is affected. Insofar as the quality of our sensation gives us information about the peculiarity of the external influence stimulating it, it can pass for a sign—but not for an image. For one requires from an image some sort of similarity with the object imaged: from a statue, similarity of form; from a drawing, equality of perspectival projection in the visual field; and from a painting, similarity of colors. A sign, however, need not have any type of similarity with what it is a sign for. The relations between the two are so restricted that the same object, taking effect under equal circumstances, produces the same sign, and hence unequal signs always correspond to unequal effects.

Compared to popular opinion, which in good faith assumes the complete truth of images that give us our sense of things, this residue of similarity which we acknowledge may seem very limited. In truth, it is not; for with it something of the very greatest significance can still be accomplished: namely, the imaging of the lawlike in the processes of the real world. Each natural law says that, given preconditions which are alike in certain respects, consequences which are alike in certain other respects will always follow. Since likeness in our world of sensation is shown by like signs, then there will also correspond to the natural-law consequence of like effects upon like causes a regular consequence in the field of our sensations.

If a certain type of berry develops (at maturity) both red pigment and sugar, then red color and sweet taste will always be found together in our sensation of berries of this type.

Hence, even though our sensations are, in their quality, only signs whose special type depends completely on our organization, they are nonetheless certainly not to be dismissed as empty appearance; rather, they are precisely signs of something, be it something enduring or occurring, and, what is most important, they can delineate for us the law of this occurring.

Hence, physiology also recognizes the qualities of sensation as a mere form of intuition. Kant, however, went further. He addressed not only the qualities of sense perception as given by the characteristics of our intuitive capability, but also space and time, since we can perceive nothing in the external world without it occurring at a definite time and in a definite place; determination in time may also be attributed, moreover, to each internal perception. He thus designates time as the given and necessary transcendental form of inner intuition, and space as the corresponding form of outer intuition. Kant thus considered spatial determinations as little belonging to the real world, "to the thing in itself," as the colors which we see belong to bodies in themselves, and which are rather brought into them through our eye. Even here the natural scientific view can, up to a certain point, go along. Namely, if we ask whether there is a marker which is common and perceptible by direct sensation, through which every perception relating to objects in space is characterized, then we find, in fact, one such marker in the circumstance that the movement of our body places us in other spatial relationships to perceived objects and thereby also changes the impression which they make on us. However, the impulse to movement that we give through innervation of our motor nerves, is something directly perceivable. We feel that we do something when we give one such impulse. We do not know directly what we do. Physiology only teaches us that we displace or innervate the motor nerves into a stimulated state; that its stimulation is conducted to the muscles; and that this contracts as a consequence and moves the limbs. Again however, without any scientific study we also know which perceptible effect follows from each different innervation that we can induce. It is demonstrable in a large number of cases that we learn this by frequently repeated experiments and observations. We can still learn in adulthood to locate the innervations necessary to enunciate letters of a foreign language or to produce a special type of voice in singing; we can learn innervations to move the ears, to squint with the eyes inward or outward, even upwards and downwards, etc. The difficulty in executing these consists only in that we must try to find by experiments the still-unknown innervations needed for such previously unexecuted movements. Moreover, we ourselves know these impulses under no other form and by no other definable feature than that they induce precisely the intended observable effect; this alone thus serves to distinguish different impulses in our own representation.

If we now make such types of impulses—take a look, move the hands, go back and forth—then we find that the sensations belonging to certain quality circles—namely, those with respect to spatial objects—can be changed; while other mental states of which we are conscious—memories, intentions, wishes, moods—cannot at all. A decisive distinction between the former and the latter is thus posited in direct perception. If we then want to call "spatial" the relation which we change directly by our will's impulse but whose type may still be quite unknown to us, then the perceptions of psychic activities do not at all enter into any such relation. All sensations of the external senses must, however, proceed through some type of innerva-tion, that is, be spatially determined. Accordingly, space will also seem physical to us, laden with the qualities of our perceptions of movement, as that through which we move ourselves, through which we are able to look. In this sense spatial intuition would thus be a subjective form of intuition, like the sensory qualities red, sweet, and cold. Naturally it would mean just as little for the former as for the latter, that the location of a certain individual object might be a mere appearance.

However, from this point of view space would appear as the necessary form of outer intuition, since we understand as the external world precisely what we perceive as spatially determined. That which has no perceptible spatial relation, we conceive as the world of inner intuition, as the world of the self-conscious.

And space would be an innate form of intuition prior to all experience insofar as its perception would be tied to the possibility of the will's motoric impulses, and for which the mental and corporeal ability must be given us through our organization before we can have spatial intuition.

Moreover, there will certainly be little doubt that the marker noted by us of the change in movement belongs to all spatial objects and perceptions, respectively.[2] On the other hand, the question remains to be answered as to whether all particular determinations of our spatial intuition are to be derived from this source. Toward that end we must consider what can be attained with the means of perception considered so far.

---

2. See Appendix I, "On the Localization of Sensations of the Inner Organs," p. 367 below.

Let us try to put ourselves in the position of a person without any experience. In order to begin without any spatial intuition, we must assume that such a human being also knows no more about the effects of his innervations beyond that by which he may have learned by putting himself in the initial state (from which he withdrew through the first impulse)—through the reduction of an initial innervation or by the carrying through of a second, opposing impulse. Since this mutual self-neutralization of different innervations is completely independent of what is perceived, the observer can discover how he has to do that without having previously reached any kind of understanding of the external world.

Let such an observer at first find himself facing an environment of objects at rest. This will initially allow itself to be recognized by him that, as long as he emits no motoric impulses, his sensations will remain unchanged. If he gives one such impulse—for example, if he moves his eyes or his hands, or if he steps forward—then the sensations change. If he returns to the previous condition through relaxation or through the appropriate counterimpulse, then all sensations again become the previous ones.

Let us call the entire group of aggregate sensations induced during the said period of time by a certain definite and finite group of the will's impulses the "current presentables"; by contrast, let us call "present" the aggregate of sensations from this group which is just coming to perception. Our observer is now bound to a certain circle of presentables from which, however, he can make each individual presentable present to himself at any moment through execution of the relevant movement. In this way it seems to him that each individual from this group of presentables exists at each moment during this period of time. He observed it in every individual instant that he chose to. The claim that he would also have been able to observe it in every other intervening moment (when he would have wanted it), is to be considered as an inductive conclusion simply drawn from each moment of a successful attempt at each moment of the concerned period of time. Thus the idea of a simultaneous and continuous existence of different things alongside one another will be achieved. "Coexistence" is a spatial designation; it is, however, justified because we have defined as "spatial" that relation changed by the will's impulses. One still does not need to think of that which is here said to coexist as substantive things. "It is bright on the right, it is dark on the left"; "resistance is ahead, behind it is not," could, for example, be said at this level of knowledge, whereby right and left are only names for definite eye movements, ahead and behind for definite hand movements.

At other times the circle of presentables becomes another circle for the same group of the will's impulses. This circle thereby confronts us with its individual contents as a given, as an "object." Those changes which can bring forth and annul by conscious impulses of the will are to be distinguished from those which are not consequences of the will's impulses and cannot be overcome by such. The latter finding is negative. Fichte's appropriate expression for it is that a *Non-ego* forces recognition of itself vis-à-vis the *Ego*.

If we inquire about the empirical conditions under which spatial intuition develops, then we mainly have to take into consideration the sense of touch, since blind people can fully form spatial intuition without help from vision. Even though the filling of space with objects for them will be sparser and less fine than for the seeing, it nonetheless seems in the highest degree improbable that the foundations of spatial intuition in both classes of persons should be completely different. Let us try to observe in the dark or with the eyes closed by touch: then we can touch very well with a finger, even with a pencil held in the hand, like a surgeon with a probe, and determine quite precisely and certainly the bodily shape of the object before us. When we want to find our way around in the dark, we usually touch larger objects with five or ten fingertips at once. We then get five to ten times as much information in the same amount of time as with one finger alone. And we also need the fingers to measure, as the tips of an open compass do, the sizes of objects. Anyway, in the case of touch the circumstance that there is a stretched-out, sensing skin surface with many sensory points retreats into the background. What we can determine by the skin's feel with a hand laid still upon an object, for example on a medal's impressions, is extraordinarily dull and poor in comparison with what we discover by feeling obtained through a motion, even if only with a pencil tip. With the sense of sight this procedure becomes much more complex in that—alongside the retina's most sensitive point, its central cavity, which in the act of seeing is simultaneously led around by the retinal image—a large quantity of still other, more sensitive points cooperate simultaneously in a much more productive manner than is the case with the sense of touch.

It is easy to appreciate that by moving the touching finger along the objects, the sequence in which the impressions of the object are presented becomes known; that this sequence shows itself to be independent of whether one feels with this or with that finger; that, furthermore, it is not a single-channelled, determined series, whose elements one must again and again traverse forward or backward in the same order so as to go from one to another—and thus is no linear series, but rather a surface-like coexistence, or, in Riemann's terminology, a manifold of the second order. The touching finger can, of course, go from one point to another of the tangible surface by means of motoric impulses other than those which displace it along the tangible surface, and different tangible surfaces require different motions in order to glide along them. A higher manifold is then required for the space in which the touching moves than in that for the tangible surface:

it will have to add a third dimension. This suffices, however, for all experiences being considered; for a closed surface fully divides the space which we know. Also, liquids and gases which are not limited to the form of human capacities for representation, cannot escape through a surface closed all the way round. And as only a surface, not a space—thus a spatial formation of two, not of three dimensions—is bounded by a closed line, so can only a space of three dimensions, not one of four, be completed by a surface.

In such a way may knowledge of the spatial ordering of things existing beside one another be acquired. Comparisons of size would be added to this by observation of the congruence of a touching hand with parts or points of body surfaces, or of retinal congruence with parts and points of the retinal image.

Hence, in the completed representation of the experienced observer it remains, finally, a wonderful consequence that this observed spatial order of things originally derives from the sequence in which the qualities of the sensation present themselves to the moved sensory organ: namely, the objects at hand in space seem to us clothed with the qualities of our sensations. They appear to us as red or green, cold or warm, to have smell or taste, etc., although these qualities of sensation belong to our nervous system alone and do not at all reach beyond into external space. Yet even when we know this, the appearance does not end, because this appearance is, in fact, the original truth; they are precisely the sensations which initially present themselves to us in the spatial order.

You see that in this way the most essential characteristics of spatial intuition can be derived. However, an intuition seems to the popular consciousness to be something simply given, that comes about without reflection and seeking, and in general is unresolvable into other mental processes. Some students of physiological optics, and also strictly observant Kantians, agree with this popular opinion, at least as far as it concerns spatial intuition. It is well known that already Kant assumed not only that the general form of spatial intuition may be transcendentally given, but that the very same form also contains from the very beginning and before all possible experience certain more specific determinations, as they are expressed in the axioms of geometry. These can be reduced to the following theorems:

1. Between two points there is only one possible shortest line. We call such a line "straight."

2. Through any three points a plane can be placed. A plane is a surface which completely includes any straight line if it coincides with any two of its points.

3. Through any point, only one line parallel to a given straight line is possible. Two straight lines are parallel if they lie in the same plane and do

not intersect with one another within any finite distance. Indeed, Kant uses the supposed fact that these geometrical theorems seem to us as necessarily correct and that we cannot even imagine any spatially deviating behavior, precisely as proof for it, that they had to be given prior to all experience and that, therefore, spatial intuition may also contain in them a transcendental form of intuition independent of experience.

I would like to emphasize at this point, on account of the controversies which have taken place in recent years concerning the issue as to whether geometrical axioms are transcendental or principles of experience, that this question is completely separate from the first question discussed (whether space in general may or may not be a transcendental form of intuition).[3]

Our eye sees all that it sees as an aggregate of color surfaces in the visual field: that is its form of intuition. Which special color appears by this or that opportunity, in which context, and in which order, is a result of external influences and is determined by no organizational law. Nothing about the facts expressed in the axioms follows from the thought that space may be a form of intuition. If such theorems should not be theorems of experience, but rather should belong to the necessary form of intuition, then this is a further special determination of the general form of space, and therefore those reasons which permit us to conclude that the form of spatial intuition may be transcendental still do not necessarily also suffice to prove that the axioms may be of transcendental origin.

Kant was influenced in his claim that spatial relations which might contradict Euclid's axioms cannot even be imagined—just as he was in his overall view of intuition as a simple, not further reducible mental process—by the then current state of mathematics and sensory physiology.

If one wants to try to imagine something never before seen, then one has to know how to imagine the series of sense impressions that would have to come about according to the known laws of the same if one observed that object and its gradual changes one after another from each point of view and with all the senses. At the same time, these impressions must be of the kind that every other interpretation is excluded. If this series of sense impressions can be given completely and clearly, then in my opinion one has to understand the object as intuitively imaginable. Since by the initial assumption the thing has never been observed, no previous experience can come to our aid and guide our fantasy in discovering the required series of impressions; rather, this can only happen through the concept of the object or relation to be imagined. Such a concept is first to be elaborated and specialized so far as is required for the given purpose. The concept of

---

3. See Appendix II, "Space Can Be Transcendental without There Being Any Axioms," p. 369 below.

spatial structures, which should not correspond to the usual intuition, can only be confidently developed by calculations using analytical geometry. For the present problem, Gauss had first (in 1828) given the analytical tool in his treatise on the curvature of surfaces, and Riemann applied this toward the invention of logically possible and consistent systems of geometry. One has not unfittingly designated these investigations as metamathematical. Moreover, it is to be noted that already in 1829 and 1840 Lobachevsky had developed a geometry without the parallel axiom in the usual, synthetic intuitive way, one which is in complete agreement with the corresponding part of the newer analytical investigations. Finally, Beltrami has given a method of modelling metamathematical spaces in part of Euclidean space, by which the determination of their manner of appearance in perspectival vision is made fairly easy. Lipschitz has proven the transformability of the general principles of mechanics to such spaces, so that the series of sense impressions which would occur in them can be given completely, whereby the intuitability of such spaces is proven in the sense of the previously given definition of this concept.[4]

Now, however, comes the contradiction. For proof of intuitability, I require only that for each path of observation the originating sense impressions be definitely and unambiguously indicated, if necessary by the use of the scientific knowledge of their laws, from which—at least for one who knows these laws—would result that the thing in question or the relation intuited may actually be at hand. The task of imagining the spatial relations in metamathematical spaces requires, in fact, some exercise in the understanding of analytical methods, perspective constructions, and optical phenomena.

This, however, contradicts the older concept of intuition, which only recognizes that as given by intuition whose representation comes to consciousness immediately with the sense impression and without recollection and effort. Our experiments for imagining mathematical spaces do not, in fact, have this ease, rapidity, and lightning-like self-evidence with which we perceive, for example, the shape of a room that we enter for the first time, the order and shape of the objects contained therein, the matter of which they consist, and much else. If this type of evidence were an originally given, necessary characteristic of all intuition, then we could not maintain as yet the intuitability of such spaces.

Now, upon further consideration there occurs to us a number of cases which show that certainty and rapidity of appearance of certain ideas in the case of certain impressions can also be achieved, even where nothing concerning such a connection is given by nature. One of the most striking examples of this sort is the understanding of our mother-tongue. The words

---

4. See my lecture "On the Origin and Significance of Geometrical Axioms," this volume.

are arbitrarily or accidentally chosen signs; each different language has different signs. Its understanding is not inherited: for a German child who has grown up among Frenchmen and who has never heard German spoken, German is a foreign language. The child becomes acquainted with the meaning of words and sentences only through examples of their use, whereby, prior to understanding the language, the child cannot even make comprehensible to itself that the sounds which it hears are signs which have a meaning. The child ultimately, when grown up, understands these words and sentences without recollection, without effort, without knowing when, where, and through which examples it has learned them; it understands the most refined changes of meaning, often such whose attempts at logical definition lag behind quite clumsily.

It will not be necessary for me to pile up examples of such processes: daily life is rich enough with them. Art—most obviously poetry and the fine arts—is virtually grounded therein. The highest type of intuition, as we find it in the vision of the artist, is such a registering of a new type of resting or moving appearance of men and nature. If the similar traces, which are often left behind in our memories by repeated perceptions, increase, then it is precisely the law-like that repeats itself most regularly in a similar manner, while fortuitous change is eliminated. By this means there develops in the loving and attentive observer an intuitive image of the typical behavior of the objects which interest him, of which he subsequently knows just as little as to how it came about as a child knows by which examples he has learnt the meaning of words. That the artist has beheld the real may be concluded from the fact that when he brings before us an example cleansed of accidental disturbances it again fills us with the conviction of truth. He is, however, superior to us in that he knew how to sift out everything accidental and confusing of the doings of the world.

So much just in order to remember how this mental process is effective from the lowest to the highest stages of development of our mental life. In my earlier works I have referred to the connection of ideas herein entering as unconscious inferences—unconscious insofar as the major premise is formed from a series of experiences which have individually disappeared from memory and have also entered into our consciousness only in the form of sentient observations, not necessarily conceived as sentences in words. In present perception, the newly appearing sense impression forms the minor premise to which is applied the rule imprinted through earlier observations. I later avoided the name of unconscious inference so as to circumvent confusion with—so it seems to me—the completely unclear and unjustified idea which Schopenhauer and his followers designate by this name. However, we are here obviously concerned with an elementary process which underlies all actual, so-called thinking, even if the critical sighting and completeness of the individual steps which enter into the scientific formation of concepts and conclusions is still also missing. Hence, as concerns, first of all, the question as to the origin of geometrical axioms: the lack of ease of representation of metamathematical spatial relations because of insufficient experience cannot be validly used as a reason against their intuitability. Moreover, the latter is completely provable. Kant's proof of the transcendental nature of geometrical axioms is thus untenable. On the other hand, the investigation of the facts of experience shows that the geometrical axioms, taken in that sense as they alone may be applied to the real world, can be tested, proven, and also eventually refuted by experience.[5]

The memory traces of earlier experiences still play an additional and highly influential role in the observation of our visual field.

An observer who is no longer completely inexperienced also receives—be it by momentary illumination through an electric discharge, be it by intentional, rigid fixation—without any eye movement a relatively rich image of the objects found before him. Yet even an adult easily convinces himself that this image becomes much richer—and, namely, much more exact—when he sweeps his gaze around the visual field, and thus applies that very type of spatial observation that I have previously described as fundamental. We are, in fact, so very much accustomed to allowing the sight of the objects we are considering to wander, that it requires a good deal of practice before we succeed in holding it steady at a certain point for a longer time while doing experiments in physiological optics. In my studies on physiological optics[6] I have sought to analyze how our knowledge of the visual field can be acquired through observation of images during eye movements, when only some sort of perceptible difference exists between otherwise qualitatively similar retinal sensations corresponding to the distinction of different places on the retina. According to Lotze's terminology, one may call one such distinction a local sign; only, that this sign may be a local sign—that is, correspond to a difference of place—and need not be known beforehand. More recent observations have again confirmed that persons who were blind from youth onwards, but subsequently regained sight through an operation, could at first not even distinguish by the eye such simple shapes as a circle and a square before they had touched them.[7] Furthermore, physiological investigation teaches that we can perform relatively exact and certain comparisons, by means of a naked-eye estimate exclusively at such lines and angles in the visual field that allow imaging through normal eye movements occurring in rapid

---

5. See my *Wissenschaftliche Abhandlungen* 2:640; and, for an extract therefrom, see Appendix III, "The Applicability of Axioms to the Physical World," p. 371 below.

6. "Vorträge über das Sehen des Menschen," in *Handbuch der Physiologischen Optik* 1:85; and in *Vorträge und Reden* 1:265.

7. Dufour (Lausanne) in the *Bulletin de la Société médicale de la Suisse Romande*, 1876.

succession on the same place of the retina. We can even estimate much more certainly the true sizes and distances of spatial objects that are not too distant than we can estimate the perspectival sizes and distances which vary with the point of view in the observer's visual field, although the former task of three-dimensional space is much more involved than the latter, which concerns only an image's surface. It is well known that one of the greatest difficulties in drawing is to free oneself from the involuntary influence exercised by the idea of the true size of an object seen. It is exactly such relations which we have to expect when we have acquired understanding of the local signs first through experience. For that which remains objectively constant, we can become well acquainted with the changing physical signs much more easily than for that which changes with each motion of our body as perspectival images do.

Moreover, for a large number of physiologists, whose opinion we can call the nativistic, in contrast to the empiricist, which I myself have sought to defend, this idea of acquired knowledge of the visual field seems unacceptable because they themselves have not clarified what is to be found so clearly in the example of language: how much accumulated memory impressions accomplish. Thus, a number of different attempts have been made to reduce at least a certain share of knowledge to an innate mechanism in the sense that certain sense impressions would cause certain, completed spatial representations. I have proven in detail[8] that until now all proposed hypotheses of this type are inadequate because ultimately one always discovers cases where our visual perception finds itself in more exact agreement with reality than those assumptions would yield. One is then forced to the additional hypothesis that the experience gained in the movements can ultimately overcome the innate intuition, and so, contrary to the latter, accomplishes what one would, following the empiricist hypothesis, accomplish without any such impediment.

Hence, the nativistic hypotheses explain, in the first place, nothing. Rather, they only assume the very fact to be explained, in that they simultaneously reject the possible reduction of the fact to the safely established mental processes, but to which they are then anyway forced to refer in other cases. Second, the assumption of all nativistic theories that complete ideas of objects are produced by the organic mechanism seems much more audacious and dubious than the assumption of the empiricist theory, that only the uncomprehended material of sensations derives from external influences, while all ideas are formed out of it according to the laws of thought.

Third, the nativist assumptions are unnecessary. The sole objection which can be brought against the empiricist explanation is the assured

8. See my *Handbuch der Physiologischen Optik*, 3. Abtheilung (Leipzig: 1867).

motion of many new-borns or even of animals hatched from eggs. The less mentally endowed they are, the more quickly do they learn what they are in general capable of learning. The narrower are the paths which their thoughts must travel, the easier do they find them. The new-born human child is extremely unskilled at seeing; it requires several days before it learns to judge the direction of the visual image towards which it must turn its head so as to reach the mother's breast. To be sure, young animals are much more independent of individual experience. What, however, is this instinct which directs it? Is direct inheritance of the parents' circles of representation possible? Does it concern only pleasure or displeasure, or a motoric drive, which attach themselves to certain aggregates of sensations? About all that we know as good as nothing. Plainly recognizable residues of the these phenomena still occur in humans. In this area, clean and critically conducted observations are highly desirable.

For facilitating the discovery of the first law-like relations the arrangements presumed by the nativist hypothesis can at best claim a certain pedagogical value. The empiricist view would also be compatible with presumptions tending in that direction: for example, that the local signs of neighboring points on the retina are as similar to one another as the more distant ones are, that corresponding positions of both retinae are more similar than disparate, etc. For our present investigation it suffices to know that spatial intuition can also originate completely in blind individuals, and that in those who can see—even if the nativist hypotheses are partially correct—the final and most precise determination of spatial relations is certainly conditioned by observations made in movement.

I return now to discussing the elementary, original facts of our perception. As we have seen, we not only have changing sense impressions which come over us without our doing anything; we also observe during our own continuing activity—and thereby attain knowledge of the existence of a lawful relationship between our innervations and the becoming present of the different impressions from the circle of the current presentables. Each of our voluntary motions by which we modify the manner of appearance of objects, is to be considered as an experiment by which we test whether we have correctly conceived the lawful behavior of the phenomenon in question, that is, its presumed existence in a definite spatial order.

The convincing force of every experiment is, however, in general so much greater than that of the observation of a process occurring without our involvement, because in the experiment the causal chain runs throughout our self-consciousness. We know one member of these causes—our will's impulse—from inner intuition, and know the motive by which it has occurred. The chain of physical causes which transpires in the course of the experiment has its initial effect from it, as from one initial member known to us and to one point in time known to us. However, an essential assumption for the conviction to be achieved is that our will's impulse has neither already been influenced by physical causes, which simultaneously determine the physical process, nor itself psychically influenced the succeeding perceptions.

In particular, this last doubt can come into consideration of our theme. The will's impulse for a definite motion is a mental act, and, so too, is the related perceived change in sensation. Now, cannot the first act bring about the second through a purely mental agency? It is not impossible. Something like this happens when we dream. Dreaming, we believe we execute a movement, and we then dream further that the very thing happens that should be its natural consequence. We dream that we climb into a canoe, push it off from the shore, slide it out onto the water, see it displace the objects lying around it, etc. Here the dreamer's expectation, that he will see the consequences of his actions occur, seems to bring about the dreamed perception in purely mental ways. Who can say how long and finely spun out, how logically executed any such dream can be? If everything therein may occur in the highest degree in a law-like manner, following the order of nature, then no other distinction from the waking state would exist than that of the possibility of awakening, the interruption of the dreamt series of intuitions.

I do not see how one could refute a system of even the most extremely subjective idealism that wanted to view life as a dream. One could explain it as so improbable, as so dissatisfying as possible—in which regard I would agree with the harshest expressions of rejection—but it may be logically feasible; and it seems to me to be very important to keep this in mind. It is well known how ingeniously Calderon developed this theme in *Life a Dream*.

Fichte, too, assumes that the Ego posits for itself the Non-Ego, i.e., the world of appearance, because it needs to develop its thinking activity. However, his idealism distinguishes itself from that just designated in that he conceives other individual human beings not as phantoms but, on the basis of the assertion of moral law, as being the same nature as his own Ego. Since, however, their images in which they imagine the Non-Ego must all harmonize again, he therefore conceived all the individual Egos as parts or emanations of the absolute Ego. The world in which those Egos then found themselves was the world of ideas, which the World Spirit posits for itself, and could again assume the concept of reality, as occurred with Hegel.

By contrast, the realistic hypothesis trusts the assertions of the normal self-observation, according to which the changes of perception following an action are in no way in any mental connection with the previously occurring impulses of the will. It regards as existing independently of our ideas, that which seems to endure in daily perception, the material world external to us. The realistic hypothesis is doubtless the simplest that we can form; tested and confirmed in extraordinarily wide circles of application; sharply defined in all individual features, and thus extraordinarily useful and fruitful as the foundation for action. We would, even from an idealistic viewpoint, hardly know how else to express the law-like in our sensations than when we say: "The acts of consciousness occurring with the character of perception take their course as if the world of material things assumed by the realistic hypothesis may really exist." However, we do not overcome this "as if"; we cannot recognize the realistic opinion as more than a superbly useful and precise hypothesis; we may not ascribe necessary truth to it, since in addition to it still other, irrefutable idealistic hypotheses are possible.

It is always good to keep this in mind so as not to conclude more from the facts than is warranted. The different shadings of idealistic and realistic opinions are metaphysical hypotheses which, so long as they are recognized as such, and however injurious they may become when represented as dogma or as supposed necessities of thought, are completely justified scientifically. Science must discuss all admissible hypotheses in order to retain a full overview of all possible attempts at explanation. Still more necessary are hypotheses for practical action, since one cannot always wait until a certain scientific decision has been reached, but instead must make a decision, be it according to probability or to aesthetic or moral feeling. In this sense, too, there is nothing objectionable in metaphysical hypotheses. It is, however, unworthy of a thinker wanting to be scientific if he forgets the hypothetical origin of his principles. The arrogance and the passion with which such hidden hypotheses are defended are the usual consequence of the dissatisfied feeling that its defender harbors in the hidden depth of his conscience about the correctness of his cause.

Yet what we find unambiguously and factually, without any hypothetical imputation, is the law-like in the phenomenon. From the first step on, where we perceive the abiding objects distributed before us in space, this perception is the recognition of a law-like connection between our movements and the sensations appearing therefrom. Thus the first elementary ideas already contain in themselves a thought and proceed according to the laws of thought. If we take a sufficiently broad concept of thought, as we did above, then everything that is added in the intuition to the raw material of sensations can be resolved in thought.

For if "conceive" means to form concepts, and if we try to summarize in the concept of a class of objects their similar characteristics, then it follows completely analogously that the concept must try to summarize a changing series of phenomena over time, one which remains the same in all its stages. The wise man, as Schiller expresses it:

Seeks the trusting law in Chance's horrifying wonders,
Seeks the resting pole in Phenomena's flight.

We call substance that which, without dependence on other things, remains the same over time; and we call the constant relationship between changeable quantities the law that binds them. It is only the latter that we perceive directly. The concept of substance can only be attained through exhaustive testing, and it always remains problematic insofar as further testing is always held in reserve. In previous times, light and heat were thought to be substances, until it later turned out that they may be transient forms of motion. Moreover, we must always be prepared for new decompositions of what are today considered to be chemical elements. The first product of the reflective understanding of a phenomenon is the law-like. If we have so far quite excluded it, so completely and certainly delimited its conditions, and, at the same time, so generally conceived it that the outcome for all possibly occurring cases that may occur is unambiguously determined, and if at the same time we become convinced that it may have maintained itself and will maintain itself in all times and in all cases, then we recognize it as something existing independently of our ideas, and we call it the cause, i.e., that which, behind the change, is the originally abiding and existing. In my opinion, in this sense alone is the use of the word justified, even if common linguistic usage generally uses it in a very vague way as antecedent or cause. Insofar, then, as we recognize the law as directing our perception and natural processes, as a power equivalent to our will, we call it "force." This concept of a power opposing us is directly conditioned by the ways and means that our simplest perceptions occur. From the very start, the changes which we ourselves make by our acts of will are separated off from those which are not made by our will and which cannot be overcome by our will. It is, namely, pain which gives us the most penetrating instruction about the power of reality. The emphasis here falls on the facts of observation, that the perceived circle of presentables is not determined by a conscious act of our idea or will. Fichte's Non-Ego is here exactly the right negative expression. To the dreamer, too, it seems that what he believes to see and feel is not produced by his will or by the conscious linking of his ideas, even if unconsciously the latter is in reality often enough the case; to him, too, it is a Non-Ego. So, too, with the idealist who looks at it as the world of ideas of the World Spirit.

We have in our language a very fortunate designation for that which, behind the change of phenomena permanently influences us, namely the "real." Herein only the effect is expressed; it lacks the additional connection to the existent as substance, which the concept of the real, i.e., the material, includes. On the other hand, the concept of the finished image of an object largely inserts itself into the concept of the objective, which does not fit into most original perceptions. We also have to designate as effective and real in the dreamer's logic those spiritual conditions or motives which, at the time, foist on him the law-like sensations corresponding to the present state of his dreamt world. On the other hand, it is clear that a separation of thought and reality first becomes possible after we know how to complete the separation of that which the Ego can and cannot change. This, however, only becomes possible after we recognize which law-like consequences the will's impulses have at that time. The law-like is thus the essential presumption for the character of the real.

I need not explain to you that it may be a contradiction in terms to want to represent the real or Kant's "thing-in-itself" through positive determinations without admitting it into the form of our representation. That has often been discussed. However, what we can attain is knowledge of the law-like order in the realm of the real; to be sure, this can only be presented in the sign system of our sense impressions.

All things transitory
But as images are sent.[9]

I consider it a good sign that here and elsewhere we find Goethe along with us on the same road. Where it concerns a broad outlook, we can well trust his clear and unprejudiced gaze for the truth. He demanded of science that it should only be an artistic ordering of facts and that beyond that it should form no abstract concepts which, to him, seem to be but empty names that only cloud the facts. In roughly the same sense, Gustav Kirchhoff has recently designated this as the task of the most abstract field among the natural sciences, mechanics, which in the completest and simplest way describes the movements occurring in nature. Concerning the "clouding," this in fact occurs when we remain in the realm of abstract concepts and do not analyze their factual sense, i.e., clarify to ourselves which observable, new, and law-like relations between the phenomena follow therefrom. Each properly developed hypothesis yields, depending on its factual sense, a more general law of phenomena than we have previously observed directly; it is an attempt to ascend to an ever-more-general and comprehensive lawfulness. Those new things that it maintains in the way of facts, must be tested and confirmed by observation and experiment. Hypotheses which do not have such a factual sense, or in general do not give clear and certain statements of the facts falling under them, are to be considered only as valueless phrases.

Every reduction of the phenomena to the underlying substances and

---

9. Johann Wolfgang von Goethe, *Faust. A Tragedy*, Two Parts, trans. Bayard Taylor (Boston: James Osgood and Company, 1871), 2:432.

forces claims to have found something unchangeable and definitive. We are never justified in making an unconditional claim of this type; for it grants neither the fragmentary nature of our knowledge nor the nature of inductive conclusions, upon which rests, from the first step on, all our perception of the real.

Every conclusion by induction is based upon trust that a previously observed law-like behavior will maintain itself in all cases which have yet to come under observation. It is a trust in the law-likeness of all that occurs. Law-likeness is, however, the condition of conceivability. Trust in law-likeness is thus simultaneously trust in the conceivability of natural phenomena. If we assume, however, that the conceptualization will be brought to completion, that we will be able to establish a final unchangeable something as the cause of the observed changes, then we call the regulative principle of our thinking, that which impels us, the causal law. We can say that it expresses the trust in the complete conceivability of the world. Conceiving, in the sense that I have described it, is the method by means of which our thinking subordinates itself to the world, orders the facts, predetermines the future. It is its right and its duty to expand the application of this method to all that occurs, and by this means it has already harvested truly great results. Still, for the causal law's applicability we have no additional security than its success. We could live in a world in which each atom might be different from every other atom, and where nothing may be at rest. There would thus be no way to discover conformity to law, and our thinking activity would have to cease.

The causal law is really an *a priori* given, a transcendental law. It is not possible to prove it by experience because, as we have seen, not even the first steps of experience are possible without the application of inductive conclusions, i.e., without the causal law; and from the completed experience, when it too taught that everything observed so far has proceeded in a law-like manner—which we are assuredly far from being justified in claiming—would always only be able to follow by an inductive conclusion, i.e., under the assumption of the causal law, that now the causal law would also be valid in the future. Only one piece of advice is valid here: trust and act!

Earth's insufficiency
Here grows to Event.[10]

This would be the response that we would have to offer to the question: what is truth in our representation? In that which has always seemed to me to constitute the most essential progress in Kant's philosophy, we still stand

10. Ibid.

on the ground of his system. In this sense I too have frequently emphasized in my previous works the agreement of modern sensory physiology with Kant's theories. To be sure, I did not, however, also thereby mean to have sworn to the master's words in all subsidiary points. I believe that one must view the modern era's most essential progress as the resolution of the concept of intuition into the elementary processes of thought, which is still lacking in Kant; this, then, also conditions his view of geometrical axioms as transcendental theorems. It was the physiological investigations on sense perception, which led us to the final elementary processes in knowledge and which were still not comprehensible in words, remained unknown and inaccessible to philosophy so long as it only investigated such knowledge that finds its expression in language.

To be sure, that which we have considered as a noteworthy deficiency, as one of the unsatisfactory developments of the specialized sciences of Kant's day, appears to those philosophers who have retained the tendency to metaphysical speculation precisely that which is most essential in Kant's philosophy. In fact, Kant's proof of the possibility of a metaphysics—about which alleged science he himself did indeed not know how to discover anything further—is based entirely on the opinion that the geometrical axioms and the applied principles of mechanics may be transcendental, *a priori* given theorems. Moreover, his entire system actually contradicts the existence of metaphysics and the obscure points of his epistemology—about whose interpretation so much has been disputed—derive from this root.

Considering all that, natural science may have its secure, well-established ground on which it can seek the laws of the real, a wonderfully rich and fruitful field of activity. So long as it limits itself to this activity, it will not be confronted by idealistic doubts. In comparison to the high-flying plans of the metaphysicians, such work may seem modest.

For never against
The immortals, a mortal
May measure himself.
Upwards aspiring,
He toucheth the stars with his forehead,
Then do his insecure feet
Stumble and totter and reel;
Then do the cloud and the tempest
Make him their pastime and sport.

Let him with sturdy,
Sinewy limbs,
Tread the enduring
Firm-seated earth;
Aiming no further, than

The oak or the vine to compare![11]

Nonetheless, the example of the one who said this may teach us how a mortal—who had indeed learned to stand even if he touched the stars with his forehead—still kept a clear eye for truth and reality. The true researcher must always have something of the artist's insight, of the insight which led Goethe, and Leonardo da Vinci, too, to great scientific thoughts. Both artist and researcher strive—even if in different ways—towards the same goal: to discover new lawfulness. One must not, however, want to propagate idle daydreams and crazy fantasies for artistic insight. Both the true artist and the true researcher know how to work properly and how to give to their work a stable form and convincing similitude.

Time and again, moreover, reality has so far revealed itself much more nobly and in a much richer way to the investigating science true to its laws than the most extreme efforts of mythical fantasy and metaphysical speculation knew how to portray them. What have all the monstrous products of Indian daydreaming, the accumulations of vast dimensions and numbers got to say in comparison to the reality of the world structure, and in comparison to the space of time into which the sun and earth formed themselves, into which life developed during geological history in ever more complete forms, adjusting itself to the quieter physical conditions of our planet?

Which metaphysics has concepts armed with effects, like those that magnets and current electricity exert upon one another—concerning whose reduction to well-determined elementary effects contemporary physics is still struggling—without yet having arrived at a clear conclusion? Even light, too, already seems to be nothing more than a form of motion of those two agents, and the space-filling aether, as a magnetizable and electrifiable medium, contains completely new characteristic properties.

And into which scheme of scholastic concepts should we classify this supply of effect-producing energy, whose constancy is expressed in the law of the conservation of force, which, like a substance, is indestructible and unincreaseable, and which is active as a motive power in each motion of inorganic and organic matter, a Proteus clothing himself in ever-new forms, operating through infinite space, and yet not without remainder divisible by space, the operative in each effect, the mover in each motion, and yet neither spirit nor matter?—Has the Poet intuited it?

In the tides of Life, in Action's storm,
A fluctuant wave,

---

11. "The Limits of Humanity," in *The Poetical Works of J. W. von Goethe*, 2 vols. (Boston: Francis A. Niccolls, 1902), 1:212–13.

A shuttle free,
Birth and the Grave,
An eternal sea,
A weaving, flowing
Life, all-glowing,
Thus at Time's humming loom 'tis my hand prepares
The garment of Life which the Deity wears![12]

We: bits of dust on the surface of our planet, itself hardly worth calling a grain of sand in the universe's infinite space; we: the most recent race among the living on earth, according to geological chronology barely out of the cradle, still in the learning stage, barely half-educated, declared of age only out of mutual respect, and yet already, through the more powerful force of the causal law, grown beyond all our fellow creatures and vanquishing them in the struggle for existence; we truly have reason enough to be proud that "the inconceivably sublime work" has been given to us to learn to understand slowly through constant work. And we need not feel in the least ashamed if this does not at once succeed in the first assault of a flight of Icarus.

---

12. *Faust* 1:28–29.

Appendixes

I. "On the Localization of Sensations of the Inner Organs." (To page 349, above.)

The question could be raised here whether the physiological and pathological sensations of the body's inner organs must not fall into the same category as the conditions of the soul, insofar as many of them likewise remain unchanged, or at least are not significantly changed, through movement. Now there are, in fact, conditions of an ambiguous character—like depression, melancholy, and anxiety—which can likewise originate from bodily as well as from psychological causes and by which there is also a lack of any idea of special localization. At best with the case of anxiety, the area around the heart is felt in some indefinite way to be the seat of the sensation, as was maintained in general by the older view (from which it clearly derives) that the heart may be the seat of many psychological feelings and that this organ is frequently altered by such feelings, whose movement one feels in part directly, and in part indirectly, when a hand is laid upon the heart. Hence, there originates in this way a sort of false bodily localization for real psychological conditions. This goes still much further in conditions of illness. I remember as a young doctor having seen a melancholy shoemaker who believed to feel his conscience press itself between his heart and his stomach.

On the other hand, there are indeed a series of bodily sensations—like hunger, thirst, satiation, and neuralgic and inflammatory pains—which we hold (if nonetheless indeterminedly) as localizable in the body and not as psychological, though they are scarcely changed by bodily movements. To be sure, the most inflammatory and rheumatic pains are considerably increased by pressure on certain bodily parts or by movement of those parts in which they are located. However, they also fall into the opposite case—similar to neuralgic pains—and can indeed be considered only as higher intensities of normally occurring feelings of pressure and tension in the parts concerned. Hence the sort of localization frequently gives a hint as to the cause by which we have experienced something about the place of sensation. Thus, almost all sensations of the abdominal intestines are displaced to definite positions of the anterior abdominal wall, even in such organs—like the duodenum, pancreas, spleen, etc., of the torso's posterior wall—which lay nearer. External pressure can, however, affect all these organs almost only through the supple anterior abdominal wall, not through the thick muscle layers between the ribs, spine, and hipbone. Furthermore, it is quite curious that with toothaches occasioned by periosteum inflammation of a tooth, the patients are at first usually uncertain as to whether (given a pair of teeth standing over one another) they suffer from the upper or the lower tooth. One must first forcefully press on the two teeth in order to discover which one is causing the pain. Does this not arise from the fact that pressure on the periosteum of the tooth's root tends in the normal state to occur only when chewing—and both pair of teeth always simultaneously undergo equally strong pressure?

The feeling of satiation is a sensation of fullness of the stomach, which is clearly increased by pressure on the epigastrium, while the feeling of hunger is, to a certain degree, lessened by the same pressure. Its localization in the epigastrium can be motivated in this way. Moreover, if we assume that the same local signs belong to nerves terminating at the same places of the body, then the distinct localization of a sensation of one such organ would also suffice for the other sensations of the same organ.

This indeed holds also for thirst, insofar as thirst is the sensation of dryness in the throat. By contrast, the more general feeling of water deficiency connected with this, which is not overcome by moistening of the mouth and throat, is not definitely localized.

The qualitatively peculiar feeling of shortness of breath, of so-called hunger for air, is diminished by breathing movements, and thereby localized. Yet sensations for breathing restrictions of the lungs and those for circulation impediments—supposing that these latter are not tied up with changes felt in heart beat—separate themselves off from one another only incompletely. Perhaps this separation is only incomplete because breathing disturbances usually also cause increased heart activity, and disturbed heart activity makes it difficult to relieve shortness of breath.

Moreover, it should be noted that we have parts capable of movement whose form and movements are extraordinarily fine and sensitive, and hence certain and well adapted—as are our soft palate, epiglottis, and larynx—yet we have no idea about them since we cannot see them without optical tools and also cannot easily touch them, and hence we have no anatomical and physiological studies of them. Indeed, all scientific investigations notwithstanding, we still do not confidently know how to describe all their movements, e.g., we cannot describe the larynx's initial movements during utterance of the falsetto. Had we innate knowledge of localization for our organs provided by the sense of touch, then we would have to expect such knowledge of the larynx as well as for the hands. In fact, however, our knowledge of the shape, size, and motion of our own organs goes only just so far as we can see and touch them.

The larynx's extraordinarily multifarious and finely executed movements also teach us about the relationship between the act of the will and its effect: namely, that what we cause is not, as we initially and immediately imagine, the innervation of a definite nerve or muscle, nor is it always the inducing of a definite position to one of our body's moveable parts; rather, it is the first observable external effect. Insofar as we can determine the position of the bodily parts by the eye or the hand, the position of these latter is the first observable effect to which a consciously intended act of the will refers. Where we cannot do this—as in the larynx and the posterior mouth parts— then these initial effects are different modifications of the voice, of breathing, of swallowing, etc.

The larynx's movements, although induced by innervations which are completely analogous to those used for the movement of the limbs, thus do not come into consideration in the observation of spatial changes. However, it is still open to question as to whether the very clear and multifarious expression of movement which music produces is perhaps not reducible to the change of pitch in singing produced through muscle innervation, and so through the same type of internal activity as the movement of the limbs.

There also exists a similar relationship for eye movements. We all know quite well how to direct our gaze to a definite position of the visual field, i.e., to cause its image to fall on the retina's central cavity. Uneducated persons do not, however, know how they move their eyes, and they do not always know to obey an optometrist's directive that they should turn their eyes somewhat towards the right, if so expressed. Indeed, even educated persons, who certainly know how to look at an object held before their nose (whereby they cast a glance inwards), do not know to obey the directive to cast a glance inwards when the corresponding object is not there.

II. "Space Can Be Transcendental without There Being Any Axioms." (To page 353, above.)

Almost all philosophical opponents of metamathematical investigations have treated both claims as identical, which they in no way are. Mr. Benno Erdmann[13] has already quite clearly explained that in the philosophers' usual mode of expression. I myself have emphasized it in an answer directed against the objections of Mr. Land in Leyden.[14] Although the author of the most recent refutation, Mr. Albrecht Krause,[15] cites both treatises, nonetheless the first five of his seven sections are again aimed at defending the transcendental nature of the form of intuition of space, while only two sections treat the axioms. To be sure, the author is not merely a Kantian, but also a follower of the most extreme nativistic theories in physiological optics, and believes that the entire content of these theories is included in Kant's system of epistemology—for which not the slightest justification may be found, even if Kant's individual opinion, which

13. *Die Axiome der Geometrie* (Leipzig, 1877), chapter III.
14. *Mind, a Quarterly Review* 3 (April 1878): 212.
15. A. Krause, *Kant und Helmholtz* (Lahr, 1878).

corresponded to the undeveloped state of physiological optics of his day, should have been approximately so. At that time the question as to whether the intuition could be more or less resolved into conceptual images had not yet been raised. Moreover, Mr. Krause ascribes to me ideas about local signs, sense memory, influence of retinal size, etc., which I have never held and never lectured on, or which I have expressly labored to refute. By sense memory I have always meant only the memory for immediate sensory impressions which have not yet been conceived in words; but I would always have vociferously protested against the claim that this sense memory may have its seat in the peripheral sense organs. I have conducted and described experiments to that end in order to show that we ourselves see with falsified retinal images (for example, in seeing through lenses, or through converging, diverging or sideward-refracting prisms), and that we quickly learn to overcome the illusion and to see correctly again. And yet Mr. Krause has imputed to me (p. 41) the claim that a child must see everything smaller than an adult because the child's eye is smaller. Perhaps the above lecture will convince the said author that he has until now completely misunderstood the sense of my empiricist theory of perception.

What Mr. Krause objects to in the sections on the axioms is in part disposed of in the present lecture, for example, the reasons why the intuitive idea of an unobserved object might be difficult. There follows a reference to my assumption, made in the lecture on the axioms of geometry,[16] of surface-like creatures living on a plane or sphere, which I made in order to demonstrate the relations between the different geometries, along with a discussion that two or even many "straightest"[17] lines can exist between two points on a sphere, but that Euclid's axiom speaks of the one "straight" line. For the surface-like creatures on the sphere, however, the straight line connecting two points on the spherical surface has, according to the assumption made, absolutely no real existence in their world. For them the "straightest" line of their world would be precisely what for us is the "straight" line. Mr. Krause indeed attempts to define a straight line as a line having only one direction. Yet how should one define direction?—surely, only again by using the straight line. We are moving here in a vicious circle. Direction is an even more specific concept, for every straight line has two opposite directions.

There follows an analysis arguing that, even if the axioms may be theorems of experience, we cannot be absolutely convinced of their correctness, as we indeed surely may be. The contention turns on precisely this very point. Mr. Krause is convinced that we would not believe measurements that might speak against the axioms' correctness. He may

16. See p. 226 of this volume.
17. This was how I designated the shortest or most geodetic lines.

well be right—at least as regards a large number of human beings who prefer to trust a theorem based on ancient authority closely tied to the rest of their knowledge rather than trust their own reflections. With a philosopher, however, it should be otherwise. Men have long behaved in a quite unbelievable manner concerning the spherical form of the Earth, its motion, and the existence of meteorites. Moreover, he is correct in his claim that it is advisable to be all the more rigorous in testing the grounds of proof against theorems of ancient authority the longer they have, in the experience of many generations, proved themselves until now as in fact correct. However, the facts, and not preconceived opinions or Kant's authority, must finally decide. If the axioms are natural laws, then it is also correct that they naturally play a part (through induction) in the (only approximate) provability of all natural laws. However, the wish to want to know exact laws is still not itself proof that there may be any such. Strange as it may nonetheless be, Mr. A. Krause, who dismisses the results of scientific measurement because of their limited exactness, sets his mind at rest (p. 62) about the transcendental intuition with the estimates made by the naked eye in order to prove that we need no measurements at all, to convince ourselves of the axioms' correctness. That is to measure friend and foe with different measures! As if any compass from the worst mathematical instruments did not measure more exactly than the best naked-eye estimate—not to mention the question, which my opponent does not at all raise, as to whether or not the latter may be innate and given *a priori*.

The expression, "measure of curvature," has given much offense to philosophical writers in its application to three-dimensional space.[18] Now the name designates a certain quantity defined by Riemann which, calculated for surfaces, converges with what Gauss has called the measure of curvature of surfaces. The geometers have retained these names as short designations for the more general case of more than two dimensions. The dispute here concerns only the name, and that, moreover, for a well-defined concept of quantity.

III. "The Applicability of Axioms to the Physical World." (To page 356, above.)

Here I want to develop the consequences to which we would be forced if Kant's hypothesis of the transcendental origins of the geometrical axioms were correct, and to discuss which values this direct knowledge of axioms

---

18. E.g., in A. Krause, l.c., p. 84.

would then have for our judgment about the relations of the objective world.[19]

§1.

In this first section I shall initially stick with the realistic hypothesis and speak its language; hence, I assume that the things we perceive objectively remain real and affect our senses. I do this initially only so as to be able to speak the simple and understandable language of everyday life and of natural science, and also thereby to express the sense of what I mean in a way that is also understandable for nonmathematicians. I reserve the right in subsequent paragraphs to abandon the realistic hypothesis and to repeat the corresponding analysis in abstract language and without any special assumption about the nature of the real.

First, we must distinguish that equality or congruency of spatial magnitudes (as it could derive from the assumption made from transcendental intuition) from that of equivalence, which is to be determined by measurement using physical means.

I call spatial magnitudes physically equivalent when, under the same conditions and in the same periods of time, the same physical processes can exist and elapse. For the determination of physically equivalent spatial magnitudes, the most frequently used process, under appropriately cautionary measures, is the transporting of rigid bodies, like the compass and the ruler, from one place to another. Furthermore, it is a completely general result of all our experience that, if the equivalence of two spatial magnitudes has been proven by any kind of sufficient method of physical measurement, then the same is also proven as equivalent vis-à-vis all other known physical processes. Physical equivalency is thus a completely definite, unequivocally objective property of spatial magnitudes, and it obviously does not prevent us from determining anything through experiments and observations as to how the physical equivalence of a definite pair of spatial magnitudes depends on the physical equivalence of another pair of such magnitudes. This would give us a type of geometry which I want to call, for the purpose of our present investigation, physical geometry, in order to distinguish it from the geometry which may be based on the hypothetically assumed transcendental intuition of space. Such a purely and intentionally realized physical geometry would obviously be possible, and would have the full character of a natural science.

---

19. Hence in order to avoid new misunderstandings, as such occur in Mr. A. Krause, l.c., p. 84: I am not the one "who knows a transcendental space with its own laws"; rather, I here seek to draw the consequences of Kant's (for me) unproven and incorrectly considered hypothesis, according to which the axioms should be given theorems through transcendental intuition so as to prove that a geometry based on such an intuition would be completely useless for objective knowledge.

Its first steps would already lead us to theorems which may correspond to the axioms, if only its physical equivalence replaces the transcendental equality of spatial magnitudes.

Namely, as soon as we may have found an appropriate method for determining whether the respective distances of two pairs of points are equal to one another (i.e., physically equivalent), then we would also be able to distinguish the special case where three points a, b, c so lie that, apart from b, there is no second point to be found that may have the same distances from a and c as b. In this case we say that the three points lie in a straight line.

We would then be in a position to seek three points A, B, C which each have the same distance from one another, thus representing the angles of an equilateral triangle. We could then seek two new points b and c equally distant from A, with b lying in a straight line with A and B, and c lying in a straight line with A and C. The question may then arise: Is the new triangle Abc also equilateral, like ABC; thus is $bc = Ab = Ac$? Euclidean geometry answers: yes, indeed; spherical geometry maintains that $bc > Ab$, if $Ab < AB$; and, under the same condition, pseudospherical maintains that $bc < Ab$. The axioms may already allow us to come near to making a factual decision. I have selected this simple example because we are here concerned only with the measurement of the equality or inequality of the distances between points (or, respectively, with the definiteness or indefiniteness of the positions of certain points) and because no complex spatial magnitudes, straight lines, or planes need in any way be constructed. The example shows that these physical geometries would have their theorems assume the role of axioms.

As far as I can see, it cannot be doubted that even for the followers of the Kantian theory it may be possible to found a purely experiential geometry in the manner described, if we did not yet have such a geometry. In this case we would only be concerned with observable, empirical facts and their laws. The science which may be achieved in this way would, to that extent, only be a theory of space independent of the constitution of the physical bodies contained in space, if the assumption proves correct that physical equivalency always appears simultaneously for all types of physical processes.

However, Kant's followers maintain that there may, in addition to one such physical geometry, also be a pure geometry which alone may be founded on transcendental intuition, and that this may in fact be the one geometry which has so far been developed scientifically. In this we may in no way be concerned with physical bodies and their types of motions. Rather, without knowing anything at all about such experience, we could form, through inner intuition, ideas about absolutely unchangeable and immoveable spatial magnitudes, bodies, surfaces, and lines which, without their ever being brought into congruence through motion belonging only to physical bodies, may nonetheless stand in the relation of equality and congruence to one another.[20]

I may be permitted to emphasize that this inner intuition of the straightness of lines and the equality of distances or angles must have absolute exactness; for otherwise, we would in no way be justified in deciding whether two straight lines, infinitely extended, intersect only once, or, perhaps just like the great circles on a sphere, twice, nor to maintain that each straight line, which intersects one of two parallel lines with which it lies in the same plane, must also intersect the other. One must not want to substitute that very incomplete naked-eye estimate of transcendental intuition, which latter requires absolute exactness.

Let us suppose that we had such a transcendental intuition of spatial structures, of their equality and congruence, and that we could convince ourselves by truly sufficient reasons that we have them: from this one could derive a system of geometry which would be independent of all properties of physical bodies, a pure, transcendental geometry. This geometry would also have its axioms. However, even according to Kantian principles it is clear that the theorems of this hypothetical, pure geometry need not necessarily agree with those of physical geometry. For the one speaks of equality of spatial magnitudes in inner intuition, the other of physical equivalency. This latter obviously depends on empirical properties of natural bodies, and not merely on the organization of our mind.

It would then have to be investigated whether the two described types of equality necessarily always coincide. This cannot be decided by experience. Does it make any sense to ask whether, according to transcendental intuition, two pairs of compass points span equal or unequal lengths? I do not know how to attach any sense to this; and insofar as I understand Kant's more recent followers, I believe that I can assume that they too they would answer with a "No." As we have said, we cannot here substitute a naked-eye estimate.

Now, could it perhaps be concluded from the theorems of pure geometry that the distances of the two compass points may be equally great? Geometrical relations between these distances and other spatial magnitudes would have to be known here, and of these one would have to know directly that they were equal in the sense of transcendental intuition. Now since one can never know this directly, one can thus also never conclude it by means of geometrical deductions.

If the claim that both types of spatial equality are identical cannot be grounded in experience, then it must be a metaphysical claim and correspond to a necessity of thought. However, such a claim would then

---

20. Land in *Mind* 2:41. — A. Krause, l.c., p. 62.

determine not only the form but also the content of empirical knowledge—for example, as in the construction of two equilateral triangles noted above—a conclusion which would virtually contradict Kant's principles. Pure intuition and thinking would then do more than Kant is inclined to grant.

Let us suppose, finally, that physical geometry may have found a series of general theorems of experience which may be identical with the axioms of pure geometry. It would then at most follow that the agreement between physical equivalence of spatial magnitudes and their equality in pure spatial intuition may be an admissible hypothesis which leads to no contradiction. It would not, however, be the only possible hypothesis. Physical space and the space of intuition could also correspond to one another as real space does to its image in a convex mirror.[21]

Since physical and transcendental geometry do not necessarily need to agree, it follows that we can in fact imagine them as not agreeing. The manner in which one such incongruence would appear emerges already from what I have analyzed in a previous essay.[22] We assume that the physical measurements may correspond to a pseudo-spherical space. The physical impression of one such space when the observer and observed objects are at rest would be the same as if we had before us Beltrami's spherical model in Euclidean space, in which the observer would find himself at its center. However, as the observer changed place, the center of the sphere of projection would have to shift with him and the entire projection displace itself. For an observer whose spatial intuitions and estimates of spatial magnitudes would be formed either from transcendental intuition or as a result of experience to date in the sense of Euclidean geometry, the impression would thus arise that, as he himself moves, all objects seen by him therefore displace themselves in a definite way and variously expand and contract themselves in different directions. In a similar way, only in quantitatively diverging relationships, we see also in our objective world how the perspectivally relative position and the apparent sizes of objects change at different distance (as the observer moves). Just as we now in fact are in a position to recognize from these changing visual images that the objects around us do not change their relative mutual position and magnitude, as long as the perspectival displacements correspond exactly to the laws which have been verified to date by experience (and which are obeyed in the case of objects at rest); and we infer, on the other hand, from each deviation of this law a motion of the objects; so would someone, as I myself, a follower of the empiricist theory of perception, believe it is permitted to assume, so would someone who

---

21. See my lecture "On the Origin and Significance of Geometrical Axioms," this volume.
22. Ibid.

made the transition from Euclidean into pseudo-spherical space initially believe he was seeing apparent motions of the objects, but would very quickly learn to adjust his estimation of the spatial relations to the new conditions.

This latter, however, is an assumption which is formed only in analogy to that which we otherwise know about sense perceptions, and cannot be proven by experiment. Let us assume then, since it may be connected with innate forms of spatial intuition, that the judgment of spatial relations by one such observer could not be further changed. The same observer would indeed quickly determine that the motions which he believes he is seeing are only apparent motions, since they always return again when he returns to his initial point of view; or a second observer would be able to determine that everything remains at rest while the first observer changes position. If, therefore, the physically constant spatial relationships are perhaps impossible to prove through unreflected intuition, one would certainly soon be able to do so through scientific investigation, perhaps just as we ourselves know through scientific investigations that the sun stands still and the Earth rotates, although the sentient appearances persist that the Earth stands still and the sun rotates around it once every twenty-four hours.

In that case, however, this completely assumed transcendental intuition *a priori* would be degraded to the rank of a sensory illusion, of an objectively false appearance, from which we must free ourselves and which we must seek to forget, just as we do with the sun's apparent motion. There would then be a contradiction between that which appears according to inner intuition as spatially equivalent and that which proves itself as such in objective phenomena. Our entire scientific and practical interest would be tied up with the latter. The transcendental form of intuition would thus only represent the physically equivalent spatial relationship as a plane map does the Earth's surface: very small pieces and strips appear correctly, while larger ones necessarily appear incorrectly. It would thus be a matter not only of the manner of appearance, which indeed necessarily conditions a modification of the content to be represented; it would also be a matter that the relations between appearance and content, which for narrower limits yield agreement between both, would yield a false appearance to broader limits.

Here is the consequence that I draw from these considerations: If there really were an innate and ineradicable form of intuition of space in us, including the axioms, then we would only be justified in their objective scientific application to the world of experience if it could be determined by observation and experiment that the parts of space which are equivalent according to the presumed transcendental intuition were also physically equivalent. This condition coincides with Riemann's requirement that the mass of curvature of space in which we live must be determined empirically through measurement.

Such measurements that have been done have produced no noticeable deviation from zero for the value of this mass of curvature. We can, then, by all means consider Euclidean geometry as in fact correct within the limits of the exactness of measurement reached so far.

§2.

The discussion in the opening paragraphs remained completely in the realm of the objective and realistic standpoint of the natural scientist, where conceptualization of natural laws is the final goal and knowledge by intuition is only a facilitating aid used to overcoming a false appearance.

Now Professor Land believes that in my analyses I may have confused the concepts of the objective and the real, that my claim that the geometrical theorems could be tested and confirmed by experience may be assumed in an unfounded manner (*Mind* II, p. 46), "that empirical knowledge is acquired by simple importation or by counterfeit, and not by peculiar operations of the mind, solicited by varied impulses from an unknown reality." Had Professor Land been familiar with my works on sensory perceptions, he would have known that I too have struggled throughout my life against the very assumption that he ascribes to me! In my essay, I have not spoken of the distinction between the objective and the real because it seemed to me that in the investigation under discussion absolutely no weight falls on this distinction. In order to substantiate my opinion, we want now to drop the hypothetical part in the realistic viewpoint and prove that the posited theorems and proofs to date are still completely correct, and that one is still justified in inquiring about the physical equivalence of spatial magnitudes and deciding on the basis of experience.

The sole assumption which we adhere to is that of the causal law: namely, that the ideas occurring in us with the character of perception occur according to enduring laws, so that, if different perceptions intrude upon us we are justified in drawing the inference to differences in the real conditions under which they have developed. Furthermore, we know nothing about these conditions themselves, about the actual real which underlies the phenomena; all opinions which we may otherwise harbor in this regard are only to be considered as more or less probable hypotheses. The above assumption, by contrast, is the fundamental law of our thinking; if we wanted to give it up, we would therewith generally renounce our ability to think conceptually about these relations.

Concerning the nature of the conditions under which ideas arise, I stress that absolutely no assumptions should be made. The hypothesis of subjective idealism may be just as admissible as that of the realistic view, whose language we have used until now. We could assume that all our perceiving may be only a dream—if nonetheless a dream which in itself is highly consistent, in which idea developed out of idea according to enduring laws. In that case, the reason that a new apparent perception entered, could only be sought in that in the dreamer's soul ideas of certain other perceptions, and perhaps also ideas of a definite type of his own will's impulse, have been present earlier. What we call natural laws in the realistic hypothesis would, in the idealistic hypothesis be laws which regulate the succession of ideas following upon one another with the character of perception.

Now, we find it is a fact of consciousness that we believe we perceive objects located at definite places in space. That an object seems at a definite, specific place and not at another will have to depend on the type of real conditions that determine the idea. We must conclude that other real conditions would have had to obtain in order for the perception to appear to be of another place of the same object. In the realm of the real, therefore, there must exist some sort of relationship or complexes of relationships which determine at which place in space an object appears to us to be. In order to characterize this in a shorthand manner, I shall call these topogenous factors. We know nothing of their nature; we know only that the occurrence of spatially different perceptions assumes a difference of topogenous factors.

At the same time, there must be other causes in the realm of the real which lead us to believe that we perceive different material things in the same place but at different times. I shall permit myself to call these by the name of hylogenous factors. I choose this new name so as to separate off any intermixing of additional meanings that could attach themselves to everyday words.

Now if we perceive anything at all and assert something that expresses a mutual dependency of spatial magnitudes, then the factual meaning of such an expression is doubtless only that a certain law-like connection occurs—whose type is unknown to us—between certain topogenous factors—whose actual nature, however, also remains unknown to us. For this very reason Schopenhauer and many of Kant's followers have come to the wrong conclusion that there may be no real content in our perceptions of spatial relations in general, that space and its relations may only be transcendental appearance, without anything real corresponding to them. We are, however, nonetheless justified in applying to our spatial perceptions the same considerations as to other sensible signs, e.g., colors. Blue is only a mode of sensation; that we, however, see blue at a certain time and in a definite direction, must have a real reason . If we see red there at one time, then this real reason must have changed.

If we observe that dissimilar types of physical processes can occur in congruent spaces during the same periods of time, then this means that in the realm of the real the same aggregates and sequences of certain hylogenous factors occur and can occur in connection with certain definite groups of different topogenous factors: namely, such as give us the perception of physically equivalent parts of space. And if experience then teaches us that each connection or each succession of hylogenous factors which can exist or occur in connection with one group of topogenous factors is also possible with each physically equivalent group of other topogenous factors, then this is a theorem which at all events has a real content, and the topogenous factors thus doubtless influence the occurrence of the real process.

The example given above of two equilateral triangles only concerns: (1) equality or inequality, i.e., physical equivalency or non-equivalency of point intervals; and (2) definiteness or indefiniteness of the topogenous factors of certain points. These concepts of definiteness and equivalence in relation to certain consequences can, however, also be applied to objects of an otherwise completely unknown nature. I therefore conclude that the science which I have called physical geometry contains theorems of real content and that its axioms are determined not by mere forms of ideas, but by relations in the real world.

This still does not justify us in declaring the assumption of a geometry founded on transcendental intuition to be impossible. One could, e.g., assume that an intuition (without any physical measurement) of equality between two spatial magnitudes is produced directly by the influence of the topogeneous factors on our consciousness; thus, that certain aggregates of topogenous factors may also be equivalent with respect to a mental, directly perceivable effect. All of Euclidean geometry can be derived from the formula which gives the distance between two points as a function of their rectangular coordinates. Let us assume that the intensity of that mental effect, whose equality appears in our representation as an equality of the distance between two points, depends in the same way on any three functions of the topogenous factors of each point (as the distance in Euclidean space of the three coordinates of each), then the system of pure geometry of one such consciousness must comply with the Euclidean axioms, since, moreover, the topogenous factors of the real world and its physical equivalence are so. It is clear that in this case the agreement between mental and physical equivalence of the spatial magnitudes cannot be decided by the form of the intuition alone. And if agreement should be reached, then this should be conceived as a law of nature, or, as I have called it in my popular lecture, as a preestablished harmony between the world of ideas and the real world, just as it is based on laws of nature that the straight line characterized by a lightray coincides with that formed by a taut string.

I believe I have thus shown that the proof that I gave in §1 using the language of the realistic hypothesis also holds without those assumptions.

If we wanted to apply geometry to the facts of experience, where it always only concerns physical equivalence, then only the theorems of that science that I have designated as physical geometry can be applied. For anyone who derives the axioms from experience, our geometry to date is in fact physical geometry, which is based only on a large amount of unplanned, accumulated experience, instead of on a system of methodically undertaken experience. It is, moreover, noteworthy that this view was already Newton's, who explains in the Introduction to the *Principia*: "Geometry itself has its foundation in mechanical practice and is, in fact, nothing other than that very part of the totality of mechanics which forms the basis of and precisely determines the art of measurement."[23]

By contrast, the assumption of knowledge of the axioms as issuing from transcendental intuition is:

1. an unproven hypothesis;
2. an unnecessary hypothesis, since it pretends to explain nothing in our factual world of representation that could not also be explained without its help; and
3. a completely unusable hypothesis for the explanation of our knowledge of the real world, since the theorems established by it may first be applied to the relations of the real world after its objective validity has been experimentally proven and determined.

Kant's theory of the *a priori* given forms of intuition is a very apt and clear expression of the relations of things; but these forms must be without content and sufficiently free to assume any content which could generally enter into the concerned form of perception. The axioms of geometry, however, limit the form of intuition of space in a way such that if geometry is to be generally applicable to the real world, then no longer can any imaginable content be included in it. If we eliminate the axioms, then the theory of the transcendality of the form of intuition of space is completely inoffensive. Kant has here, in his *Critique*, not been critical enough; but, of course, here it concerned theorems from mathematics, and this piece of critical work must be completed by the mathematicians.

Speech held at the Commemoration-Day Celebration of the Frederick William University in Berlin, August 3, 1878. Revised and expanded version.

---

23. Fundatur igitur Geometria in praxi Mechanica, et nihil aliud est quam Mechanicae universalis pars illa, quae artem mensurandi accurate proponit ac demonstrat.

14

Hermann von Helmholtz. An Autobiographical Sketch

IN the course of the past year, and most recently on the occasion of the celebration of my seventieth birthday, and the subsequent festivities, I have been overloaded with honours, with marks of respect and of goodwill in a way which could never have been expected. My own sovereign, his Majesty the German Emperor, has raised me to the highest rank in the Civil Service; the Kings of Sweden and of Italy, my former sovereign, the Grand Duke of Baden, and the President of the French Republic, have conferred Grand Crosses on me; many academies, not only of science, but also of the fine arts, faculties, and learned societies spread over the whole world, from Tomsk to Melbourne, have sent me diplomas, and richly illuminated addresses, expressing in elevated language their recognition of my scientific endeavours, and their thanks for those endeavours, in terms which I cannot read without a feeling of shame. My native town, Potsdam, has conferred its freedom on me. To all this must be added countless individuals, scientific and personal friends, pupils, and others personally unknown to me, who have sent their congratulations in telegrams and in letters.

But this is not all. You desire to make my name the banner, as it were, of a magnificent institution which, founded by lovers of science of all nations, is to encourage and promote scientific inquiry in all countries. Science and art are, indeed, at the present time the only remaining bond of peace between civilised nations. Their ever-increasing development is a common aim of all; is effected by the common work of all, and for the common good of all. A great and a sacred work! The founders even wish to devote their gift to the promotion of those branches of science which all my life I have pursued, and thus bring me, with my shortcomings, before future generations almost as an exemplar of scientific investigation. This is the proudest honour which you could confer upon me, in so much as you thereby show that I possess your unqualified favourable opinion. But it would border on presumption were I to accept it without a quiet expectation on my part that the judges of future centuries will not be influenced by considerations of personal favour.

My personal appearance even, you have had represented in marble by a master of the first rank, so that I shall appear to the present and to future generations in a more ideal form; and another master of the etching needle has ensured that faithful portraits of me shall be distributed among my contemporaries.

I cannot fail to remember that all you have done is an expression of the sincerest and warmest goodwill on your part, and that I am most deeply indebted to you for it.

I must, however, be excused if the first effect of these abundant honours is rather surprising and confusing to me than intelligible. My own consciousness does not justify me in putting a measure of the value of what I have tried to do, which would leave such a balance in my favour as you have drawn. I know how simply everything I have done has been brought about; how scientific methods worked out by my predecessors have naturally led to certain results, and how frequently a fortunate circumstance or a lucky accident has helped me. But the chief difference is this—that which I have seen slowly growing from small beginnings through months and years of toilsome and tentative work, all that suddenly starts before you like Pallas fully equipped from the head of Jupiter. A feeling of surprise has entered into your estimate, but not into mine. At times, and perhaps even frequently, my own estimate may possibly have been unduly lowered by the fatigue of the work, and by vexation about all kinds of futile steps which I had taken. My colleagues, as well as the public at large, estimate a scientific or artistic work according to the utility, the instruction, or the pleasure which it has afforded. An author is usually disposed to base his estimate on the labour it has cost him, and it is but seldom that both kinds of judgment agree. It can, on the other hand, be seen from incidental expressions of some of the most celebrated men, especially of artists, that they lay but small weight on productions which seem to us inimitable, compared with others which have been difficult, and yet which appear to readers and observers as much less successful. I need only mention Goethe, who once stated to Eckermann that he did not estimate his poetical works so highly as what he had done in the theory of colours.

The same may have happened to me, though in a more modest degree, if I may accept your assurances and those of the authors of the addresses which have reached me. Permit me, therefore, to give you a short account of the manner in which I have been led to the special direction of my work.

In my first seven years I was a delicate boy, for long confined to my room, and often even to bed; but, nevertheless, I had a strong inclination towards occupation and mental activity. My parents busied themselves a good deal with me; picture books and games, especially with wooden blocks, filled up the rest of the time. Reading came pretty early, which, of course, greatly increased the range of my occupations. But a defect of my mental organisation showed itself almost as early, in that I had a bad memory for disconnected things. The first indication of this I consider to be the difficulty I had in distinguishing between left and right; afterwards, when at school I began with languages, I had greater difficulties than others in learning words, irregular grammatical forms, and peculiar terms of expression. History as then taught to us I could scarcely master. To learn prose by heart was martyrdom. This defect has, of course, only increased, and is a vexation of my mature age.

But when I possessed small mnemotechnical methods, or merely such as are afforded by the metre and rhyme of poetry, learning by heart, and the retention of what I had learnt, went on better. I easily remembered poems by great authors, but by no means so easily the somewhat artificial verses of authors of the second rank. I think that is probably due to the natural flow of thought in good poems, and I am inclined to think that in this connection is to be found an essential basis of aesthetic beauty. In the higher classes of the Gymnasium I could repeat some books of the Odyssey, a considerable number of the odes of Horace, and large stores of German poetry. In other directions I was just in the position of our older ancestors, who were not able to write, and hence expressed their laws and their history in verse, so as to learn them by heart.

That which a man does easily he usually does willingly; hence I was first of all a great admirer and lover of poetry. This inclination was encouraged by my father, who, while he had a strict sense of duty, was also of an enthusiastic disposition, impassioned for poetry, and particularly for the classic period of German Literature. He taught German in the upper classes of the Gymnasium, and read Homer with us. Under his guidance we did, alternately, themes in German prose and metrical exercises—poems as we called them. But even if most of us remained indifferent poets, we learned better in this way, than in any other I know of, how to express what we had to say in the most varied manner.

But the most perfect mnemotechnical help is a knowledge of the laws of phenomena. This I first got to know in geometry. From the time of my childish playing with wooden blocks, the relations of special proportions to each other were well known to me from actual perception. What sort of figures were produced when bodies of regular shape were laid against each other I knew well without much consideration. When I began the scientific study of geometry, all the facts which I had to learn were perfectly well known and familiar to me, much to the astonishment of my teachers. So far as I recollect, that came out incidentally in the elementary school attached to the Potsdam Training College, which I attended up to my eighth year. Strict scientific methods, on the contrary, were new to me, and with their help I saw the difficulties disappear which had hindered me in other regions.

One thing was wanting in geometry; it dealt exclusively with abstract forms of space, and I delighted in complete reality. As I became bigger and stronger I went about with my father and my schoolfellows a great deal in the neighbourhood of my native town, Potsdam, and I acquired a great love of Nature. This is perhaps the reason why the first fragments of physics which I learned in the Gymnasium engrossed me much more closely than purely geometrical and algebraical studies. Here there was a copious and multifarious region, with the mighty fulness of Nature, to be brought under the dominion of a mentally apprehended law. And, in fact, that which first fascinated me was the intellectual mastery over Nature, which at first confronts us as so unfamiliar, by the logical force of law. But this, of course, soon led to the recognition that knowledge of natural processes was the magical key which places ascendency over Nature in the hands of its possessor. In this order of ideas I felt myself at home.

I plunged then with great zeal and pleasure into the study of all the books on physics I found in my father's library. They were very old-fashioned; phlogiston still held sway, and galvanism had not grown beyond the voltaic pile. A young friend and myself tried, with our small means, all sorts of experiments about which we had read. The action of acids on our mothers' stores of linen we investigated thoroughly; we had otherwise but little success. Most successful was, perhaps, the construction of optical instruments by means of spectacle glasses, which were to be had in Potsdam, and a small botanical lens belonging to my father. The limitation of our means had at that time the value that I was compelled always to vary in all possible ways my plans for experiments, until I got them in a form in which I could carry them out. I must confess that many a time when the class was reading Cicero or Virgil, both of which I found very tedious, I was calculating under the desk the path of rays in a telescope, and I discovered, even at that time, some optical theorems, not ordinarily met with in text-books, but which I afterwards found useful in the construction of the ophthalmoscope.

Thus it happened that I entered upon that special line of study to which I have subsequently adhered, and which, in the conditions I have mentioned, grew into an absorbing impulse, amounting even to a passion. This impulse to dominate the actual world by acquiring an understanding of it, or what, I think, is only another expression for the same thing, to discover the causal connection of phenomena, has guided me through my whole life, and the strength of this impulse is possibly the reason why I found no satisfaction in apparent solutions of problems so long as I felt there were still obscure points in them.

And now I was to go to the university. Physics was at that time looked upon as an art by which a living could not be made. My parents were compelled to be very economical, and my father explained to me that he knew of no other way of helping me to the study of Physics, than by taking up the study of medicine into the bargain. I was by no means averse from the study of living Nature, and assented to this without much difficulty. Moreover, the only influential person in our family had been a medical man, the late Surgeon-General Mursinna; and this relationship was a recommendation in my favour among other applicants for admission to our Army Medical School, the Friedrich Wilhelms Institut, which very materially helped the poorer students in passing through their medical course.

In this study I came at once under the influence of a profound teacher—Johannes Müller; he who at the same time introduced E. Du Bois Reymond, E. Brücke, C. Ludwig, and Virchow to the study of anatomy and physiology. As respects the critical questions about the nature of life, Müller still struggled between the older—essentially the metaphysical—view and the naturalistic one, which was then being developed; but the conviction that nothing could replace the knowledge of facts forced itself upon him with increasing certainty, and it may be that his influence over his students was the greater because he still so struggled.

Young people are ready at once to attack the deepest problems, and thus I attacked the perplexing question of the nature of the vital force. Most physiologists had at that time adopted G. E. Stahl's way out of the difficulty, that while it is the physical and chemical forces of the organs and substances of the living body which act on it, there is an indwelling vital soul or vital force which could bind and loose the activity of these forces; that after death the free action of these forces produces decomposition, while during life their action is continually being controlled by the soul of life. I had a misgiving that there was something against nature in this explanation; but it took me a good deal of trouble to state my misgiving in the form of a definite question. I found ultimately, in the latter years of my career as a student, that Stahl's theory ascribed to every living body the nature of a *perpetuum mobile*. I was tolerably well acquainted with the controversies on this latter subject. In my school days I had heard it discussed by my father and our mathematical teachers, and while still a pupil of the Friedrich Wilhelms Institut I had helped in the library, and in my spare moments had looked through the works of Daniell, Bernoulli, D'Alembert, and other mathematicians of the last century. I thus came upon the question, 'What relations must exist between the various kinds of natural forces for a perpetual motion to be possible?' and the further one, 'Do those relations actually exist?' In my essay, 'On the Conservation of Force,' my aim was merely to give a critical investigation and arrangement of the facts for the benefit of physiologists.

I should have been quite prepared if the experts had ultimately said, 'We know all that. What is this young doctor thinking about, in considering himself called upon to explain it all to us so fully?' But, to my astonishment, the physical authorities with whom I came in contact took up the matter quite differently. They were inclined to deny the correctness of the law, and in the eager contest in which they were engaged against Hegel's Natural Philosophy were disposed to declare my essay to be a fantastical speculation. Jacobi, the mathematician, who recognised the connection of my line of thought with that of the mathematicians of the last century, was the only one who took an interest in my attempt, and protected me from being misconceived. On the other hand, I met with enthusiastic applause and practical help from my younger friends, and especially from E. Du Bois Reymond. These, then, soon brought over to my side the members of the recently formed Physical Society of Berlin. About Joule's researches on the same subject I knew at that time but little, and nothing at all of those of Robert Mayer.

Connected with this were a few smaller experimental researches on putrefaction and fermentation, in which I was able to furnish a proof, in opposition to Liebig's contention, that both were by no means purely chemical decompositions, spontaneously occurring, or brought about by the aid of the atmospheric oxygen; that alcoholic fermentation more especially was bound up with the presence of yeast spores which are only formed by reproduction. There was, further, my work on metabolism in muscular action, which afterwards was connected with that on the development of heat in muscular action; these being processes which were to be expected from the law of the conservation of force.

These researches were sufficient to direct upon me the attention of Johannes Müller as well as of the Prussian Ministry of Instruction, and to lead to my being called to Berlin as Brücke's successor, and immediately thereupon to the University of Königsberg. The Army medical authorities, with thankworthy liberality, very readily agreed to relieve me from the obligation to further military service, and thus made it possible for me to take up a scientific position.

In Königsberg I had to lecture on general pathology and physiology. A university professor undergoes a very valuable training in being compelled to lecture every year, on the whole range of his science, in such a manner that he convinces and satisfies the intelligent among his hearers—the leading men of the next generation. This necessity yielded me, first of all, two valuable results.

For in preparing my course of lectures, I hit directly on the possibility of the ophthalmoscope, and then on the plan of measuring the rate of propagation of excitation in the nerves.

The ophthalmoscope is, perhaps, the most popular of my scientific performances, but I have already related to the oculists how luck really played a comparatively more important part than my own merit. I had to explain to my hearers Brücke's theory of ocular illumination. In this, Brücke was actually within a hair's breadth of the invention of the ophthalmoscope. He had merely neglected to put the question, To what optical image do the rays belong, which come from the illuminated eye? For the purpose he then had in view it was not necessary to propound this question. If he had put it, he was quite the man to answer it as quickly as I could, and the plan of the ophthalmoscope would have been given. I turned the problem about in various ways, to see how I could best explain it to my hearers, and I thereby hit upon the question I have mentioned. I knew well, from my medical studies, the difficulties which oculists had about the conditions then comprised under the name of Amaurosis, and I at once set about constructing the instrument by means of spectacle glasses and the glass used for microscope purposes. The instrument was at first difficult to use, and without an assured theoretical conviction that it must work, I might, perhaps, not have persevered. But in about a week I had the great joy of being the first who saw clearly before him a living human retina.

The construction of the ophthalmoscope had a very decisive influence on my position in the eyes of the world. From this time forward I met with the most willing recognition and readiness to meet my wishes on the part of the authorities and of my colleagues, so that for the future I was able to pursue far more freely the secret impulses of my desire for knowledge. I must, however, say that I ascribed my success in great measure to the circumstance that, possessing some geometrical capacity, and equipped with a knowledge of physics, I had, by good fortune, been thrown among medical men, where I found in physiology a virgin soil of great fertility; while, on the other hand, I was led by the consideration of the vital processes to questions and points of view which are usually foreign to pure mathematicians and physicists. Up to that time I had only been able to compare my mathematical abilities with those of my fellow-pupils and of my medical colleagues; that I was for the most part superior to them in this respect did not, perhaps, say very much. Moreover, mathematics was always regarded in the school as a branch of secondary rank. In Latin composition, on the contrary, which then decided the palm of victory, more than half my fellow-pupils were ahead of me.

In my own consciousness, my researches were simple logical applications of the experimental and mathematical methods developed in science, which by slight modifications could be easily adapted to the particular object in view. My colleagues and friends, who, like myself, had devoted themselves to the physical aspect of physiology, furnished results no less surprising.

But in the course of time matters could not remain in that stage. Problems which might be solved by known methods I had gradually to hand over to the pupils in my laboratory, and for my own part turn to more difficult researches, where success was uncertain, where general methods left the investigator in the lurch, or where the method itself had to be worked out.

In those regions also which come nearer the boundaries of our knowledge I have succeeded in many things experimental and mechanical—I do not know if I may add philosophical. In respect of the former, like any one who has attacked many experimental problems, I had become a person of experience, who was acquainted with many plans and devices, and I had changed my youthful habit of considering things geometrically into a kind of mechanical mode of view. I felt, intuitively as it were, how strains and stresses were distributed in any mechanical arrangement, a faculty also met with in experienced mechanicians and machine constructors. But I had the advantage over them of being able to make complicated and specially important relations perspicuous, by means of theoretical analysis.

I have also been in a position to solve several mathematical physical problems, and some, indeed, on which the great mathematicians, since the time of Euler, had in vain occupied themselves; for example, questions as to vortex motion and the discontinuity of motion in liquids, the question as to the motion of sound at the open ends of organ pipes, &c. &c. But the pride which I might have felt about the final result in these cases was considerably lowered by my consciousness that I had only succeeded in solving such problems after many devious ways, by the gradually increasing generalisation of favourable examples, and by a series of fortunate guesses. I had to compare myself with an Alpine climber, who, not knowing the way, ascends slowly and with toil, and is often compelled to retrace his steps because his progress is stopped; sometimes by reasoning, and sometimes by accident, he hits upon traces of a fresh path, which again leads him a little further; and finally, when he has reached the goal, he finds to his annoyance a royal road on which he might have ridden up if he had been clever enough to find the right starting-point at the outset. In my memoirs I have, of course, not given the reader an account of my wanderings, but I have described the beaten path on which he can now reach the summit without trouble.

There are many people of narrow views, who greatly admire themselves, if once in a way, they have had a happy idea, or believe they have had one. An investigator, or an artist, who is continually having a great number of happy ideas, is undoubtedly a privileged being, and is recognized as a benefactor of humanity. But who can count or measure such mental flashes? Who can follow the hidden tracts by which conceptions are connected?

> That which man had never known,
> Or had not thought out,
> Through the labyrinth of mind
> Wanders in the night.

I must say that those regions, in which we have not to rely on lucky accidents and ideas, have always been most agreeable to me, as fields of work.

But, as I have often been in the unpleasant position of having to wait for lucky ideas, I have had some experience as to when and where they came to me, which will perhaps be useful to others. They often steal into the line of thought without their importance being at first understood; then afterwards some accidental circumstance shows how and under what conditions they have originated; they are present, otherwise, without our knowing whence they came. In other cases they occur suddenly, without exertion, like an inspiration. As far as my experience goes, they never came at the desk or to a tired brain. I have always so turned my problem about in all directions that I could see in my mind its turns and complications, and run through them freely without writing them down. But to reach that stage was not usually possible without long preliminary work. Then, after the fatigue from this had passed away, an hour of perfect bodily repose and quiet comfort was necessary before the good ideas came. They often came actually in the morning on waking, as expressed in Goethe's words which I have quoted, and as Gauss also has remarked.[1] But, as I have stated in Heidelberg, they were usually apt to come when comfortably ascending woody hills in sunny weather. The smallest quantity of alcoholic drink seemed to frighten them away.

Such moments of fruitful thought were indeed very delightful, but not so the reverse, when the redeeming ideas did not come. For weeks or months I was gnawing at such a question until in my mind I was

> Like to a beast upon a barren heath
> Dragged in a circle by an evil spirit,
> While all around are pleasant pastures green.

And, lastly, it was often a sharp attack of headache which released me from this strain, and set me free for other interests.

I have entered upon still another region to which I was led by investigation on perception and observation of the senses, namely, the theory of cognition. Just as a physicist has to examine the telescope and galvanometer with which he is working; has to get a clear conception of what he can attain with them, and how they may deceive him; so, too, it seemed to me necessary to investigate likewise the capabilities of our power of thought. Here, also, we were concerned only with a series of questions of fact about which definite answers could and must be given. We have distinct impressions of the senses, in consequence of which we know how to act.

---

1. Gauss, *Werke*, vol. v. p. 609. 'The law of induction discovered Jan. 23, 1835, at 7 A.M., before rising.'

The success of the action usually agrees with that which was to have been anticipated, but sometimes also not, in what are called subjective impressions. These are all objective facts, the laws regulating which it will be possible to find. My principal result was that the impressions of the senses are only signs for the constitution of the external world, the interpretation of which must be learned by experience. The interest for questions of the theory of cognition, had been implanted in me in my youth, when I had often heard my father, who had retained a strong impression from Fichte's idealism, dispute with his colleagues who believed in Kant or Hegel. Hitherto I have had but little reason to be proud about those investigations. For each one in my favour, I have had about ten opponents; and I have in particular aroused all the metaphysicians, even the materialistic ones, and all people of hidden metaphysical tendencies. But the addresses of the last few days have revealed a host of friends whom as yet I did not know; so that in this respect also I am indebted to this festivity for pleasure and for fresh hope. Philosophy, it is true, has been for nearly three thousand years the battle-ground for the most violent differences of opinion, and it is not to be expected that these can be settled in the course of a single life.

I have wished to explain to you how the history of my scientific endeavours and successes, so far as they go, appears when looked at from my own point of view, and you will perhaps understand that I am surprised at the universal profusion of praise which you have poured out upon me. My successes have had primarily this value for my own estimate of myself, that they furnished a standard of what I might further attempt; but they have not, I hope, led me to self-admiration. I have often enough seen how injurious an exaggerated sense of self-importance may be for a scholar, and hence I have always taken great care not to fall a prey to this enemy. I well knew that a rigid self-criticism of my own work and my own capabilities was the protection and palladium against this fate. But it is only needful to keep the eyes open for what others can do, and what one cannot do oneself, to find there is no great danger; and, as regards my own work, I do not think I have ever corrected the last proof of a memoir without finding in the course of twenty-four hours a few points which I could have done better or more carefully.

As regards the thanks which you consider you owe me, I should be unjust if I said that the good of humanity appeared to me, from the outset, as the conscious object of my labours. It was, in fact, the special form of my desire for knowledge which impelled me and determined me, to employ in scientific research all the time which was not required by my official duties and by the care for my family. These two restrictions did not, indeed, require any essential deviation from the aims I was striving for. My office required me to make myself capable of delivering lectures in the University; my family, that I should establish and maintain my reputation as an investigator. The State, which provided my maintenance, scientific appliances, and a great share of my free time, had, in my opinion, acquired thereby the right that I should communicate faithfully and completely to my fellow-citizens, and in a suitable form, that which I had discovered by its help.

The writing out of scientific investigations is usually a troublesome affair; at any rate it has been so to me. Many parts of my memoirs I have rewritten five or six times, and have changed the order about until I was fairly satisfied. But the author has a great advantage in such a careful wording of his work. It compels him to make the severest criticism of each sentence and each conclusion, more thoroughly even than the lectures at the University which I have mentioned. I have never considered an investigation finished until it was formulated in writing, completely and without any logical deficiencies.

Those among my friends who were most conversant with the matter represented to my mind, my conscience as it were. I asked myself whether they would approve of it. They hovered before me as the embodiment of the scientific spirit of an ideal humanity, and furnished me with a standard.

In the first half of my life, when I had still to work for my external position, I will not say that, along with a desire for knowledge and a feeling of duty as servant of the State, higher ethical motives were not also at work; it was, however, in any case difficult to be certain of the reality of their existence so long as selfish motives were still existent. This is, perhaps, the case with all investigators. But afterwards, when an assured position has been attained, when those who have no inner impulse towards science may quite cease their labours, a higher conception of their relation to humanity does influence those who continue to work. They gradually learn from their own experience how the thoughts which they have uttered, whether through literature or through oral instruction, continue to act on their fellow-men, and possess, as it were, an independent life; how these thoughts, further worked out by their pupils, acquire a deeper significance and a more definite form, and, reacting on their originators, furnish them with fresh instruction. The ideas of an individual, which he himself has conceived, are of course more closely connected with his mental field of view than extraneous ones, and he feels more encouragement and satisfaction when he sees the latter more abundantly developed than the former. A kind of parental affection for such a mental child ultimately springs up, which leads him to care and to struggle for the furtherance of his mental offspring as he does for his real children.

But, at the same time, the whole intellectual world of civilised humanity presents itself to him as a continuous and spontaneously developing whole, the duration of which seems infinite as compared with that of a single individual. With his small contributions to the building up of science, he sees that he is in the service of something everlastingly sacred, with which he is connected by close bands of affection. His work thereby appears to him more sanctified. Anyone can, perhaps, apprehend this theoretically, but actual personal experience is doubtless necessary to develop this idea into a strong feeling.

The world, which is not apt to believe in ideal motives, calls this feeling love of fame. But there is a decisive criterion by which both kinds of sentiment can be discriminated. Ask the question if it is the same thing to you whether the results of investigation which you have obtained are recognised as belonging to you or not when there are no considerations of external advantage bound up with the answer to this question. The reply to it is easiest in the case of chiefs of laboratories. The teacher must usually furnish the fundamental idea of the research as well as a number of proposals for overcoming experimental difficulties, in which more or less ingenuity comes into play. All this passes as the work of the student, and ultimately appears in his name when the research is finished. Who can afterwards decide what one or the other has done? And how many teachers are there not who in this respect are devoid of any jealousy?

Thus, gentlemen, I have been in the happy position that, in freely following my own inclination, I have been led to researches for which you praise me, as having been useful and instructive. I am extremely fortunate that I am praised and honoured by my contemporaries, in so high a degree, for a course of work which is to me the most interesting I could pursue. But my contemporaries have afforded me great and essential help. Apart from the care for my own existence and that of my family, of which they have relieved me, and apart from the external means with which they have provided me, I have found in them a standard of the intellectual capacity of man; and by their sympathy for my work they have evoked in me a vivid conception of the universal mental life of humanity which has enabled me to see the value of my own researches in a higher light. In these circumstances, I can only regard as a free gift the thanks which you desire to accord to me, given unconditionally and without counting on any return.

An Address delivered on the occasion of his Jubilee, 1891.

15

Goethe's Presentiments of Coming Scientific Ideas

IT IS A BEAUTIFUL CUSTOM of the Goethe Society that it allows representatives of the most diverse forms of scientific and literary activity to relate their own ways of thought to those of this incomparable man whose surviving signs it seeks to locate, and to whom it seeks to remain faithful. Men who, like him, had absorbed all aspects of the culture of their day, without limiting the freshness and natural independence of their capacity to experience, and who, as morally free men in the noblest sense of the word, needed to follow only their warm, innate interest for all variations of the human soul so as to find the right path between the crags of life, have become quite rare; and will probably become rarer still. The impartiality and health of the Goethean spirit stands out as all the more worthy of wonder because it originated in a deeply artificial era, when even the longing to return to nature assumed the most unnatural forms. His example has thus bequeathed to us an inestimably valuable standard for the authentic and the original in man's spiritual nature, by which we should not fail to measure our own efforts, even with their more limited goals.

I myself undertook, at the beginning of my scientific career, to give an account of Goethe's natural scientific work. This was, for the most part, concerned with a defense of the physicists' scientific standpoint against the reproaches of the Poet. At that time, he found much greater acceptance among the nation's cultured classes than did the young natural science, whose warrant for joining the association of the other venerable sciences, with their old traditions, was viewed not completely without mistrust.

Since then, forty years of fruitful, scientific development have passed over Europe. Through the transformation of all practical relations of life issuing from them, the natural sciences have proven the trustworthiness and fruitfulness of their principles; moreover, they have acquired broad perspectives from which the complete picture of nature, both animate and inanimate, appears deeply altered: one has only to think of Darwin's *Origin of Species* and of the law of the conservation of energy. That alone would have given sufficient motivation to recall old thoughts and subject them to new testing.

For me, however, there is an additional special interest. Early on, my course of study had involved physiological problems: namely, the laws of nervous activity, where the question of the origin of the sense perceptions could not be avoided. Just as the chemist must investigate the correctness and trustworthiness of his scale before starting his own professional work, and the astronomer that of his telescope, so, too, the natural sciences as a whole must test the mode of operation of their instruments that are the source of all our knowledge: namely, the human sense organs. One already knew that so-called sensory illusions occur; one now had to try to learn as much as necessary about the way in which they originated so as to be safely able to avoid them. Contemporary philosophy gave virtually no help. For even Kant—who for us, his successors, had drawn together the results of the previous epistemological efforts—condensed into one act, which he named intuition, all the connecting links between pure sense perception and the formation of ideas of the perceived, spatially extended object. This plays a role for him and his followers as if it were merely the result of a natural mechanism that could not be an object of further philosophical and psychological investigations—apart from his final result, which is precisely a representation, and thus can fall, under certain formal conditions of representation that Kant had sought.

As soon as it was permitted to assume that true perceptions may be acquired by means of our senses, the path of future investigation was basically prescribed by the inductive methods of the natural sciences. The main emphasis here is that natural laws of appearances must be found and successfully expressed in clearly defined terms. One can only describe the initial, still insufficiently tested attempts to establish a natural law as hypotheses. The consequences of such hypotheses open to observation are pursued and compared, under the widest possible manifold changed conditions, with the facts. The possibility of formulating a conjectured law in words has the great and decisive advantage that it is communicated to many, and that many participate in testing it; that this can be undertaken over an indefinitely long period of time and in an unlimited number of cases; that, as the number of confirmations grows, so too does attention to the true or apparent exceptions increase, until an overwhelming amount of observational material is finally brought together such that it is no longer possible to doubt the law's correctness, at least not within the range of conditions thoroughly tested.

This is a long and laborious road, whose success, as I stress once again, essentially depends on the possibility of expressing the law in question in precisely defined words. Nevertheless, we can now in fact already completely reduce large fields of natural processes—namely, those within the simpler relations of inorganic nature—to well-known and rigorously defined laws.

Yet whoever knows the law of phenomena gains not only knowledge; he also gains the power of intervening, at appropriate moments, in the course of nature and enabling it to continue working according to his will and utility. He gains insight into the future course of these very same phenomena. He gains, in truth, the abilities once sought in mystical times by prophets and magicians.

Nonetheless, we find that insight can be gained into the complicated mechanism of nature and of the human mind in yet another way than that of science, and that it can be communicated to others such that it still retains full conviction of the truth of what is communicated. Such a way is given in artistic representation. You will not find it difficult to convince yourselves that, at least in individual branches of art, something like this is done. We will later have to discuss the problem as to whether such effects are limited to individual branches of art or whether something similar occurs in all of them.

Consider any masterpiece of tragedy. You see human feelings and passions develop and intensify, until noble or terrible actions finally issue therefrom. You understand completely that, under the given conditions and events, the result must occur precisely as it is presented to you by the writer. You feel that in the same situation you yourself would have the same instinct to act in just this way. You become acquainted with the depth and power of sensations that are never awakened in quiet daily life, and leave with a deep conviction of the truth and correctness of the movements presented to the soul, although you were, at the same time, at no moment in doubt that everything you saw was only metaphorical appearance.

This truth that you recognize is thus only the inner truth of the soul's events that are presented, their logical structure, their agreement with what you yourself know of the development of such moods, i.e., it is the rightness of the representation of the natural course of these conditions. The artist must have had this knowledge; so, too, must the listener, at least insofar as he recognizes it again when it is presented to him.

Now, where does such knowledge, which shows itself particularly clearly in precisely those fields where the efforts of science to date have had the least success—namely, in the fields of the soul's workings, the properties of character, and individual decision making—come from? It has certainly not been gained through the laborious road of science, through reflective thinking. On the contrary: where the author starts to reflect and wants to help by means of philosophical insights, the listener is almost instantaneously disenchanted and critically disposed. He feels that a surrogate has entered in place of the artist's lively fantasy.

Artists themselves have little to say about where these images originate; indeed, precisely the most able among them only slowly learn, through the success of their works, that they do something which the majority of other human beings is not in a position to emulate. They clearly give less importance to the type of activity they perform (for the most part so easily)—achieving the incomprehensible—than to matters of secondary importance which have given them trouble. In this regard, Goethe once said to Eckermann that with his theory of colors he believed to have accomplished something more important than with his poetry. And I myself once heard Richard Wagner declare that he valued his verses more highly than his music.

Now we know that this type of intellectual activity which occurs so easily, quickly, and without a second thought, can only be described by the name intuition, in particular artistic intuition. The concept of intuition is, however, almost entirely negative in character. According to philosophical terminology, it constitutes the antithesis to thought, i.e., to the conscious comparison of ideas already acquired through similarities to concepts. Sensory intuition is present without reflection, without intellectual effort, immediate, just as the corresponding sense impression affects us. No arbitrariness occurs with respect to it; it seems to us that the perception of the object corresponding to it is completely determined by the sense impression, so that the same impression always calls forth the same idea.

To be sure, the artistic imagination at times functions not only in response to the present sense impression, but often enough—namely, in poetry—also with recollected images of such impressions, which, however, in the relations just discussed, do not distinguish themselves from the sense images directly present.

The prevailing conceptual definition of sensory intuition has, as I already noted, attempted no analysis of the same; it thus does not help us understand artistic intuition.

We have, therefore, sufficient reasons to raise objections to the view that both types of intuition might be fully free from the influence of experience. Experience, however, is a result of processes which fall within the field of thinking.

It should be considered, first, that often enough—namely, in suddenly arising dangers but also in the case of propitious opportunities demanding a quick response—lightning-like decisions shoot through the brain even though these are not due solely to the nature of the present sense impressions. In general, all cases where we praise the actor's presence of mind belong here; knowledge of the danger thereby normally rests not on extraordinarily shocking sense impressions but rather only on judgment based upon previous experiences. It cannot be doubted, therefore, that the speed with which an idea occurs determines nothing about its physiological-mechanical origin and its independence from results of previous thought.

The other characteristic of sensory intuition cited above—that the idea of the object which originates through intuition should depend only on the type of sense impression which is present—excludes, to be sure, the involvement of experiences about changed relations in the external world (but not those experiences based on unchanging relations) and which thus always repeat themselves over and over in the same way; and thus, if they combine in a new sense impression, this can only be completed again and again in the same way as all its predecessors.— All relations regulated by a firm law of nature obviously belong here.

To give an example: A cast shadow can fall on an illuminated surface only when the shadow-throwing body lies in front of that side of the surface on which the light falls. It is for this reason that one of the most important aids in artistic representations of the mutual positions of non-transparent bodies in space is to be able to give the shadow side correctly. Indeed, stereoscopic pictures can present us with the case where ideas depending on the active sense impression of the position of perceived outlines in the picture's depth and in varying distances from the eye can be repressed through a wrongly placed cast shadow, so that the correct spatial intuition cannot become contrasted to it.

In general, the influence that the laws of perspective, shadowing, hiding of outlines of more distant bodies, aerial perspective, &c. have on the spatial interpretation of our visual images is extraordinarily great, and yet this influence can be reduced solely to the activity of past experiences, even though it makes itself felt just as certainly and unhesitatingly in the picture as the colors and outlines do.

Hence, in my view there can be no doubt that factors derived from experience participate in the immediate perceptions (by means of our senses) towards the formation of our ideas of objects. The specific physiological investigation of the dependency of our perceptions on the underlying sensations provides hundreds of examples. Of course, it is often difficult in an individual to separate definitely what belongs to the physiological mechanics of the nerves and what trained experience about the unchanging laws of space and of nature has provided. I myself am inclined to grant the greatest sphere of operation to the latter.

Furthermore, the little that we know so far about the laws of our memory allows us to guess how such effects may come about.

We all well know how repetition of the same consequences from the same impressions strengthens the traces of the same that remain behind in the memory; already in school this was the means that we often used to learn poems, sayings, and grammatical rules by heart. Intentional repetition works more certainly; but a strengthening of the remembered image also occurs when the repetition is accomplished without our help. We have already discussed that what must necessarily and without exception be repeated in the same way—and thus become fixed through repetition—are consequences of events united with one another by means of a natural law, by means of the necessary linkage of cause and effect. We may at the same time expect that all those characteristics of any such process that are conditioned by accidental, changing circumstances will mutually disturb one another in their effect on the memory and that most will cease to exist. However, precisely these accidents are the means by which the individual examples of a law-like process presented to us distinguish themselves individually from one another. If the memory of these disappears, then we also thereby lose the means of distinguishing the individual cases from one another in our memory, and the ability to enumerate these again individually. We retain knowledge of the law-like, but lose sight of the individuality of the cases from which our knowledge of the law derives; and, ultimately, are therefore unable to give an account, to others or to ourselves, as to how we have come to such knowledge. In the end, we only know that it is always so and that we have never seen it otherwise.

We can obtain such knowledge of the law-like from the most diverse things and relations: beginning in childhood, with the simplest spatial relations and effects of gravity; clearly increasing in adults; and for attentive observers with fine senses extensible without limits—insofar as law and order rule in nature and in the mind.

These considerations, which I have here given initially for the example of the sensory intuitions, can also be completely transferred to the artistic intuitions. Although they occur without effort, flashing up suddenly, and although the possessor does not know where they come from, it does not follow that they contain no results taken from experience and include recollections of the lawfulness of experience. Here we are directed toward a positive source of the artistic imagination, which is also completely fitting for justifying the strict consistency of the great works of art, in contrast to the free play of fantasy formerly extolled by the poets of the Romantic School.

Since artistic intuitions are not found by means of conceptual thought, they also cannot be defined in words; and one designates the type of knowledge of regular behavior acquired by the coalescence of intuitions as knowledge of the relevant type of appearance, in order to emphasize the contrast.

Just as the manifold nature of sensory perception is richer than the word descriptions that one can give of its objects, so the artistic presentation (as against the scientific) can naturally turn out to be all the richer, finer, and full of life. In addition, there is the sudden surfacing of remembered images that become associated at apt points of contact, so that it becomes possible for the artist to give the listener or viewer an extraordinarily large amount of content in a short period time or in a limited pictorial space.

When I initially wanted to remind you that art, like science, can represent and transmit truth, I limited myself at first to the most outstanding example, that of tragic art. You will perhaps ask whether this should also be the case for other branches of art. It seems to me indubitable that an artist's work can only succeed if the artist bears within himself a fine knowledge of the law-like behavior of the presented phenomena, as well as with their effect on the listener or viewer. Whoever has yet to become acquainted with the finer effects of art, easily lets himself be misled—namely, vis-à-vis the works of pictorial art—by considering absolute truth-to-nature as the essential criterion for a picture or bust. In this regard, every well-made photograph would obviously be superior to all drawings, etchings, or engravings of the greatest masters—and yet we quickly recognize how much more expressive these latter are.

This fact is also a clear sign that artistic representation cannot be a copy of an individual instance, but rather must be a representation of the type of phenomenon concerned.

Here we get closer to the much-disputed question concerning the nature and secret of the beauty of art. We do not want to attempt to answer this question completely today; instead, we want only to touch upon it just insofar as it concerns our theme of the representation of truth in art.

First of all, with respect to beauty and depth of expression it is clear that if still other demands are made upon the artist than the copying of the individual case would involve, then he will be able to fulfill these demands only by transforming the individual case, yet without going beyond the lawfulness of the type. Hence, the more exact his intuitive picture of the latter is, the more easily will he be able to meet the demands of beauty and expression.

If an increase in beauty or expression can be accomplished by means of more important factors, then this transformation of artistic form often goes so far that truth-to-nature will be intentionally relegated to a subsidiary level.

As examples I want only to adduce meter and rhyme in poetry, and the addition of music to the text of a drama or song.

The given uses of words of language are, vis-à-vis the content of poetry, an external, indifferent, or even ugly accessory, or arbitrary human construction; they already change in translation into another language. Rhythm and rhyme give them a type of external order, but also something of musical movement, whose retardation, acceleration or interruption can make an impression. If we raise the language to that of singing on stage, we destroy even more the truth-to-nature, though we thereby gain the advantage of expressing the emotions of the acting persons' souls in the much richer, finer, and more expressive movements of sound.

Since consideration of expressiveness of representation goes together in the broadest realms of art with requirements of beauty and of the purest representation of the type, and has already been discussed often and thoroughly, I believe that I here need only remind you about it.

In my book on the sensations of tone, I endeavored to prove that in music, too, the more or less harmonic effect of the interval in melody and harmony goes together with special sensory-perceptible phenomena (the overtones), and that the simpler and purer these are, the more clearly and exactly are the harmonic intervals demarcated.

Investigations on the sensations of the sense of sight teach us that certain mid-range brightnesses, which for seeing are the most pleasant to us, at the same time favor the finest discriminations in modelling of spatial shapes and of the smallest objects, and that, if the eye is not to be disturbed by color after-images, a certain equilibrium of colors is also necessary.

In general, we ought not to scorn the pleasant sensations as elements of beauty; for in the long work of generations nature has so formed our body that we find satisfaction in an environment where the perceiving activities of our soul can develop in the freest and surest way.

Furthermore, I consider the prominent influence of the beautiful on the memory of man as an outward sign of what I have here called easily understandable or comprehensible. Poetry is remembered much more easily than prose. It is clear that peoples who did not yet have writing, or among whom only a few could write, preserved their stories, their legends, their laws, and their moral rules in verses. One can never again forget a beautiful building or picture or song; a melody can become so solidly implanted, that one has trouble getting free of it again.

Now, I believe that an essential part of the effect of the beautiful rests in this, its effect on the memory. Even when we first begin to consider it, we quickly come to a definite idea of the whole, one that puts us in a position to continue an overview and consideration of particulars in a quieter and more comfortable way, while we feel continuously well-oriented on the relation to the whole.

We come now to the point where the ways of the researcher and the artist begin to separate. That the artist's memory for those phenomena that interest him—namely, in relation to the details of the phenomenon—is finer and truer than with the majority of other human beings, is seen in innumerable examples. A landscape painter must be able to retain in faithful memory the picture of rapidly disappearing illuminations, of transient phenomena of thunder; similarly, he conjures up in his picture the moon's illumination—by which he cannot paint—and the rolling waves of the sea—which do not hold still even for a moment—through innumerable details on the canvas. What he can retain in a moment through fleeting sketches of some details is very scanty. For the most part, he will have to rely throughout on his pictorial memory of what happened.

To us, the most astounding thing seems to be the memory of those musicians who, without having notes before them, know how to play innumerable compositions with their instruments; and still more astounding are the conductors who, without a musical score, are in a position to conduct innumerable symphonies whose individual musical notes would number in the millions. But I believe not to be mistaken when I assume that what they have in their heads are certainly not the notes and the number of intervals, but rather only the musical phrases of the masterpieces, their succession and linkage, including changing timbres; and that they are only in a position to obtain what they want to hear (with greater security and speed) by retranslating as necessary into the picture of the musical score by giving their musicians the right hints.

For scientific work, in contrast to artistic, a wide-ranging, accurate memory does not have the same importance. For, what we can conceive in words, we can also fix in writing. Only the first inventive thought, which must precede conceptualization in words, will become formed and must always emerge in both types of activity in the same way. In fact, that happens only and always in an analogous way to artistic intuition, as a presentiment of a new law-likeness. One such consists in the discovery of a previously unknown similarity of type, as when certain phenomena conform to a group of typically agreeing cases. We call the ability to discover previously unimagined similarities "wit." Our ancestors used this word in the original sense. It always designates a suddenly emerging insight that one cannot obtain through methodical reflection but that, instead, appears like a sudden piece of luck.

In the oldest Latin designation the poet's name is therefore identical with that of the seer. The suddenly emerging insight gets designated as divination, as a type of godly inspiration.

A favorable accident can occasionally also be of help, and an unknown relation be revealed; but the accident will be used only with difficulty if the one to whom it happens has not already collected in his mind sufficient material from insights so as to convince him of the presentiment's correctness. Goethe's story of the discovery of the vertebral structure of the skull in the course of finding a crumbling sheep's skull in the sand of the Lido in Venice, seems to me typical for this type of discovery. He also mentions it in one version of his story as an initial discovery, in another only as a confirmation of an earlier-known truth.

I have now given you the reasons for my belief in the affinity of science and art, and we want now to turn to Goethe's special activity.

Goethe was not the only artist who simultaneously conducted scientific research; just to mention one other, I name Leonardo da Vinci, who, however, dedicated himself more to practical questions of engineering science and optics—and thereby displayed much farseeing insight.

The field in which Goethe won the greatest fame, and wherein his merits are seen most easily and clearly, is that of animal and plant morphology.

Here he became convinced that a common structural plan provides the foundation for the bodily formation of the different animal and plant forms—and does so with thorough consistency right down into the seemingly unimportant details. This was a task which was particularly closely related to the artistic way of looking at things, and one in which it was already a gain simply to attain and to hold on to the standpoint corresponding to artistic intuition. The scientific anatomists and zoologists of that time were hindered by a prejudice—namely, in the belief in the unchanging nature of organic types—in looking in the direction opened up by Goethe and, as he proposed to them, to follow up on his insights. At the same time, he himself could say equally little about what meaning or origin these agreements of forms might have. He says revealingly about this:

> None resembleth another, yet all their forms have a likeness;
> Therefore, a mystical law is by the chorus proclaim'd;
> Yes, a sacred enigma! Oh, dearest friend, could I only
> Happily teach thee the word, which may the mystery solve![1]

Darwin was the first to find the resolving idea in that he freed himself from the above-mentioned prejudice of his predecessors, and drew attention to man's ability—already long known through numerous examples—to transform types when he breeds races. He then further showed that conditions of a similar type have an effect on animals living in the wild, just as the breeder intentionally fixes them, and that over a series of generations considerable transformation of the animal forms can be produced. I believe that I do not need to explain this further to this Society; it is part of one of the greatest revolutions in biology which has excited the most general attention, and has precisely thus been much and thoroughly discussed among all educated people. I refrain therefrom all the more so as in the university of this state there is a representative of the theory of development, one who is among the most active and rich in ideas. Moreover, Goethe's activity in this regard has been discussed often and in much detail. In the most recent volume of the Goethe Yearbook, Professor K. Bardeleben has given us a description of the poet's industry in this regard.

His efforts in another direction of natural scientific research—namely, in the theory of color—were less fortunate. I have already given an extensive report on the reasons for his failure in my older essay on Goethe's natural scientific works. They lay essentially in his inability to observe the decisive facts by means of the relatively inadequate apparatus that he had in hand. He never obtained completely pure, simple, colored light before his

---

1. "The Metamorphosis of Plants," in *The Poems of Goethe*, trans. Edgar Alfred Bowring (London: George Bell, 1880), pp. 269–71, on 269.

eyes, and so he did not want to believe in its existence. Men like Sir David Brewster, who had the best instruments and who was much more experienced and skilled in optical experiments than Goethe, have failed at the point of this difficulty of obtaining complete purity of simple spectral colors. Brewster, furthermore, put forward a false theory of colors in which he, like Goethe, maintained that it is not the different refractivity of light rays that determines the colors of the prismatic images, but rather that there are three different types of light—red, yellow, and blue—each of which may occur with each degree of refractivity. Brewster was misled in that he was not acquainted with the real and invariable darkening of transparent bodies—upon which Goethe had built his entire theory—and that on account of the darkening he missed the false light diffused in the observer's visual field.

Precisely because I investigated the phenomena described by Brewster, which seemed to stand in contradiction to Newton's theory, I had occasion to obtain an even more precise purification of the colored light than Newton, Goethe, and Brewster had ever known. I finally reached my goal, though not without effort. I thus know from my own experience how inappropriate it would be if I wanted to give you here an extensive analysis of the deficiencies of Goethe's experiments, the neglected sources of error, the misunderstandings of Newtonian principles, &c.—all the more so, since there also lies hidden in this unfortunate effort of the poet a highly important kernel of new insight.

He declares it as his firm conviction that in all branches of physics one must seek an "Ur-phenomenon," an ultimate event, to which the multiplicity of phenomena may be reduced. The opposing approach that so repelled him was the abstraction of intuitionless concepts with which theoretical physics then commonly calculated. Matter was supposedly—according to its pure concept without forces, and thus also without properties—and was yet again in each special case the bearer of forces residing within it. The forces themselves, if one wants to consider them as detached from matter, were imagined to have a capacity to operate, yet without any point of attachment for any kind of effect. He wanted nothing to do with such supernatural, inconceivable abstractions. And one must admit that his opposition was not unjustified; and that these abstractions, even if they were also employed meaningfully and without any contradiction by the great theoretical physicists of the seventeenth and eighteenth centuries, nonetheless contained the germ of the most confused misunderstandings that also at times made themselves loudly known in confused and mystical minds: namely, in the followers of animal magnetism as well as in the theory of the life force, wherein the forces detached from matter have played a fateful role.

Nonetheless, in this regard contemporary physics has already completely followed the roads that Goethe wanted to follow. The immediate historical connection with the stimulus given by him has, unfortunately, been broken due to his incorrect interpretation of the example he selected and the embittered polemic against the physicists that followed thereupon. It is quite regrettable that at that time he did not know the wave theory of light which Huygens had already advanced; this would have put in his hands a much more correct and insightful "Ur-phenomenon" than the scarcely appropriate and very convoluted process which, towards this end, he selected in the colors of transparent media. In external nature, these are of course important since the blue of heaven and the sunset belong to them.

Newton's corpuscular theory of light had in fact to make many a cumbersome and artificial assumption: namely, for the explanation of the recently discovered polarization and interference of light. It has thus now been entirely abandoned by physicists, who have turned much more to Huygen's wave theory.

Mathematical physics received its impulse to the progress just discussed without any recognizable influence from Goethe. It came mainly from Faraday, who was an untaught autodidact and, like Goethe, an enemy of abstract concepts, which he did not know how to handle. His entire view of physics rested on insight into phenomena; and he, too, sought to avoid explanations of everything that was not immediately an expression of observable facts. Perhaps Faraday's wonderful power of following clues that led to the discovery of new phenomena went together with this naiveté and freedom from theoretical prejudices in traditional science. In any case, the number and importance of his discoveries was well suited to lead others—at first the most able among his countrymen—down this road. German researchers also soon followed the same direction. Gustav Kirchhoff begins his *Textbook of Mechanics* with the explanation: The task of mechanics is "to describe completely and in the simplest ways the movements taking place before one in nature." What Kirchhoff here understands by the "simplest ways" of description may, in my opinion, not lie so far from the Goethean "Ur-phenomenon."

It may be noted that precisely the most outstanding among the older mathematical physicists were also not so far from the same view. Newton and his contemporaries found great difficulty in imagining actions at a distance that operated through empty space, just as Faraday and his students objected to the same idea and have really removed the electric-magnetic distance forces from physics.

On the other hand, it is certainly not difficult to express the basic law of planetary motion in the way called for by Goethe, as an Ur-phenomenon, so that only observable facts are spoken of, namely, "When gravitational masses are simultaneously present in space, each one of them continually sustains an acceleration of its motion towards every other, with the value depending on the masses and their mutual distances in the way given by Newton." It is thereby assumed that the concept of acceleration is already explained, and also the meaning one may have to attribute to the simultaneous existence of several accelerations and speeds in different directions. Masses, their speeds, and accelerations are observable and measurable phenomena. Only these are referred to in the expressed law. And yet that very law contains within itself the germ from which the entire branch of astronomy which calculates the movements of the heavenly bodies may be fully developed. However, you also see at the same time how cumbersome and far-reaching one such form usually turns out to be.

Newton himself expressed his fundamental conception of the law of gravitation in a form which introduces that which goes beyond the phenomena only as "similitude." According to it, the heavenly bodies move as if they were attracted towards one another by a force of attraction of the quantity adduced. Goethe uses the word "similitude" in a similar way, and indeed in a sense of praise, when he analyzes the opinions of the English monk, Roger Bacon, in the history of the theory of colors. Some weight still thereby falls, to be sure, on the old scholastic assumption of a certain similarity between cause and effect, which modern natural science no longer recognizes.

In Schiller, the insight is given that it is a matter of law:

The wise man
Sees the trusting law in Chance's horrifying wonders,[2]
Seeks the resting pole in Phenomena's flight.

Now, of course, natural law has still another meaning vis-à-vis us humans. It is not only a guide for our observing intelligence; it also rules over the operating of all processes in nature without our having to pay attention to, wish, or want it; indeed, often enough, unfortunately also against our wishes and wills. We thus have to recognize it as a means of expression of a power that is ready to take effect in every moment where the conditions for its efficacity occur. In this sense, we call it force; and since this force maintains itself as ready and able to take effect in every moment, we ascribe a continuous existence to it. In my opinion, thereupon also rests the designation of force as the cause of changes that occur under its influence; it is the permanent being behind the change of phenomena. The meaning of the term "thing" corresponds to the Latin *res*, from which the terms "real" and "reality" are derived; here they designate the lasting, the effective.

---

2. In the poem "The Walk." The middle stanza doubtless refers to Loder's collection of anatomical monsters at Jena. Loder proved its connection with the normal type.

All these transformations of the concept are fully justified insofar as they designate definite relations of facts open to observation. When used correctly, the abstract form of designation provides the great advantage of being a much shorter linguistic expression than the description of the Ur-phenomenon in statements about conditions. It is, incidentally, certainly not peculiar to theoretical physics that the use of abstract concepts in the mouths of foolish people, who no longer know a concept's original sense, can lead to the wildest nonsense.

Naturally, it may be an illusion to believe that through these abstract transformations a deeper insight can be won into the nature of the problem. Goethe says in his prose sayings: "If I finally stop with the Ur-phenomena, then it is only due to resignation. But there is a big difference as to whether I resign myself to human limits or to a hypothetical limitation of my ignorant individuality." And again: "The immediate awareness of the Ur-phenomena gives us a type of anxiety. We feel our inadequacy; only enlivened through the eternal game of empirical phenomena do they delight us."

To this point we must give the highest recognition to the Poet's healthy feeling and his deep insight. On the other hand, we should not overlook that what the Poet sought to attain in the theory of colors displays certain deficiencies which the scientific treatment of this field could not let stand. In his theory of colors, he analyzes often and extensively how, in his opinion, blue or yellow light originates. In this way, the images of brighter or darker surfaces, with which he operates, are always there. These images have, in his opinion, displaced themselves towards one another: the light of the one is supposed to penetrate through that of the other, the latter operating as a transparent medium upon the transmitted light (which, incidentally, is asking a great deal of the reader's fantasy). But nowhere does he analyze how, in his view, first blue and then yellow light should be distinguished from one another. He is simply satisfied with the assertion that both may have contained something shaded in their passage through the bodies; yet he clearly does not consider it his duty to give an account of how the shaded in the blue distinguishes itself from that in the yellow, and how both are distinguished from that in the mixture of the two, which he considers to be green. And precisely in this relation Newton's and, still more, Huygens's wave theory gives exact definitions that have been completely confirmed by the most rigorous measurements, and, ultimately, have led to astronomical determinations of the orbital elements of the most distant double stars—something that one had never believed could have been hoped for. It is the number of light vibrations in equal times that determines the color, just as, on the other hand, the number of sound vibrations in equal times determines the pitch.

Obviously for him, the optical image is what the intuition of a definitely formed bodily object or field occasions, the latter intuitively imaginable and thus the limit of his interest. On the other hand, the means by which such a sensory intuition is achieved, recedes; similarly, he says just as little of a definite nature about how he conceives of the relations of sensations which are brought forth in the seeing eye to the objective agent (to the light), whose presence and type is indicated by the sensation.

Yet these questions have been addressed to him by his friends. He reports[3] that, on their urging, he has been studying Kant; and that in the *Critique of Judgment* he has found much to stimulate him, that brings him close to Schiller, while he obviously could not befriend the *Critique of Pure Reason*. "I applauded completely all friends who maintained, with Kant, that if all our knowledge commences with experience, it does not therefore all originate from the experience." "It was the entrance, which pleased me; I could not consider entering into the labyrinth; the poetic gift soon stopped me, soon common sense; and I felt myself nowhere improved." Using the occasion of Faust's trip to the "Mothers," he has unforgettably, and with light irony, described the aesthetic impression that "Kant's world of things in themselves" made on him:

No Space around them, Place and Time still less;
Only to speak of them embarrasses.

Naught shalt thou see in endless Void afar,—
Not hear thy footstep fall, nor meet
A stable spot to rest thy feet.[4]

Now, the physiological investigation of the sense organs and their activity has shown results that, ultimately, and in the most essential points (so far, at least, as I myself consider them essential), agree with Kant; indeed, already in the physiological field there are the clearest analogies to Kant's transcendental aesthetic. However, an objection coming from the natural scientific point of view had to be raised against the borderline that Kant had drawn between the facts of experience and the forms of intuition given *a priori*. And with the required redrawing of the border—whereby, namely, the fundamental principles of spatial theory are subsumed under the facts of experience—we ought perhaps to expect that Goethe, too, would no longer feel prevented in attaching himself to what he calls "human understanding."

There may even be such forms of intuition for the perceptions of the individual senses, just as Kant seeks to prove them for the entire range of our field of representation.

3. *Zur Naturwissenschaft im Allgemeinen. Einwirkung der neueren Philosophie.*
4. Johann Wolfgang von Goethe, *Faust: A Tragedy.* Two Parts, trans. Bayard Taylor (Boston: James Osgood and Company, 1871), 2:89 and 91, resp.

The optic nerve senses everything that it, in general, senses in the form of light phenomena within the visual field. It does not have to be external light which stimulates it. A push or pressure on the eye; a stretching of the retina through the eye's rapid motion; electricity, which circulates through the head; or altered blood pressure can also stimulate sensation in it. However, in all these cases the stimulated sensation is always only light sensation, and makes completely the same impression in the visual field as if it originated from external light. Pushing, pressure, stretching, electric current can, however, also stimulate the skin; we feel it, in that case, as sensations of touch. Indeed, the same sun rays that appear to the eye as light, stimulate the sensation of heat radiation in the skin. By means of electric currents we also stimulate sensations of taste or hearing, depending on whether they strike the tongue or the eye.

From this follows the much-discussed principle of recent times that precisely the most radical differences of our sensations do not depend at all on the means of stimulation but rather only on the sense organ that has been stimulated. When we speak of man's five different senses we recognize the deeply decisive nature of this designated distinction. A comparison between the sensations of the different senses is not even possible; nor is a relation of similarity or dissimilarity. That we see an object as a colored visual image depends only on the eye; the particular colors in which we see it, however, also depends, to be sure, on the type of light that it emits towards us. This law has been proven by Johannes Müller, the physiologist, and designated as the law of specific sense energies. However, further detailed comparisons of the qualities of sensations with the qualities of the effective means of stimulation reveal that equality of color impressions can occur with quite different light mixtures and in no way coincides with the equality of any other physical effect of light.

I have thus believed it necessary so to formulate the relationship between the sensation and its object such that I would interpret the sensation only as a sign of the object's effect. To the nature of a sign belongs only the property that for the same object the same sign will always be given. Moreover, no type of similarity is necessary between it and its object, just as little as that between the spoken word and the object that we designate thereby.

We cannot even call the sense impressions images, for an image images the same by means of the same. In a statue we give bodily form by means of bodily form; in a drawing the object's perspectival view by means of the same of the image; in a picture, again, color by means of color.

Only in reference to the temporal process can the sensations be images of the course of events (corrections reserved). Number falls among the determinants of the temporal process. In these relations they thus produce, in fact, more than mere signs would do.

Goethe knew a good deal about the subjective sensations of the eye; discovered some himself; and has learned the theory of the specific sense energies, at best in incomplete form, through Schopenhauer. He has rejected whatever in Kant or the elder Fichte could have led to it, because it was tied up with other claims that were unacceptable to him. How surprised we must be when we find, at the end of *Faust,* the condition of the blessed spirits, who look at eternal truth face to face, described as follows in the words of the Chorus mysticus:

> All things transitory
> But as images are sent,

that is, what happens in time, and what we perceive through the senses, we know only as an image. I would hardly know how to express that end result of our physiological epistemology more suggestively.

> Earth's insufficiency
> Here grows to Event.

All knowledge of natural laws is inductive; and no induction is ever absolutely finished. After the Poet's confession as noted above, we feel our insufficiency penetrate more deeply into a type of anxiety. The result that occurs justifies for the first time the results of earthly thought.

> The Indescribable,
> Here it is done.[5]

The indescribable, that is, that which cannot be captured in words, we know only in the form of artistic presentation, only in the image. For the blessed, it becomes reality.

Our epistemological considerations are here at an end. The final stanza turns itself towards a higher sphere. It aims at the promotion of all spiritual activity in the service of humanity and of the moral ideal, which is symbolized through the Eternal Feminine.

The deeper we try to go into the innermost workshop of the Poet's thoughts, the weaker become the traces left by him that we may follow. In the meantime, if our own way has finally led us to the same goal as his, then we must indeed note it well—and also where the connecting links are missing and where the context appears doubtful.

Faust saves himself—for action—from the unsatisfactory condition of the knowledge and brooding going on inside himself, from which he may

---

5. Ibid., 2:432.

not hope to arrive at the secure possession of truth and by which he may not know how to grasp reality. Before he makes the pact with Mephistopheles, Goethe presents him—obviously with the intent of preparing the subsequent development of the second part—in the scene (added later) where he undertakes to translate the Gospel according to Saint John. He runs up against the much-discussed concept of the *Logos*: "In the beginning was the word." The word is only a sign of its meaning; this must be meant; the meaning of a word is a concept, or, if it refers to something that happens, a natural law, which, as we saw, when it is conceived as continuous and effective is designated as force. There thus lies in this transition from word to meaning, and then to force, which Faust makes in his attempt at translation, above all a continuous, further development of the concept. However, even force does not satisfy him. He now makes a decisive intellectual leap:

> The Spirit aids me: now I see the light!
> "In the Beginning was the *Act*," I write.[6]

To be sure, the Gospel's position refers to the original conditions of the Creative Spirit, but Faust seeks his own spiritual comfort and finds a hope for it in this thought, which fills the devilish poodle with increased discomfort because he sees his victim is looking to discover a means of rescue. Hence, I do not believe that Goethe wanted to present us with Faust moved here only by the theoretical interest in the act of the world's creation, but rather, and still more, by his subjective thirst after the paths to truth.

The epistemological counterpart to this scene lies in the efforts of the philosophical schools to establish belief in the existence of reality, efforts that must remain unsuccessful so long as they proceed only from passive observations of the external world. They could not get beyond their world of images; they did not recognize that human actions, which are posited by the will, form an indispensable part of our sources of knowledge. We have seen that our sense impressions are only a sign language which reports to us about the external world. We humans must first learn to understand this sign system, and that happens when we observe the success of our actions and so learn to distinguish which changes in our sense impressions follow from our acts of will, and which others enter independently of the will.

I have elsewhere analyzed that and how we thereby attain knowledge of reality.[7] It would here lead us too far into the most abstract areas of thought; the fact may suffice that, in order to become sure of reality, even an

---

6. Ibid., 1:69.
7. See "The Facts in Perception," this volume.

epistemology based on the physiology of the senses has to instruct humans how to proceed to act.

I must mention yet another of Goethe's allegorical figures, namely, the Earth Spirit in *Faust*, to which I have already previously referred. His words, in which he describes his own nature, so well fit another, newest conception of natural science, that only with difficulty can one tear oneself away from the thought that it is intentional. The Spirit says:

> In the tides of Life, in Action's storm,
> A fluctuant wave,
> A shuttle free,
> Birth and the Grave,
> An eternal sea,
> A weaving, flowing
> Life, all-glowing,
> Thus at Time's humming loom 'tis my hand prepares
> The garment of Life which the Deity wears![8]

Now, we know today that there resides in the world an indestructible and unincreasable supply of energy or effective motor power that can appear in the most manifold, ever-changing forms—now as a raised weight, now in oscillation with accelerated masses, now as heat or chemical affinity, &c.—which, in this exchange, constitutes the active force in each effect, both in the realm of living nature and in inanimate bodies.

The germs of this insight into the constancy of the value of energy were already at hand in the previous century, and could well have been known to Goethe. The comparison with contemporaneous essays by him (*Nature*, 1780) perhaps brings one closer to the idea that the Earth Spirit should be the representative of organic life on Earth, although with this, admittedly, the words "Life, all-glowing," fit badly. The two versions do not necessarily contradict one another, since Robert Mayer, as well as myself, have been led to the generalization of the law of constancy of energy precisely through considerations of the general character of the life processes.

We can, of course, now no longer limit the constant supply of energy to the Earth, but must at least include the sun. However, a presentiment of the Poet does not have to be exact in all details.

As a concluding result, we may well summarize the outcome of our considerations as follows: Where it is a matter of problems that can be solved by poetic divinations proceeding in intuitive images, the Poet has shown himself able to the highest achievements; where only the consciously applied inductive method may have been able to help, he has failed.

---

8. *Faust*, 1:28–29.

However, once again, where it is a matter of the highest questions concerning the relationship of reason to reality, his healthy adherence to reality protects him from aberrations and leads him securely to insights that reach the limits of human understanding.

Speech held in the General Assembly of the Goethe Society, Weimar, 1892.

Index

DATE DUE